Particle Technology Series

Volume 20

Series editor

José Manuel Valverde Millán, University of Sevilla, Spain

Many materials exist in the form of a disperse system, for example powders, pastes, slurries, emulsions and aerosols, with size ranging from granular all the way down to the nanoscale. The study of such systems necessarily underlies many technologies/products and it can be regarded as a separate subject concerned with the manufacture, characterization and manipulation of such systems. The series does not aspire to define and confine the subject without duplication, but rather to provide a good home for any book which has a contribution to make to the record of both the theory and applications of the subject. We hope that engineers and scientists who concern themselves with disperse systems will use these books and that those who become expert will contribute further to the series.

The Springer Particle Technology Series is a continuation of the Kluwer Particle Technology Series, and the successor to the Chapman & Hall Powder Technology Series.

More information about this series at http://www.springer.com/series/6433

Frank Babick

Suspensions of Colloidal Particles and Aggregates

Springer

Frank Babick
Institute of Process Engineering and
Environmental Technology
Technische Universität Dresden
Dresden
Germany

ISSN 1567-827X
Particle Technology Series
ISBN 978-3-319-30661-2   ISBN 978-3-319-30663-6 (eBook)
DOI 10.1007/978-3-319-30663-6

Library of Congress Control Number: 2016934674

Printed on acid-free paper

This Springer imprint is published by Springer Nature
The registered company is Springer International Publishing AG Switzerland

... the responsibility for this indiscriminate "wrapping" of ten centuries lies to some extent on the medieval culture itself, which [...] is perpetually commenting comments and citing authoritative formulae without giving the impression of saying anything new. This is wrong. The medieval culture has a sense of innovation, but also the talent for concealing it behind the robe of repetitions (in contrast to modern culture, which pretends to innovate even when repeating).

... la colpa di questo "impacco" indiscriminato di dieci secoli è anche un poco della cultura medievale la quale [...] lavora commentando commenti e citando formule autorevoli, con l'aria di non dire mai nulla die nuovo. Non è vero, la cultura medievale ha il senso dell'innovazione, ma si ingegna a nasconderla sotto le spoglie della ripetizione (al contrario della cultura moderna, che finge di innovare anche quando ripete).

Umberto Eco, *Arte e bellezza nell'estetica medievale.*
Bompiani, Milano, 1987

Preface

Why colloidal, why not nano?

This question, recently raised after a presentation of my research, suddenly reminded me of my academic beginnings and that my relation with colloids had lasted much longer than I was aware of before. Certainly, there are objective reasons for my persistent use of the term *colloidal*, even though the term *nano* may sometimes attract a larger audience. Most of all, the both terms are not congruent. With regard to particle size *colloidal* covers a wider range than *nano*, whereas *nano* refers to a wider spectrum of products and applications (e.g. surface structures or nano-electronics). Moreover, *nano* has also non-technical connotations, which may provoke expectations beyond the scope of my research. Yet, in principle the question was justified. Regarding the material systems I usually deal with both terms could have been equally employed. In the instant of that question I realised that I have been fond of colloidal systems since my first scientific steps.

My first encounter with the field of colloid science occurred during my undergraduate studies, when I worked in a project on the electroacoustic zeta-potential measurement of colloidal suspensions. I probably did not understand much of the measurement technique, but I became fascinated with colloids for several reasons. First, they behave "strangely"—at least for a student of process engineering whose education in physics stopped right after mechanics and thermodynamics. Second, in our library I found some of the historical papers by Smoluchowski and others, which impressed by their age and the clear and comprehensible explanation of the physical phenomena in colloidal suspensions. Further on, I felt attracted by the interrelation of physics, chemistry and engineering. Last but not least, there was this mystery that our curriculum completely ignored the subject of colloids.

I remained stuck to the field of particle characterisation when I started my scientific career at the Technische Universität Dresden. In particular, I dealt with the ultrasonic attenuation spectroscopy and its potential use for process characterisation. The performance of this sizing technique was to be examined for colloidal systems. I soon learnt that there is no meaningful particle measurement without a profound understanding of the material's physico-chemical properties. This holds particularly true for colloidal systems, which are prone to aggregation and agglomeration. A proper characterisation, therefore, needs to address the questions of sample stability, signal interpretation in the case of aggregated suspensions, or the description of aggregate morphology. These are not just academic issues, but were and are still addressed in several projects that our research group has conducted for industry partners (e.g. in the semiconductor industry, or in pigment manufacturing) or legal authorities (e.g. with regard to nanotoxicology). For me it was, hence, a logical step to expand my research focus from the characterisation techniques to the physical behaviour of colloidal particles and aggregates. This book reflects a major part of my research in this field.

I have been working in the university for more than 15 years. I would like to acknowledge the opportunities that this environment has offered for my scientific work. On the one hand, there are the manifold contacts to industry or other research groups, which ensure a certain variety of scientific topics as well as the confrontation with different views and new aspects. On the other hand, there is enough room to deeply explore scientific problems, which is essential for really moving forward in science. Besides this, the university life offers the chance to go on sabbaticals, which I was able to use for short-term stays at the Ecole Nationale Supérieure des Mines in St.-Etienne/F and at the Doshisha University in Kyoto/JP. Both stays clearly widened my views (not only on colloids) and gave substantial input to my research. At this occasion I would like to thank Frédéric Gruy and Yasushige Mori for their logistic support and our scientific discussions.

Of course, I also took advantage of the university environment while writing this book. Its content is definitely influenced by the experience I gained in several research projects and, naturally, I could profit from the expertise of my colleagues at the Research Group of Mechanical Process Engineering. I gratefully acknowledge their scientific as well as non-scientific support. Indeed, I really appreciate the family atmosphere I can work in. Special thanks are dedicated to Michael Stintz, the head of our group, for his principal support, his helpfulness and his professional advice. Last but not least, I sincerely thank Margrit Hansel for being our "firewall" against bureaucracy, and even more for her respectful manners and her extraordinary kindness.

In addition, I want to acknowledge all the students and Ph.D. students who contributed to my research through literature research, experiments, programming, discussions or publications. Especially I would like to mention Uwe Kätzel, who wrote an exceptionally good diploma thesis on the scattering of aggregates, subsequently worked in our research group, and eventually submitted an excellent Ph.D. thesis on the characterisation of pyrogenic silica with dynamic light scattering. Additionally, I would like to appreciate the outstanding theses of Karin Schwarzenberger (née Schießl) on the subjects of aggregate scattering and aggregate interaction. Even though she conducted these works as a student, they were of remarkably high quality and provided a sound basis for future investigations. Last but not least, I want to acknowledge the input of Karina Paciejewska, who received a doctorate on the stability of colloidal suspensions under my supervision. I was able to benefit from her intuitive understanding of colloidal phenomena, her experimental work, and all the inspiring discussions. Most of all, I thank her for our close friendship.

Further thanks go to Lee-Ellen Reed who subedited the manuscript and gave valuable assistance in finding the best English expressions.

Finally, I would like to thank my family, which has always shown its support for me and my scientific work and which provides me with the positive emotions that are needed for accomplishing a monograph like this.

Dresden
November 2014

Frank Babick

Contents

Nomenclature

n.d. = not (uniformly) defined

Latin Letters

A	Area (m^2)
A_{123}	Hamaker function, Hamaker coefficient (J)
$A_{2/3}$	Second/third virial coefficient (–)
B	Slope of the scattering function (n.d.)
a	Particle radius (m)
a_c	Centrifugal acceleration (m/s^2)
C_{abs}	Absorption cross section (m^2)
C_{ext}	Extinction cross section (m^2)
C_{sca}	Scattering cross section (m^2)
c_m	Mass concentration (kg/m^3)
c_N	Number concentration ($1/m^3$)
c_n	Molar concentration (mol/m^3)
c_V	Volume concentration (–)
$\mathbf{D}$	Rate of strain tensor (1/s)
D	Diffusion coefficient (m^2/s)
d_f	Fractal dimension (–)
d_S	Surface fractal dimension (–)
$\mathbf{E}$	Rate of strain tensor of imposed flow (1/s)
$\mathbf{E}$	Electric field (V/m)
E	Extinction (–)
E_{ij}	Sticking probability between clusters of relative mass i and j (–)
$\mathbf{e}$	Standard basis vector (–)
e_0	Elementary electric charge, $e_0 = 1.602 \times 10^{-19}$ C
F	Faraday constant, F = 96485 C/mol
$\mathbf{F}_{ad}$	Adhesion force (N)
$\mathbf{F}_D$	Drag force (N)

Fa	Fragmentation number (–)
f	Frequency (Hz)
f	Polarisation factor (–)
G'	Storage modulus (Pa)
G''	Loss modulus (Pa)
g	Gravitational acceleration, g = 9.81 m^2/s
$g^{(1)/(2)}$	Normalised field/intensity correlation function (–)
h	(minimum) Surface distance (m)
$\mathbf{I}$	Identity tensor (–)
I_0	Incident radiation intensity (W/m^2)
I_{sca}	Scattered radiation intensity ($W/m^2/sr$)
$\mathbf{J}$	Oseen tensor (–)
K_a	Acid dissociation constant (–)
K_{ij}	Kinetic coefficient for collisions between clusters of relative mass i and j (m^3/s)
K_{reg}	Double layer capacity (F)
k	Wave number (1/m)
k	Permeability (m^2)
ka	Dimensionless wave number (–)
k_B	Boltzmann's constant, $k_B = 1.381\times10^{-19}$ J/K
k_t	Kinetic constant ($mol/s/m^2$)
L	Length (m)
L	Truncation level (–)
l_g	Gravitational length (m)
$\mathbf{M}$	Mobility matrix (n.d.)
$\mathbf{M}_{mn}^{(k)}$	Vector spherical wave functions (–)
M	Molar mass (kg/mol)
M	Torque (N·m)
m	Refractive index (–)
$\mathbf{N}_{mn}^{(k)}$	Vector spherical wave functions (–)
N	Aggregation number (–)
N_C	Coordination number (–)
N_A	Avogadro's number, $N_A = 6.022 \times 10^{23}$ 1/mol
$\mathbf{n}$	Normal vector (–)
n	Number, absolute frequency (–)
P	Form factor for particle scattering (–)
P	Power (W)
P_V	Volume specific power (W/m^3)
PDI	Polydispersity index from DLS experiments (–)
Pe	Peclet number (–)
p	Probability, relative frequency (–)
p_n^j	Harmonic function (–)
Q	Porod invariant (–)

Q_r	Sum or cumulative function; r indicates the type of quantity (–)
Q_{sca}	Scattering efficiency (–)
q	Electric charge volume density (C/m^3)
q_r	Density function; r indicates the type of quantity (n.d.)
q_{lat}	Latent heat (J/kg)
$\mathbf{R}$	Rotation matrix (–)
R	Molar gas constant, R = 8.3145 J/mol/K (J/mol/K)
R	(aggregate) Radius (m)
Re	Reynolds number (–)
$\mathbf{r}$	Space coordinate (–)
r	Radial coordinate (–)
r_{AB}	Centre-distance between aggregate A and B (–)
r_{ij}	Centre-distance between particles i and j (–)
$\mathbf{S}$	Tensor of stresslets (n.d.)
S	Entropy (–)
S	Surface area (m^2)
$S_{\mathrm{agg/sus}}$	Static structure factor of the aggregate/of the suspension (–)
S_m	Mass specific surface area (m^2/kg)
S_V	Volume-specific surface area (1/m)
S_{reg}	Parameter of double layer regulation (C/m^2)
s_r	Standard deviation; r indicates the type of quantity (n.d.)
$\mathbf{T}$	Translation matrix (–)
T	Transmission (–)
T	Thermodynamic temperature (K)
T_{m}	Melting point (K)
t	Time (s)
$\mathbf{u}$	Fluid velocity vector (m/s)
u_{mn}	Solid spherical harmonics (–)
V	Volume (m^3)
$V_{\mathrm{B/hyd}}$	Energy potential for Born and hydration interaction (J)
$V_{\mathrm{DL/vdW}}$	Energy potential for double layer and van der Waals interaction (J)
V_{t}	Total interaction energy potential (J)
V_{m}	Molar volume (m^3/mol)
$\mathbf{v}$	(particle) Velocity vector (m/s)
v_{sett}	(terminal) Settling velocity (m/s)
x	Particle size, equivalent diameter (m)
x_{cum}	Intensity weighted harmonic mean size from DLS experiments (m)
$x_{\mathrm{h,t/r}}$	Hydrodynamic (equivalent) diameter of translation/rotation (m)
x_{Stokes}	Stokes diameter or sedimentation equivalent diameter (–)
x_V	Volume equivalent diameter (m)
z_i	Valency of ionic species i (–)

Greek Letters

α	Attenuation coefficient (dB/m)
α	Atomic polarisability (F·m^2)
α'	Polarisability volume (m^3)
Γ	Interfacial concentration (n.d.)
γ	Interfacial tension (N/m)
ε	Porosity (–)
ε	Relative permittivity (–)
$\varepsilon(i\omega)$	Dielectric function (–)
ε_0	Electric constant, $\varepsilon_0 = 8.854 \times 10^{-12}$ C/V/m
ζ	Zeta-potential (V)
η	Dynamic viscosity (Pa·s)
η_{sp}	Specific viscosity (–)
θ	Scattering angle (–)
κ	Debye–Hückel parameter (1/m)
κa	Dimensionless Debye–Hückel parameter (1/m)
λ	Loss coefficient (–)
λ	Wavelength (m)
μ_{ep}	Electrophoretic mobility (A•s^2/kg)
μ_{hd}	Hydrodynamic mobility (s/kg)
$\mathbf{\Pi}$	Stress tensor (N/m^2)
ρ	Mass density (kg/m^3)
σ	Electric surface charge density (C/m^2)
σ	Mechanical strength (N/m^2)
σ_T	Tensile strength (N/m^2)
σ_{ln}	Standard deviation of log-normal distribution (–)
τ	Turbidity (1/m)
τ	Decay time (s)
τ	Time scale (s)
τ_0	Yield stress (N/m^2)
Φ_n^j	Harmonic function (–)
φ_m	Mass fraction (–)
φ_N	Number fraction (–)
φ_V	Volume fraction, packing density (–)
ϕ	Azimuthal angle, angle of polarisation (–)
χ_n^j	Harmonic function (–)
ψ	Electric potential (–)
ω	Angular frequency (s^{-1})

German Letters

$\mathfrak{R}$	Resistance matrix (n.d.)

Subindices

0	Zero, surface
agg	Aggregate, agglomerate
av	Average (i.e. number weighted arithmetic mean)
cr	Critical
C	Convex hull/circumscribing spheres
c	Centre
cav	Cavitation
cb	Collapsing bubbles
DL	Double layer
disp	Dispersion
GM	Grinding media
g	Gyration
m	Suspension medium, solvent
max	Maximum
mean	Mean value
meas	Measurement
min	Minimum
p	(primary) particle
proj	(orthogonal) projection
S	Surface
s	Solid
sat	Saturation
sca	Scattering
sett	Settling
slip	Slip
stoch	Stochastic
stress	Stress
sus	Suspension
US	Ultrasonic
vdW	Van der Waals
μe	Micro-eddy

Superindices

diff	Diffusion
E	Excess
eq	Equilibrium
if	Interfacial

iso	Isolated particle
sh	Shear
tu	Turbulent

Mathematical Symbols and Operators

e	Euler's number, e = 2.71828
i	Imaginary unit, $i^2 = -1$
π	Archimedes' constant, Ludolphian number, $\pi = 3.14159$
δ_{ij}	Kronecker delta
Δ	Difference
Δ	Laplace operator
∇	Nabla operator
$\langle \cdot \rangle$	Volume or time average
$\approx$	Approximately
$\propto$	Proportional
∞	Infinity
$\prod$	Product
$\sum$	Sum
log	Logarithm to an arbitrary base
Im	Imaginary part
Re	Real part

Abbreviations

AFM	Atomic force microscope
APPI	Approximation of primary particle interaction
BET	Brunauer–Emmett–Teller model for gas adsorption
CCC	Critical coagulation concentration
CMC	Critical micelle concentration
CMD	Cluster mass distribution
CV	Coefficient of variation
CVP	Colloid vibration potential
DLCA	Diffusion limited cluster–cluster aggregation
DLS	Dynamic light scattering
DLVO	Derjaguin, Landau, Verwey, Overbeek
DUM	Dynamic ultramicroscopy
EDL	Electric double layer
EMS	Electroacoustic mobility spectroscopy
ESA	Electrokinetic sonic amplitude
FFF	Field-flow fractionation
HP	High pressure
HPPF	High pressure post feeding
IEP	Isoelectric point

IHP	Inner Helmholtz plane
LD	Laser diffraction
MADLS	Multi-angle dynamic light scattering
NIR	Near infrared
OCA	Optical centrifugation analysis
OHP	Outer Helmholtz plane
OPC	Optical particle counting
OS	Optical spectroscopy
PBE	Poisson–Boltzmann equation
PBE	Population balance equation
PCR	Point of charge reversal
PZC	Point of zero charge
RLCA	Reaction limited cluster–cluster aggregation
rms	Root mean square value (i.e. quadratic mean)
SANS	Small angle (static) neutron scattering
SAXS	Small angle (static) X-ray scattering
SEM	Scanning electron microscope
SLS	Static light scattering
SOP	Standard operation procedure
SPLS	Single particle light scattering
SVWF	Spherical vector wave function
TEM	Transmission electron microscope
UV	Ultraviolet

Chapter 1
Introduction

1.1 Colloidal Suspensions

Colloidal suspensions are heterogeneous materials which consist of a liquid continuum and solid, submicrometre particles. They form a large group of dispersed systems which comprises, e.g., black ink as used by the ancient Egyptians, ruby glass as employed for decorative glassware in the Victorian era, paints with white or coloured pigments below 1 μm, sun lotions with UV-absorbing colloidal titania, or polishing slurries as used for the planarisation of silicon wafers. Due to their fineness, colloidal particles are not visible to the human eye, not even with conventional light microscopes. Moreover, if the particles are much smaller than the wavelength of light, colloidal suspensions may appear transparent. This means the dispersed state is not obvious and the distinction from true solutions relies on physical phenomena like scattering (Tyndall effect) or membrane penetration. The latter effect was used by Graham as a criterion to classify all transparent, non-settling liquid materials. He gave the name "colloids" to those materials which behaved similarly to glue (Latin: *colla*).

The small size of colloidal particles does not only affect their visibility, but coincides with extremely low settling velocities and relatively high diffusibilities. Furthermore, it increases the significance of particle–particle interactions, in particular those related to electric surface charges, and it means that a certain solid mass is dispersed into a comparatively huge number of particles, which, e.g., promotes their mutual collisions. Besides this, we can frequently simplify the mathematical treatment of physical phenomena in colloidal suspensions: The hydrodynamic drag, for instance, is dominated by viscous friction, and the scattering on very small colloids obeys the Rayleigh scattering limit. In other instances, however, the mathematical treatment of very fine colloids might become more complicated (e.g., because the curvature of the surface cannot be neglected).

The physico-chemical properties of colloidal particles and their interfaces to the dispersion medium, the interaction between these particles, the macroscopic

F. Babick, *Suspensions of Colloidal Particles and Aggregates*,
Particle Technology Series 20, DOI 10.1007/978-3-319-30663-6_1

properties of the colloidal suspensions, as well as the various characterisation techniques are subjects of colloid science. This scientific discipline on the border of chemistry and physics has several roots in the nineteenth century but was not really acknowledged until the advent of the ultramicroscope by Siedentopf and Zsigmondy (1903). The ultramicroscope did not only reveal the existence of very fine, "invisible" particles, but was soon able to be used by Perrin (1909) to demonstrate the validity of Einstein's theory on Brownian motion. The first half of the twentieth century experienced a rapid development of colloid science—with regard to theory (e.g. Smoluchowski's contribution on aggregation and electrokinetic phenomena or the electric double layer model by Gouy and Chapman) as well as with regard to instrumentation (e.g. Svedberg's invention of the ultracentrifuge or Guiniers application of X-rays to the characterisation of colloids). The scientific achievements in the field of colloid science received a lot of recognition and were explicitly acknowledged by the Nobel prizes to Zsigmondy (in 1925), Perrin (in 1926), and Svedberg (in 1926). An outstanding landmark in that period was the DLVO (Derjaguin, Landau, Verwey, and Overbeek) theory, which quantifies the interaction between suspended particles with electrically charged surfaces and allows the prediction of suspension stability based on Fuchs' approach. In the second half of that century, a lot of work was done to refine our understanding of the electric double layer (e.g. by Hunter, Lyklema), to enhance the theory of electrokinetic phenomena (e.g. by Booth, Dukhin, O'Brien & White), and to elaborate the calculation of particle aggregation and suspension stability (e.g. by Hogg & Healy). Colloid science obtained major theoretical input through the concept of fractality, which was introduced by Mandelbrot (1977) and which clearly enhanced the quantification of aggregate morphology and the distinction between different types of aggregation. Additionally, a large variety of characterisation techniques was newly developed or became widely available to the scientific community, such as electron microscopy, dynamic light scattering, field-flow fractionation, or scanning force microscopy. This technical development has not ceased yet and is even being accelerated by the ongoing advances in computation technique. The latter have additionally acted as catalysts for a variety of numerical tools that allow the calculation of physical properties of irregularly shaped bodies or that facilitate the simulation of multi-particle systems.

Today, colloid science is characterised by a wide diversity of subjects and narrow relations to other scientific disciplines (e.g. polymer chemistry, process engineering, particle sizing). It has gained additional significance by the commercial success of the nanotechnologies, which i.a. generate, process and employ nanomaterials. Such materials contain particles (isolated, aggregated or agglomerated) with at least one outer dimension below 100 nm—a size range, which is fully covered by the term *colloidal*. Since the handling of nanomaterials may considerably differ from that of coarse particle systems, there is a clear demand from industry as well as from legal authorities for a comprehensive understanding of their physical and physiological behaviour and for appropriate characterisation. Colloid science provides major input for solving these tasks.

1.2 Scope of the Book

This book addresses the properties of colloidal particles and suspensions, in particular when the particles are aggregates[1] of even finer, often nano-sized (i.e. ≤ 100 nm) particles. Such particle aggregates evolve during the particle synthesis by precipitation or crystallisation, and may result from the aggregation (coagulation) of isolated primary particles or can already be present in the powder which is dispersed in the liquid medium. It is therefore rather likely to encounter such aggregates in colloidal suspensions. The aggregates influence optical and rheological suspension properties, affect the sedimentation or filtration of the dispersed phase, reduce the homogeneity within composite materials, and may show other physiological effects than the primary particles. In any case, there is a specific impact of the aggregate morphology on the physical aggregate properties and on the related macroscopic properties. This refers i.a. to optical suspension properties (e.g. scattering, turbidity), to the particles' dynamic properties (e.g. diffusion coefficient, acoustophoretic mobility), or to electrokinetic properties (e.g. electrophoretic mobility). The knowledge of the governing structure-property relations supports the tailored adjustment of product qualities and is required for the interpretation of measurement results that are used to evaluate if a desired suspension quality (e.g. a fractional concentration, a certain size distribution, a required stability) is achieved.

The objectives of this book are to review the theories and methodological tools available for calculating structure-property relations, to show the corresponding results for selected aggregate types and to apply this knowledge to practical engineering tasks. It is intended to enable readers to independently conduct their own investigations with the appropriate models, methods and techniques. Consequently, the book does not only consider the physical properties of colloidal aggregates, but also discusses characterisation techniques, fundamentals of colloid science and the preparation of colloidal suspensions. The book does not claim to completely cover all of these aspects in full detail; yet, it provides references to numerous textbooks and scientific publications that may answer the reader's questions instead. Last but not least, it should be noted that the relationships between the properties of individual colloidal aggregates with those of a suspension of such aggregates, i.e. the coupling between microscopic and macroscopic suspension properties, are beyond the scope of this book.

---

[1]The terms *aggregation* and *aggregates* are commonly used in colloid science for any type of particle cluster. Note that in other scientific communities they imply the existence of strong interparticle bonds, whereas for weak bonds or for the general case, *agglomeration* and *agglomerate* would be preferred (cf. glossary A.1).

1.3 On the Structure and Thinking of This Book

The structure-property relations of colloidal aggregates concern very different technical and scientific issues. Accordingly, this book aims at a widespread readership (e.g. colloidal scientists, process engineers, developers of particle measurement techniques, or toxicologists), whose scientific expertise and specific interest on this subject may vary considerably. For this reason, the book is divided into four main chapters:

- characterisation techniques (Chap. 2), which reveal morphology parameters of aggregates and quantify their physical properties;
- physical fundamentals of colloidal suspensions (Chap. 3), on which the more sophisticated treatment of aggregates is based;
- morphological and physical properties of aggregates (Chap. 4), including the derivation of structure-property relations and the interactions between aggregates;
- two fundamental aspects of processing colloidal suspensions: dispersion of powders and suspensions stability (Chap. 5).

Though these chapters refer to each other they can be read independently. For the sake of convenience the references are listed at the end of each chapter, sorted by topic. Due to its broad scope, the book can only discuss the general ideas and procedures of physical theories, mathematical models or numerical tools, while details have to be elicited from the cited literature. Only a few topics with major relevance (e.g. the models on light scattering) are described more comprehensively in the appendices.

A general methodological approach of this book is the comparison of sophisticated and simplified models for the computation of aggregate properties. A further particularity might be the discussion on the significance of measurement results when characterising colloidal suspensions (containing aggregates) for various purposes. The former aims at a clear understanding of the limits of approximate solutions, which are frequently favoured in practice because of ease, lack of model parameters or computational costs. A typical example is the "spherical particle" assumption which is standard in particle sizing instruments. If, however, this assumption is, e.g., applied to a centrifugation analysis of colloidal aggregates >100 nm, one may wrongly measure much finer (even nano-sized) particles, which might have severe implications for the (regulatory) classification of the material. A further failure of sphere models is related to the light scattering properties, which show a oscillating dependency on size for submicrometre spheres >100 nm and a monotonic dependency for aggregates in that size range. Optical sizing instruments may, therefore, calculate multimodal size distributions, where a monomodal aggregate size distribution prevails. Another example for the need of sophisticated models is the decrease of suspension viscosity during the dispersion of highly porous aggregates, which cannot be explained without acknowledging their fractal structure.

The particular attention to characterisation techniques does not only result from the author's biography, but is due to the specific sensitivity of a given characterisation technique to aggregate morphology (sensitivity with regard to aggregate property, size, or concentration range). Hence, the evaluation of product quality and process changes requires a clear understanding of the relevant parameters of the dispersed system (e.g. mean or maximum aggregate size, or concentration of micrometre particles) and a basic knowledge of the applicability and limits of measurement techniques.

Although this book focuses on the properties of individual aggregates and dilute colloidal suspensions, it cannot ignore the impact of single particles and aggregates on the macroscopic suspension properties, since we seldom quantify the properties of an individual particle, but rather those of a particle ensemble. Frequently, a linear correlation between the microscopic properties and macroscopic behaviour is assumed, e.g. between particle settling and suspension sedimentation; particle scattering and Tyndall effect; light extinction at the particle and turbidity; particle self-diffusion and the fluctuation of scattered light; or between the zeta-potential of a particle and the sedimentation potential of a suspension. However, non-linear effects may result from steric, viscous, optical or other interactions between the particles and from the corresponding suspension structuring. Such effects—e.g. hindered or zone sedimentation, multiple scattering, gradient diffusion or secondary electro-viscous effects—are usually significant at moderate or high particle concentrations and then need adequate consideration.

Eventually, the notation and terminology of this book needs to be commented upon. This is necessary because its subject bears relation to several scientific communities (chemists, process engineers, physicists, measurement engineers) with their own traditions regarding terms and notation, which may be further distinguished by national and/or continental schools. Differences among such systems are not just arbitrary, but result from different objectives and examined objects and most often they have pragmatic origins. Consequently, they can be only partly resolved. Attempts to completely harmonise the terminology and notation in colloid science by a uniform system of definitions will fail in practice and for principal reasons. Even though this book seeks a uniform and consistent terminology, it avoids strictly unambiguous definitions and is aware of possible inconsistencies as well as of differences to other books and authors. Important terms, such as *aggregation* and *agglomeration*, are explained in the glossary A.1.

References

B. Mandelbrot, *Fractals, form, chance, and dimension*, 1st edn. (Freeman & Co, San Fransisco, 1977). ISBN 0-716-70473-0

J. Perrin, Mouvement brownien et realité moléculaire. Ann. Chim. Phys. **8**(18), 1–114 (1909)

H. Siedentopf, R. Zsigmondy, Über Sichtbarmachung und Größenbestimmung ultramikroskopischer Teilchen, mit besonderer Anwendung auf Goldrubingläser. Ann. Phys., IV **10**(1), 1-39 (1903). doi:10.1002/andp.19023150102

Bibliography

K. Edelmann, Lehrbuch der Kolloidchemie, vol. I (Verlag der Wissenschaften, Berlin, 1962), pp. 1–7

W. Ostwald, Über die andere geschichtliche Wurzel der Kolloidwissenschaft. Kolloid Z 84(3), 258–265 (1938). doi:10.1007/BF01518685

B. Vincent, Early (pre-DLVO) studies of particle aggregation. Adv. Colloid interface Sci. **170**(1–2), 56–67 (2012). doi:10.1016/j.cis.2011.12.003

S. Wall, The history of electrokinetic phenomena.Curr. Opin. Colloid Interface Sci. **15**(3), 119–124 (2010). doi:10.1016/j.cocis.2009.12.005

Additionally, the reader is referred to the biographic papers cited in Sect. A.2.

Chapter 2
Characterisation of Colloidal Suspensions

Working with colloidal suspensions means to adjust or to monitor certain properties of the single particles and of the whole disperse system. This may concern the generation and modification of colloidal particles, the preparation and processing of suspensions from such particles, or the observation of the colloids' behaviour in physiological and environmental media. Each of these tasks requires appropriate techniques for the determination of the relevant particle or suspension properties. There are various potentially interesting properties which may be classified as follows (Polke et al. 2003; Stintz et al. 2010):

- particle morphology (size, shape, structure)
- particle properties based on hydrodynamics or specific interactions with external fields that relate to morphology (e.g. diffusion, settling, scattering)
- surface area of the dispersed phase
- interfacial properties which depend on solvents (e.g. surface charge, zeta-potential)
- concentration of particles (e.g. mass or number concentration, total or per size fraction)
- suspension properties (e.g. turbidity, viscosity, stability)
- bulk properties of the dispersed phase and its interaction with the solvent (e.g. solubility and dissolution kinetics)

Each point corresponds to numerous measurands and even more characterisation techniques. In practice, however, only a few parameters are important for evaluating process performance or product quality. This chapter will focus on characterisation techniques that allow for a quantification of particle size and aggregate structure, which are of fundamental importance in describing any colloidal suspension. Besides this, relevant techniques for the quantification of interfacial properties are presented.

F. Babick, *Suspensions of Colloidal Particles and Aggregates*,
Particle Technology Series 20, DOI 10.1007/978-3-319-30663-6_2

2.1 Particle Size Distribution

Particle size is one of the most important parameters when describing colloidal suspensions. It is the only criterion for classifying a particle as a *colloidal particle* or even *nanoparticle*, it changes during aggregation processes or during dispersion and comminution and it is related to other properties like surface area or settling velocity. For this reason, particle size serves as one of the fundamental parameters when characterising colloidal suspensions.

However, the term *particle size* does not describe an unambiguous quantity, but rather a variety of measurands which are related to the outer particle dimensions. Indeed, particle size is always derived from a geometrical or physical property. If the property is not a length, it is usually converted to a diameter of a sphere being equivalent to the particle with regard to this property. The corresponding diameter is called equivalent diameter (e.g. equivalency in volume V leads to volume equivalent diameter x_V). Table 2.1 lists some of these properties and the associated equivalent diameters.

In general, particle properties vary within a particle system and the particle size x should be regarded as a distributed parameter. Size distributions can be depicted by the cumulative function Q_r, the density function q_r, or the transformed density function q_r^*:

$$\mathrm{d}Q_r = q_r \mathrm{d}x = q_r^* \mathrm{d}\ln x = \frac{\text{amount of particles in } x \ldots x + \mathrm{d}x}{\text{total amount of particles}}, \tag{2.1}$$

where the index r characterises the type of quantity[1] which is employed to weigh the individually size fractions (Rumpf and Ebert 1964; Leschonski 1984; ISO 9276-1 2004).

The details of size distributions are frequently summarised in a few parameters that reflect the "average size" or the polydispersity of the particle system:

- modal value(s) $x_{\mathrm{mod},r}$ or $x_{\mathrm{mod}*,r}$ of the density functions q_r and q_r^*, respectively
- median $x_{50,r}$: of Q_r
- mean size $\bar{x}_{k,r} = \left(\int x^k \mathrm{d}Q_r\right)^{1/k}$ requires complete knowledge of q_r or Q_r (important: $k = -1, 0, 1, 2 \rightarrow$ harmonic, geometric, arithmetic mean and rms value)
- percentile (or: quantile) $x_{p,r}$: inverted cumulative distribution, $Q_r(x_{p,r}) = p\%$
- standard deviation: $s_r = \int (x - \bar{x}_{1,r})^2 \mathrm{d}Q_r$

All three types of location parameters—modes, medians and means—are frequently used in practice; yet their meaning and robustness is not identical.

The *modal values* are characteristic values of the density functions q and q^*, which appear to be the immediate result of most sizing techniques (e.g. line-start disc centrifugation, spectroscopic techniques). Modal values represent the most

[1] A cumulable quantity like volume, number, or scattering intensity of a dilute particle system.

Table 2.1 Particle properties, in brackets the associated equivalent diameters (if existing)

Geometric properties	Properties related to mobility	Interaction with external fields
Volume (x_V)	Settling velocity (x_{Stokes})	Weight/inertia (x_m)
Surface (x_S)	Translational diffusion coefficient ($x_{\text{h,t}}$)	Extinction cross section with regard to light or sound
Characteristic length at 2D projection (e.g. x_{Feret})	Acoustophoretic mobility (x_{ap})	Scattering cross section with regard to light or sound

frequent particle size(s) and, thus, bear a relatively low statistical uncertainty (provided the sizing technique allows to resolve modal values). Moreover, they are usually not affected by the relatively large measurement uncertainties of rare size fractions or by a lacking sensitivity to very fine or very coarse particles. Note, that $x_{\text{mod},r}$ and $x_{\text{mod}*,r}$ are not identical and that the latter—derived from q_r^*—is usually employed for very broad size distributions. Last but not least, modal values do not offer an unambiguous characterisation, since a size distribution can consists of several modes.

Median values split the particle population into a fine and a coarse fraction of equal weight (in the selected type of quantity). They are obtained from the cumulative function Q_r, which commonly has to be computed from the originally measured density functions q or $q*$. Only a few methods directly reflect the cumulatively function Q_r in their results (e.g. cuvette centrifugation). Unlike modal values, the median is adversely affected if the weights of certain size classes, in particular at the lower and upper end of the size distribution, are not accurately quantified. However, the exact knowledge of, e.g., x_{min} or x_{max} is not required. Medians are unequivocally defined and reliably reflect global changes in the size distribution.

A *mean* particle size corresponds to the expectation value of a certain power k of x, i.e. to a certain moment of the distribution function. It is affected by the main modes of distribution as well as by the rare fractions at the ends of the size distribution. In the case of arithmetic means all size classes are equally weighted (with given type of quantity), whereas quadratic means (rms values) are more sensitive to coarse particles, and harmonic means are more affected by the fine ones. The accurate calculation of means values from a distribution function, therefore, requires the exact knowledge of q_r for the total size range including the size and frequency of the finest and coarsest size fractions. The major advantage of mean values is that they correspond to effective properties of the dispersed phase (e.g. to its specific surface area, or to the suspension turbidity). They can, thus, be measured directly without the necessity of resolving the size distribution. Such values (e.g. the intensity weighted harmonic mean $x_{-1,\text{int}}$ or x_{cum} in the case of dynamic light scattering) can be measured with higher reliability and lower uncertainty than even modal or median values.

Other characteristic size parameters of the distribution function are the *percentiles* (also: *quantiles*) other than $x_{50,r}$. They are commonly used to quantify the polydispersity, either as indicators for the minimum or maximum particle size (e.g. by $x_{10,r}$ and $x_{90,r}$, respectively) or as the difference between high and low order percentiles (e.g. $x_{90,r}$–$x_{10,r}$). In contrast, the standard deviation s_r is rarely computed for a size distribution q_r. Yet, it may be employed as a regression parameter when describing a given measurement result by means of an analytical distribution function.

Note that all mentioned distribution parameters are subject to the specific type of quantity in which the size distribution is weighted. Conversion between different types of quantity requires (i) models that correlate the detected particle properties with the types of quantity and (ii) the accurate quantification of q_r over the entire size range. Solving the two points is prerequisite for comparing measured size distributions. They can be, therefore, considered as fundamental challenges to any particle sizing technique.

2.2 Particle Sizing

The strict differentiation of particle size distributions with regard to the particle property and the type of quantity is not just an academic issue but indispensable when coping with the huge variety of particle sizing techniques. There are several means and physical phenomena that allow for the determination of size distributions (e.g. imaging, sedimentation, extinction). A more general distinction of particle sizing techniques is based on how the weights of the individual size fractions are determined (Fig. 2.1; cf. Stintz 2005; Stintz et al. 2010):

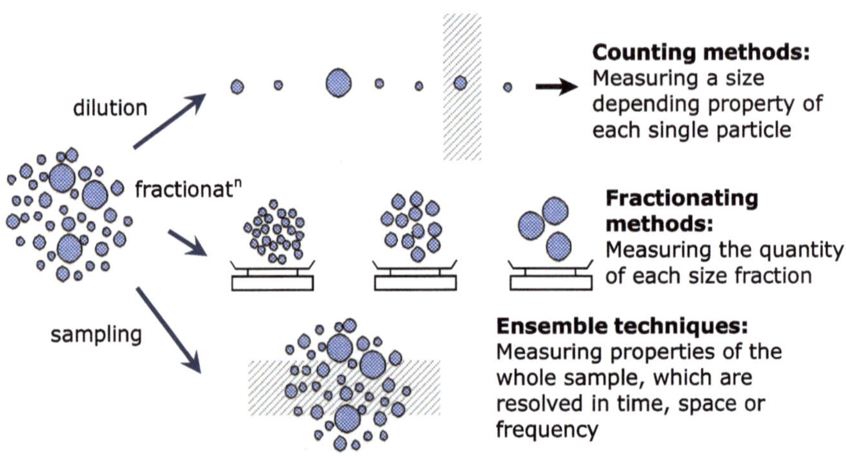

Fig. 2.1 Classification of particle sizing techniques

- counting techniques (measuring particle properties at individual particles)
- fractionating techniques (measuring the amount or concentration of size/property classes after fractionating the particle system)
- spectroscopic ensemble techniques (measuring the spectral or parametric response of a representative particle ensemble of the total particle system)

Counting techniques inherently yield number weighted distributions (Q_0) of a certain particle property or of a physical quantity that is related to a certain particle property (e.g. the average displacement as a measure of the diffusion coefficient). They rely on the individualisation of the particle sample, which can be either achieved by analysing microscopy images (e.g. from electron microscopes) or by sufficient sample dilution.[2] The number of detected particles per each size class is a stochastic quantity that obeys a Poisson distribution. That means that the relative uncertainty in class frequency is inversely proportional to the square root of particle counts. The probed particle property may be either geometric (in particular for image analysis), optical (e.g. scattering cross section), or related to mobility (diffusion coefficient).

Fractionating (ensemble) techniques include the two steps of fractionation and detection. The former can either result in a physical separation of the different size classes or in the depletion of coarse or fine particles in the measurement zone. In the case of colloidal suspensions, the fractionating effect is usually related to the mobility of the particles (e.g. settling velocity). The detection system monitors the fractionation process and, thus, serves for evaluating the class frequencies. It frequently employs the phase shift, extinction, or scattering of some radiation (e.g. X-rays). The applied detection system determines the type of quantity in which the size fractions are intrinsically weighted (e.g. extinction of X-rays is mass proportional → Q_3).

The immediate result of a spectroscopic ensemble technique is a signal spectrum, i.e. the variation of the measured signal g over the spectral parameter s (time, space or frequency). Each size fraction x possesses a characteristic spectrum $k_r(s,x)$, which in general covers the whole spectral range. Assuming that each size fraction contributes independently and linearly to the measured signal spectrum, the determination of the size distribution requires the inversion of a linear integral equation (Fredholm type):

$$g(s) = b(s, c_r) + c_r \cdot \int k_r(s, x) q_r(x) \mathrm{d}x, \tag{2.2}$$

where c_r is the particle concentration for the type of quantity r and $b(s, c_r)$ is a size-independent baseline signal. When s is varied over a sufficiently wide range, the spectrum can be normalised, which eliminates the concentration influence.[3]

---

[2] "Sufficient" means a vanishing likelihood of having two or more particles in the measurement volume.

[3] Note that the impact of concentration c_r is lost if $g(s)$ refers to the phase shift rather than to the magnitude of a measured quantity.

Unlike counting or fractionating techniques, the *intrinsic type of quantity* is not necessarily obvious. Indeed, Eq. (2.2) can be expressed and solved for any type of quantity r. Yet, the intrinsic type of quantity of spectroscopic techniques refers to the impact of a single particle to the integrated signal spectrum.

If the integral over the signal spectrum of one particle is proportional to the particle volume, the intrinsic type of property is the volume ($r = 3$); proportionality to the partial scattering cross section implies that the signal is intrinsically weighted by scattering intensity.

Ideally, if the value of the intrinsic type of quantity is kept constant, a variation in size will only shift the spectrum $g(s)$ along the s-axis without changing its height or shape (Babick and Ripperger 2002). Note that in measurements only a limited range of the spectral parameter s is covered, which may affect the real sensitivity to certain size fractions. The probed particle property of the spectroscopic technique can be easily described by the terms of Table 2.1. However, a profound correlation to morphological parameters is much more complicated.

The inversion of Eq. (2.2) is an essential feature of spectroscopic techniques, which requires appropriate numerical algorithms (e.g. Twomey 1977; Kandlikar and Ramachandran 1999). Moreover, any measured spectrum bears only limited information on the size distribution, i.e. only a few details of size distributions are accessible by the measurement (Fig. 2.2). This results from the discreteness and the restricted range of the spectral parameter s as well as from the uncertainty of the measured signal g. The information content is further affected by the physics of the measurement principle, which is expressed by the kernel function $k_r(s, x)$: the more structured the shape of this function, the higher the achievable resolution of the size distribution $q_r(x)$. The information content can be quantified, but just for a given set of instrument parameters. Such data have only been published for a few techniques (dynamic light scattering: Finsy et al. 1992; electroacoustic mobility spectroscopy: Knösche 2001; ultrasonic spectroscopy: Babick and Ripperger 2002).

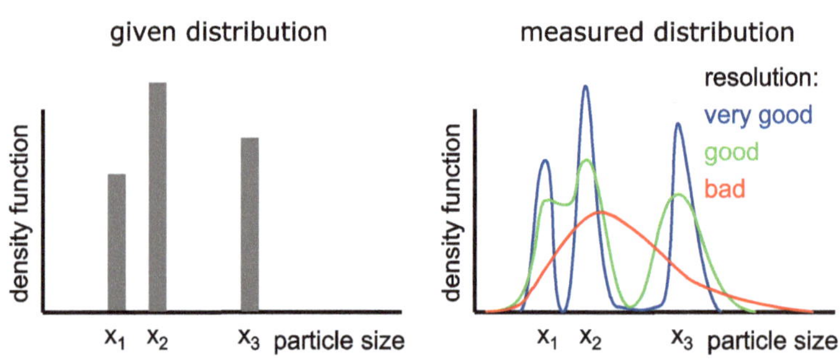

Fig. 2.2 Schematic representation of the meaning of resolution at the example of a tridisperse size distribution: truth and possible measurement results; low resolution results from low information content in the case of spectroscopic techniques

Inversion is not an issue for most counting and fractionating techniques, where the immediate measurement results can be ideally considered as scaled density and sum functions of the size distribution. However, when counting techniques measure a physical property that does not unambiguously correspond to size or when a fractionation is not perfect, the integral Eq. (2.2) also has to be applied.

In summary, every particle sizing technique is characterised by the particle property probed and its intrinsic type of quantity. Experimental size distributions, which are commonly derived for the assumption of spherical particles, may bear very different physical meanings and most often disagree (e.g. the volume weighted distribution Q_3 of the Stokes diameter x_{Stokes} differs from the number weighted distribution Q_0 of the hydrodynamic diameter x_{h}). For the purpose of comparison, it is in theory possible to recalculate measured distribution functions by converting the particle properties and the type of quantity. However, such a conversion requires models on the correlation between the different particle properties and the different types of quantities, which are available only for few ideal particle shapes. Moreover, the limited resolution of experimental size distributions and the uncertainty in class frequencies affect the quality of such a conversion as well (e.g. the uncertainty about the coarse particle fraction in a number weighted distribution is over-proportionally magnified when converted to volume weighted size distributions, Witt et al. 2007). Last but not least, the comparability of differently measured size distributions may be impeded by specific limitations of the covered size ranges. Such limitations occur when signals of very coarse or very fine particles cannot be separated from noise (e.g. negligible turbidity of non-absorbing nanoparticles may set a lower limit of detection for particle sizing with analytical photocentrifugation). Additionally, they can be related to a rapid migration of particles out of the measurement zone (e.g. due to sedimentation in light scattering experiments). Hence, a perfect agreement of experimental size distributions is illusory even after conversion. Yet, the intrinsic differences between the measurement methods offer the opportunity to study different aspects of particle systems (e.g. selective analysis of coarse particles) or to derive quantitative information on the shape of particles and structure of particle aggregates.

The inevitable imperfection of particle sizing compels a cautious interpretation of measured size distributions. Any computation or any conversion increases the uncertainty. Hence, data should be presented in a way that best resembles the original measurement with regard to the particle property, the type of quantity, and the distribution function (Q_r, q_r or q_r^*). Highest accuracy can be expected for the characteristic location parameters, like $x_{\mathrm{mod},r}$ or $x_{50,r}$, whereas the values of minimum and maximum particle sizes are least reliable. The polydispersity is generally less accurately accessible than the location parameters. This holds particularly true for spectroscopic (ensemble) techniques, where the details of the size distribution (e.g. number of modal values/sizes) significantly depend on the employed inversion algorithm and its parameterisation (Stock and Ray 1985; Riebel and Löffler 1989; Koo and Hirleman 1992; Babick and Ripperger 2002).

2.3 Selected Characterisation Techniques

This section gives a brief survey on characterisation techniques that can be used to determine size distributions and interfacial properties of colloidal suspensions. They are grouped according to the classification scheme in Fig. 2.1 (counting, fractionating, and spectroscopy) and the main physical principle (e.g. imaging, scattering). Some methods that are considered most relevant for colloidal suspensions are discussed more extensively than others. The survey starts with the ultramicroscopy; not only because it belongs to the counting methods, but because of its historical relevance for colloid science. The section is concluded by a comparison of the major characterisation techniques.

2.3.1 Ultramicroscopy

Fine colloidal particles are usually smaller than the spatial resolution of an ordinary light microscope, which means that they are invisible with regard to an affine projection. However, when colloidal particles are laterally illuminated by very intense light against a dark background (dark field microscopy), it is possible to see the scattering patterns with an optical microscope (Fig. 2.3). Since such an instrument facilitates the indirect visualisation of fine particles beyond the optical limit, it was named *ultramicroscope*.

The first ultramicroscope was developed by the Austrian-Hungarian physicist Richard Zsigmondy and the German technician Henry Siedentopf. It was based on an illumination with sunlight, which was focussed and collimated by using a pair of fine slits (Siedentopf and Zsigmondy 1903; cf. Fig. 2.4). Later, Zsigmondy proposed several optical modifications (e.g. immersion objectives) which increased the illuminance and, thus, improved the resolution (Zsigmondy 1913; Zsigmondy und Bachmann 1914).

The invention of the ultramicroscope finished a long debate of the nineteenth century on the true composition of "colloidal solutions" as it eventually disclosed their heterogeneous, disperse nature. For this reason, this technique attracted

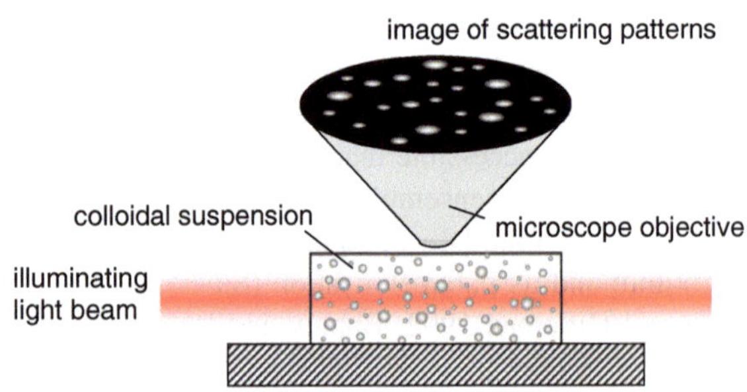

Fig. 2.3 Principal set-up of the ultramicroscope

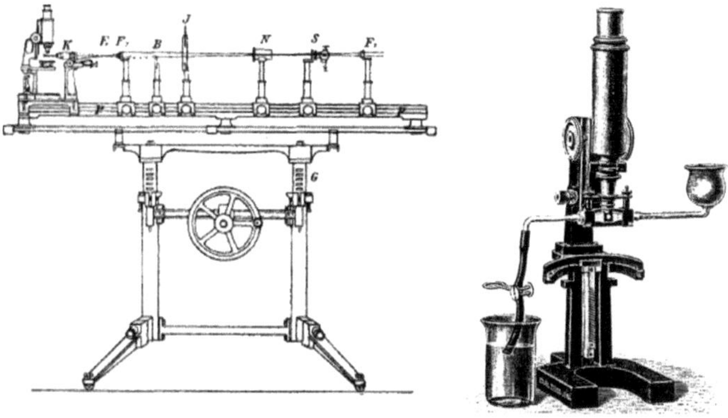

Fig. 2.4 Slit ultramicroscope (original drawing from Siedentopf and Zsigmondy 1903) and advertisement for an commercial instrument by ZEISSWERK Jena from 1907 (© Zeiss Archiv, Jena)

attention by many scientists, in particular since the discontinuity of matter and the statistical interpretation of thermodynamic phenomena had not yet been universally accepted at that time (Bigg 2008).

The instrument soon became a basic tool for colloidal scientists; it raised colloid science from a speculative theory to an independent academic discipline. The Nobel prize that was awarded to Zsigmondy in 1925 explicitly acknowledges his achievements "for elucidating the heterogeneous nature of colloidal solutions as well as for the thus employed methods, which are fundamental to modern colloid chemistry".

Characterisation of Colloidal Suspensions by Ultramicroscopy

Already in the first publication on the ultramicroscope, Siedentopf and Zsigmondy (1903) described its application to the characterisation of colloidal systems, specifically of gold ruby, i.e. glasses in which colloidal gold particles are embedded. They showed that neither the diameter nor the colour of the diffraction patterns correlated to particle size. Instead, they estimated the size of the gold particles from the particle number concentration (obtained by counting the illuminated particles within a defined volume) based on the known mass concentration of gold. Reissig (1908), however, pointed to the fact that there is a non-linear correlation between the gold content and the detected particle number concentration. This results from the fact that the scattered light of coarse gold particles completely surpasses the scattering signals of neighbouring fine particles. For this principal reason it is difficult to accurately determine the particle number concentration by this measurement technique.

When ultramicroscopy is used for particle sizing nowadays, one evaluates the Brownian motion of the scattering centres (i.e. particles). According to Einstein (1905), the mean square displacement $\overline{\Delta r^2}$ is proportional to the translational diffusion coefficient D_t:

$$\overline{\Delta r^2} \sim D_t \cdot t. \tag{2.3}$$

The diffusion coefficient, in turn, is inversely proportional to the translational hydrodynamic diameter $x_{h,t}$ (Stokes-Einstein equation):

$$D_t = \frac{k_B T}{3\pi \eta x_{h,t}}. \tag{2.4}$$

Particle sizing, thus, involves the quantification of the trajectory lengths (Δr) of the individual scattering centres for a given time step (Δt). This measurement principle is called *dynamic ultramicroscopy* (DUM) or *particle tracking analysis(PTA)*. It was first employed by Perrin (1908, 1909) for the quantification of emulsion droplets.

While at Perrin's time the particle trajectories were recorded by means of a *camera lucida*, contemporary ultramicroscopes produce digital video sequences and evaluate them by image analysis software. The immediate results of such measurements are the average displacements $\overline{\Delta r}$ of the individual scattering objects per time step (i.e. per frame). If averaging was conducted for an infinite number of time steps, the mean displacement of a particle would be identical to the collective average. In practice, however, only a finite number of time steps is evaluated and the average displacement of particles of size $x_{h,t}$ is a stochastic quantity. The measured frequency distribution of the average displacement is, therefore, a "smeared" projection of the number weighted size distribution $q_0(x_{h,t})$:

$$p\left(\overline{\Delta r}\right) = \int P\left(\overline{\Delta r}, x_{h,t}\right) \cdot q_0\left(x_{h,t}\right) \mathrm{d}x_{h,t}, \tag{2.5}$$

where the kernel function P depends on the number of time steps. Its shape can usually be approximated by a normal distribution (Saveyn et al. 2010). The size distribution $q_0(x_{h,t})$ can be derived from the distribution of mean displacement $p\left(\overline{\Delta r}\right))$ by inversion of Eq. (2.5).

The application limits of ultramicroscopy result from the requirements that the particle distances should be much larger than the optical resolution limit of the microscope and that the scattered light of all individual particles is sufficiently strong for detection. While the former can be achieved by appropriate dilution, the latter requirement is missed for particles below a material-specific size limit.

Additionally, there are principal difficulties in detecting weak scatterers in the presence of strong scatterers. That means, though providing number weighted distributions (similar to ordinary microscopy), the dynamic ultramicroscopy has a bias to strongly scattering particles (similar to dynamic light scattering; cf. Domingos et al. 2009). This concerns very broad size distributions as well as multi-component particle systems. On the other hand, the method's high sensitivity to strong scatterers can be used to evaluate the homogeneity or the possible contamination of a colloidal suspension.

2.3.2 Imaging Methods

Images of the particle systems give a clear idea about their morphological characteristic and are of the highest persuasiveness. Apart from a qualitative morphology description, they can be employed for quantitative analyses of size, shape, and structure as well. Particle images may be obtained from light, electron, or scanning force microscopy.

Light microscopy is the oldest and the simplest imaging method. However, the lower resolution limit for conventional light microscopy lies above 200 nm. Its application to colloidal particles is, therefore, restricted to rather large colloids and aggregates of them. Even though recent developments, like stimulated emission depletion (STED) microscopy, have shifted the optical resolution limit below 100 nm (Hell 2007), light microscopy is not really relevant for the characterisation of colloidal suspensions.

With regard to resolution, *electron microscopy* provides a powerful alternative which even allows the detection of single atoms. First instruments were developed in the 1930s by Ruska (e.g. Knoll and Ruska 1932), who later received the Nobel Prize in physics for this achievement, and von Ardenne (1938). Two fundamental types are distinguished:

- the *transmission electron microscope* (TEM), which is operated in wide-field irradiation mode and for which the image results from the local disturbance of the electron field when it passes through an ultra thin specimen, and
- the *scanning electron microscope* (SEM), which employs a very fine electron beam that scans the surface of the specimen and creates the image from backscattered primary electrons and emitted secondary electrons.

In principle, the TEM can be operated in a scanning mode, too—this modification is called transmission scanning electron microscopy (TSEM).

Traditional electron microscopy is conducted in high vacuum, which imposes specific efforts to sample preparation. Particles from colloidal suspensions have to be deposited onto an appropriate substrate (e.g. on carbon or silica films) and dried. Alternatively, the suspensions can be shock-freezed and particles are subsequently excavated from the continuous phase by special cryo-preparation techniques (Schmidt et al. 1994, pp. 694–705). The sample preparation can be considerably reduced with environmental scanning electron microscopes (ESEM), which are operated up to 1000 Pa and, thus, even facilitate the analysis of wet surfaces. However, the ease in operation is at the expense of resolution (Danilatos 1993).

A further technique able to image colloidal particles is the *scanning force microscopy*, which was developed by Binnig and his co-workers in the 1980s (e.g. Binnig et al. 1986). The scanning force microscopy (or: *atomic force microscopy*) and its related techniques are based on the interaction between a very fine probe tip with the atoms or molecules at the surface of the sample (Giessibl 2003; Danzebrink et al. 2006; cf. Fig. 2.5). This can be used to resolve surface morphologies or particles on a substrate with lateral resolution of 0.1–10 nm (axial resolution: 10 nm).

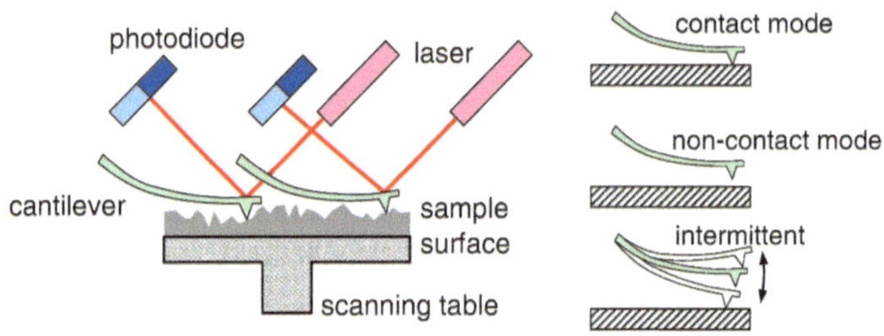

Fig. 2.5 Schematic set-up of scanning force microscopy and modes of operation; variation in surface properties are detected via the deflection of the cantilever with probing tip

The SFM is usually employed for the characterisation of films and surfaces (e.g. roughness), whereas the morphological characterisation of particles is of minor relevance. Its real strength is the sensitivity to the forces between probe and sample, which allows an evaluation of surface chemistry (e.g. functional groups, hydrophobicity) and the quantification particle interactions, or interactions between particles and surfaces (e.g. adhesion, friction; Heim et al. 1999; Butt et al. 2007).

The results of imaging methods are (mainly) number weighted distributions. That means that the sample size (number of probed particles) should be sufficiently high for ensuring low uncertainty in class frequencies. Moreover, the sample size required to achieve a certain confidence level increases with polydispersity. The accuracy of the measured particle properties depends on a variety of factors (e.g. magnification or spatial resolution of the scanning mode, or image processing). Most crucial, however, is the representativity of the imaged particles for the whole particle system. That requires that the particle deposition on the substrate is neither size-selective nor inhomogeneous (Fiala et al. 2011). In general, sample preparation is a key issue for imaging methods.

2.3.3 Fractionating Sizing Techniques

Fractionating sizing techniques combine a size-related classification process with the measurement of particle quantities. The classification may yield a physical separation of differently sized particles (e.g. by sieving, cf. Fig. 2.1) or it may successively deplete the disperse system of the coarsest or finest particles (e.g. in a sedimentation column). Accordingly, the measured quantities, which can be absolute amounts or concentration values, represent either a density or a cumulative function of the size distribution. Anyhow, the type of quantification (e.g. weighing) determines the type of quantity of the measured size distribution (e.g. mass), whereas the classification defines the probed particle property. The classification process should be ideal (i.e. with maximum selectivity) and well-defined by a monotone correlation between the particle property (e.g. settling velocity) and the parameter(s) of the classification process (e.g. settling time). This relationship is either known from first principle or has to be determined by means of calibration. If both approaches fail to provide sufficiently accurate size information, the concentration measurement can be supplemented with a spectroscopic sizing technique. This notably increases the costs of analysis, yet finally provides accurate and highly resolved particle size distributions. Even more, such a measurement setup—classification followed by particle sizing technique(s)—facilitates, in principle, the experimental correlation between different particle properties and, thus, offers an opportunity to quantify particle shape and structure.

This section addresses two basic principles of classification—*sedimentation* and *field-flow fractionation*—and the corresponding sizing techniques. Additionally, a chromatographic technique is briefly introduced. The focus lies on sedimentation or centrifugation analysis, which corresponds to its practical relevance for the characterisation of colloidal suspensions.

2.3.3.1 Analytical Sedimentation and Centrifugation

Analytical sedimentation counts among the most popular techniques of particle sizing. To a major extent, this can be attributed to the good comprehensibility of the employed classification principle: Isolated particles settle in a quiescent liquid with a stationary velocity, which solely depends on their individual size, shape, and density. The corresponding changes in local particle concentration, which are observed along the settling path and over time, therefore reflect the size distribution of the particle system (Edelmann 1962, pp. 74–86; Bernhardt 2010).

This basic concept can be varied with respect to the force field, the concentration measurement, or the mode of operation. For that reason, several types of sedimentation instruments have been developed (Leschonski 1982); those that are based on centrifugation are of particular relevance for colloidal suspensions. The set-up of an analytical centrifuge for colloids was first described by Svedberg and co-workers (Svedberg and Nichols 1923; Svedberg and Rinde 1924). They called the instrument "ultra-centrifuge" and successfully tested its performance for the size analysis of colloidal gold sols. Nowadays, two fundamental types of centrifuges are distinguished: disc centrifuges and cuvette centrifuges.

Disc centrifuges consist of a hollow disc ("tyre") which contains the suspension medium. Upon rotation, the liquid forms a stagnant layer on which a thin layer of the particle system is injected (line-start technique, cf. Fig, 2.6). The particles migrate according to their settling velocity to the bottom of the disc.[4] All particles of a certain size (or settling velocity) move in a narrow band with growing distance from the initial position. In the case of multidisperse particle systems, one can observe several of such bands in analogy to chromatographic techniques. The radial

[4]For the sake of convenience only the case $\rho_p > \rho_m$ is considered throughout this section.

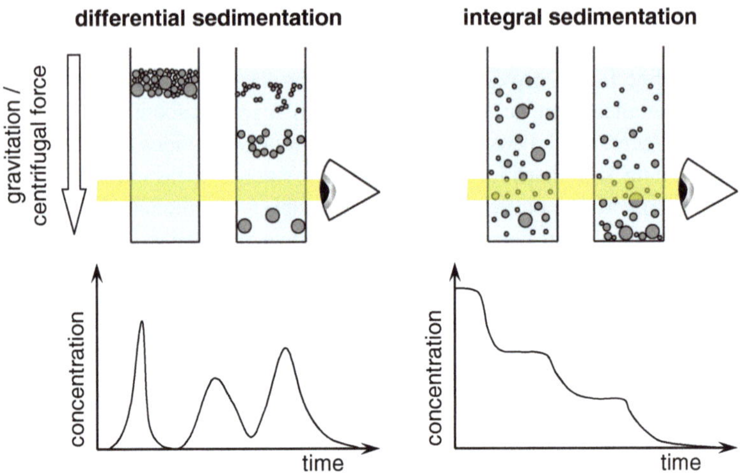

Fig. 2.6 Principles of differential sedimentation (line-start technique) and integral sedimentation (homogeneous technique); the former is employed for disc centrifuges, the latter for cuvette centrifuges; below: the corresponding time-curves of local particle concentration

concentration profile is, hence, a distorted projection of the density function of the size distribution ($q(x_{\text{Stokes}})$). The line start technique requires a density gradient in the suspension medium (e.g. by sugar) before the particles are injected; otherwise there was a convective transport of particles within strands of the (heavy) suspension layer. The density gradient sets some practical limits to the measurement (e.g. duration) and has to be calibrated before conducting size measurements.

A different set-up and even mode of operation is found in *cuvette centrifuges*, where the particle sedimentation is observed in small cuvettes that are fixed on a rotating table. In this case, the particles are homogeneously suspended in the continuous phase before the centrifugation starts (homogeneous technique). During the centrifugation, all particles migrate towards the bottom of the cuvette, which results in the formation of a sediment, in a steady decline of local particle concentrations above the sediment, and in a monotone decrease of particle concentration in the direction from the sediment to the meniscus. These two types of variation in particle concentration, the temporal evolution, and the radial profile, can be considered as distorted projections of the cumulative function of the particle size distribution ($Q(x_{\text{Stokes}})$).

In order to quantify the changes in the local particle concentration, several techniques can be employed. For instance, one can quantify the growth rate of the sediment by means of a sedimentation balance. This indirect way of concentration measurement was proposed by Odén (1916) in one of the earliest papers on analytical sedimentation. In contrast, modern analytical centrifuges directly probe the concentration in the suspension phase. Four important principles of detection are discussed below.

(a) Optical extinction

The extinction of light is a characteristic behaviour of particle systems, which is governed by particle size and concentration—apart from material properties. For dilute, weakly scattering particle systems, the extinction obeys Lambert-Beer law, which states a linear dependency between turbidity τ and particle concentration:

$$\tau = E/L = -\ln T = c_N C_{\text{ext}}, \tag{2.6}$$

where T denotes the transmission of the suspensions, E the natural extinction, c_N the particle number concentration, and C_{ext} the (average) extinction cross section of the particles. The latter is a function of particle size, shape, and optical properties. In general, it is not possible to unambiguously relate the light extinction in suspensions to a geometrical particle property. However, in the case of very fine colloids at the nanoscale, one can apply the scattering rules of the Rayleigh limit (Appendix B.2.2). When such particles absorb light (e.g. soot particles or metal colloids), the extinction cross section C_{ext} is proportional to the particle volume and the turbidity is a multiple of the particle volume concentration. Non-absorbing materials, on the other hand, extinguish light according to the squared particle volume:

$$\text{Rayleigh limit, absorbing particles}: E \sim c_N x^3 \sim c_V, \tag{2.7}$$

$$\text{Rayleigh limit, non-absorbing particles}: E \sim c_N x^6 \sim c_V x^3. \tag{2.8}$$

That means, an optical detection of sedimenting Rayleigh scatterers leads to either volume weighted (q_3 or Q_3) or squared volume weighted distributions (q_6 or Q_6) of the Stokes diameter x_{Stokes}. For the general case, one should prefer the term extinction weighted size distribution (e.g. $Q_{\text{ext}}(x_{\text{Stokes}})$), even though this term is meaningless without additional pieces of information on the employed radiation and the optical properties. Optical extinction is probably the most popular quantification principle for analytical centrifugation. However, its sensitivity to very fine particles is rather low for non-absorbing materials, which means that these size fractions may be disregarded in the case of large polydispersity.

(b) Extinction of X-rays

The extinction of X-rays in a colloidal suspension follows the same fundamental rules as the extinction of light. Nonetheless, its separate treatment is quite common, since the sensor techniques differ considerably. Besides this, X-rays have much shorter wavelengths (approx. 0.01–0.1 nm) than visible light and hardly interact with matter ($|m\text{-}1| \ll 1$, cf. Sect. 2.3.4). The scattering of X-rays with colloidal particles can, therefore, be considered as Rayleigh-Debye-Gans scattering with significant absorption (van de Hulst 1981, p. 85), for which the extinction is proportional to volume concentration:

$$\text{extinction of X-rays at colloids}: E \sim c_N x^3 \sim c_V. \tag{2.9}$$

Accordingly, X-ray sedimentometers yield volume weighted size distributions (q_3 or Q_3). However, X-ray absorption is strongly affected by the subatomic structure of the involved elements. The mass specific absorption cross section grows approximately with the third power of the atomic number for light elements (up to potassium), when typical X-ray sources (i.e. with emission energies of several keV) are employed.[5] An X-ray detection may, therefore, be inappropriate for the characterisation of organic substances. Instrumental aspects are, e.g., discussed by Conlin et al. (1967) and Enomoto et al. (1979).

(c) (Interferometric) determination of the suspension's refractive index
The interaction of light with colloidal particles does not only affect the intensity of the transmitted beam, but also its phase velocity, and, thus, the refractive index of the suspension (m_{sus}). This quantity depends linearly on the particle concentration for dilute suspensions (appendix B.2):

$$m_{\text{sus}} - m_{\text{m}} \approx c_V \cdot \frac{\partial m_{\text{sus}}}{\partial c_V} \approx c_m \cdot \frac{\partial m_{\text{sus}}}{\partial c_m}, \tag{2.10}$$

where m_{m} is the refractive index of the suspension medium. The increment $\partial m_{\text{sus}}/\partial c_V$ ($\partial m_{\text{sus}}/\partial c_m$) is, in general, a function of the particle size. Yet, the size dependency vanishes for very fine, nanosized colloidal particles, in which case the shift in refractive index is a measure of the change in particle volume concentration and the detected size distribution becomes, therefore, volume weighted (q_3 or Q_3). Interferometric detection systems are mainly employed for analytical ultracentrifuges (e.g. Schilling 1999).

(d) Manometric determination of the hydrostatic pressure
The hydrostatic pressure results from the weights of the continuous and dispersed phase and can, thus, serve as a measure for the particle mass or volume concentration. In the context of analytical sedimentation, it was already utilised by Ostwald and Hahn (1922), who quantified the rate of sedimentation of flocculated suspensions by means of a hydrostatic pressure gauge. More recent papers report on the manometric determination of the hydrostatic pressure in analytical cuvettes centrifuges with electronic pressure transmitters (Bickert 1997; Beiser 2005). In contrast to the detection systems portrayed above, these *manometer centrifuges* do not measure a local particle concentration, but the total mass of all particles that are suspended above the point of measurement. The cumulative function of the volume weighted size distribution ($Q_3(x_{\text{Stokes}})$) can be, thus, computed from the time derivative of the hydrostatic pressure. In that regard, the manometric detection shows similarity to the sedimentation balance.

[5]Comprehensive tables on X-ray absorption are e.g. provided by Henke et al. (1993).

The various analytical sedimentation techniques differ not only with respect to the immediate measurement results (scaled density or cumulative function, type of quantity), but also with regard to their practical limits of application. Cuvette centrifuges, for instance, require sufficiently high particle concentrations, which yield extinction signals well above the noise for finely resolving the steady decrease in concentration. On the other hand, the particle concentration should be low enough to exclude multiple scattering and hydrodynamic concentration effects.

A reasonable value for the initial sample transmission lies typically in the range between 5 and 95 %. Let us assume an illumination with red light (670 nm), a cuvette of 10 mm thickness, and a suspension of silica particles. An initial particle concentration of 0.1 wt% would then yield a size-dependent turbidity which is too high for particles >400 nm and which is too low for particles <70 nm. While high turbidity can be easily dealt with by dilution or utilisation of thinner cuvettes; analogous measures are restricted for the fine particles. In particular, an increase in particle concentration at this level would inevitably raise the significance of viscous coupling for the settling motion and eventually affect the mean size and the width of the measured distribution function. Note that such limits are material-specific; for titania suspensions, a lower size limit of 15 nm would be obtained for the assumed conditions and concentration, while a colloidal gold sol would be completely opaque then—irrespective of size.

The immediate result of centrifugation analysis is the temporal change or spatial distribution of the particle concentration.[6] Time and space then correspond to values of the stationary settling velocity, which can be calculated via:

$$v_{\mathrm{sett}} = \frac{a_{\mathrm{c}}\Delta\rho}{18\eta} x_{\mathrm{Stokes}}^2 \approx \frac{a_{\mathrm{c}}\Delta\rho}{18\eta} \frac{x_V^3}{x_{\mathrm{h,t}}} (\text{with}\, a_{\mathrm{c}} = \omega^2 r) \tag{2.11}$$

for friction-dominated settling (i.e. Re < 1). This condition holds true for the centrifugation of virtually all colloidal suspensions. The equivalent diameter of settling is called Stokes diameter x_{Stokes} and is related to the volume and hydrodynamic equivalent diameters. Values of the settling velocity are meaningful only when reference is made to the centrifugal acceleration. Alternatively, the settling behaviour of particles can be described by the sedimentation coefficient according to Svedberg:

$$s = \frac{v_{\mathrm{sett}}}{\omega^2 r} = \frac{\Delta\rho}{18\eta} x_{\mathrm{Stokes}}^2. \tag{2.12}$$

During centrifugation, the settling particles are steadily accelerated in the centrifugal field. As a result, the sedimentation distance Δr grows exponentially with time:

[6] An alternative to the analysis of time curves or radial profiles is to evaluate the shift of the radial profile with time (e.g. Salinas-Salas 2007, pp.76–80; Paciejewska 2010, pp. 62–64).

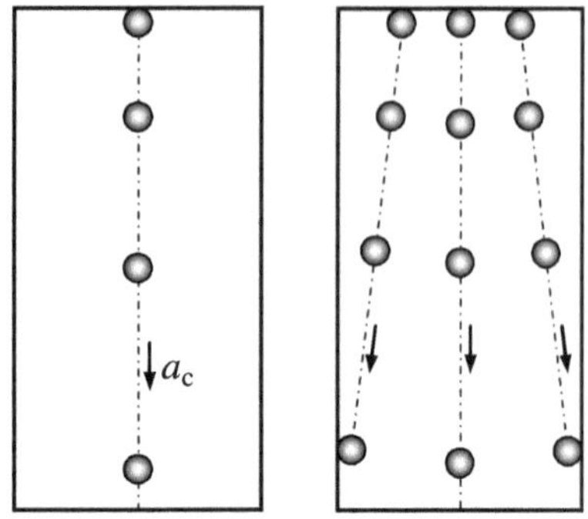

Fig. 2.7 Particle depletion in centrifugal fields due to radial acceleration and radial trajectories

$$\ln\frac{r_0+\Delta r}{r_0} = s\cdot\int\omega^2\mathrm{d}t \propto x^2 t \tag{2.13}$$

and the fractional particle concentration at a given location steadily decreases before the suspension level has reached this point. Moreover, the radial trajectories of the settling particles result in depletion as well (Fig. 2.7).

The depletion effect occurs uniformly along the radial coordinate. It is a function of the sedimentation coefficient (i.e. particle size)

$$\ln\frac{c}{c_0} = -2s\cdot\int\omega^2\mathrm{d}t \tag{2.14}$$

and has to be considered, when converting concentration values to class frequencies. Details can be taken from the literature (Kamack 1972; Leschonski 1982; Detloff et al. 2006).

When dealing with the sedimentation of colloidal particles, it is principally necessary to regard the Brownian motion of the particles, which results in diffusive particle transport and, thus, acts against the migration in the gravitational or centrifugal field. The relevance of the Brownian motion can be roughly estimated by means of a Peclet-number:

$$Pe = \frac{D_t}{v_{\mathrm{sett}}x} = \frac{6}{\pi}\frac{\mathrm{k_B}T}{a_c\Delta\rho x^4} = \frac{\mathrm{k_B}T}{a_c\Delta\rho x V}. \tag{2.15}$$

It is easily shown that for gravitational settling this Pe-number lies well above 1 for almost the complete colloidal size range (i.e. ≤ 1 μm), which means that diffusion does matter. In centrifugal fields, the critical particle size is shifted to lower values, yet even for centrifugal accelerations of $10^6\times$g this value would remain above 10 nm. The particle diffusion broadens the sedimentation profiles and leads to an overestimation of the distribution width, if not properly accounted for in data analysis. However, its influence on the concentration profiles is decreasing with ongoing sedimentation (Fig. 2.8) because the diffusive particle displacement

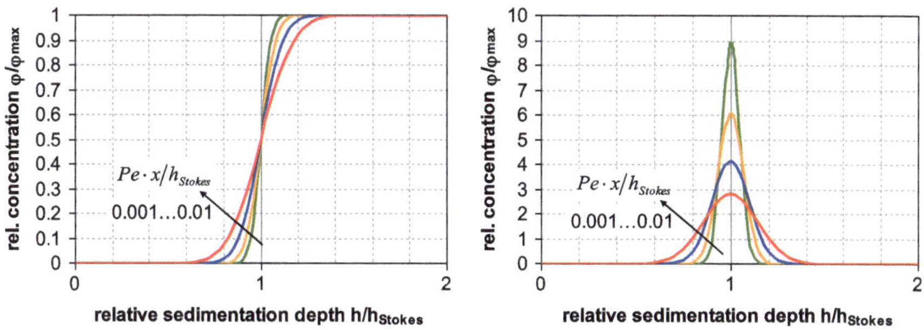

Fig. 2.8 Impact of diffusion on the concentration profiles in sedimentation experiments for a monodisperse particle system; *left* homogeneous technique, *right* line-start technique

grows with the square root of the sedimentation time, whereas the settling distance is linearly related to time.

After the sedimentation process has finished, the diffusion is still present and leads to a fuzzy interface between the sediment and the particle-free supernatant. The thickness of this layer depends on particle size. It may even extend over the whole vessel (cuvette), which means that the particles are not separated, but obey an sedimentation-diffusion equilibrium, which is characterised by an exponential concentration profile (cf. Sect. 3.1.2):

$$c(z) = c_{\mathrm{b}} \cdot \exp\left(-z/l_{\mathrm{g}}\right) \quad \text{with} \quad l_{\mathrm{g}} = \frac{\mathrm{k_B}T}{\Delta\rho V \bar{a}_{\mathrm{c}}}. \tag{2.16}$$

where c_{b} denotes the particle concentration at the bottom just above the sediment, z the distance from the sediment and $\bar{a}_c$ the average centrifugal acceleration within this distance. The evaluation of such concentration profiles offers a further opportunity to quantify the (average) particle size via centrifugation.

2.3.3.2 Field-Flow Fractionation

The term *field-flow fractionation* (FFF) refers to a class of preparation techniques which classify particle systems with regard to size in a laminar flow channel. A subsequent quantification of the particle concentrations in the eluent transforms FFF into an analytical tool for particle sizing.

Common to all FFF techniques is the application of an external "field" that acts perpendicular to the direction of flow. Particles that are injected into the channel migrate in this field and are, thus, accumulated at one wall of the channel. At the same time, they are subject to Brownian motion, which counteracts the migration. The two effects eventually cause size-dependent, stationary concentration profiles. Typically, coarse particles are less evenly distributed across the channel than fine particles. They accumulate near the channel wall, where the flow velocities are

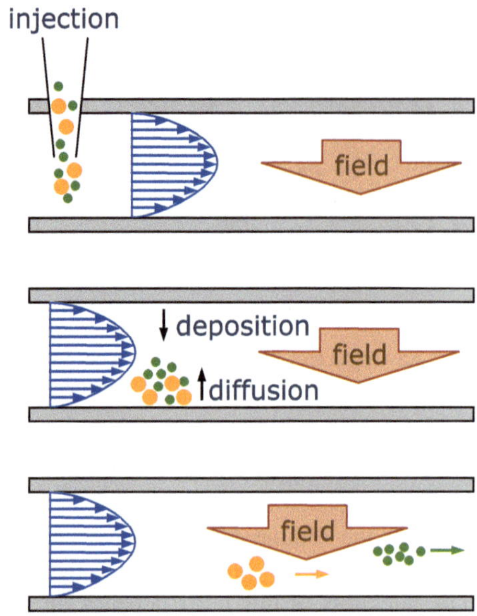

Fig. 2.9 Classification of particle systems by field-flow fractionation (FFF)

lower than at the channel centre. Hence, coarse particle need more time to leave the channel than fine ones. Due to the size-dependency of the residence time distribution, it is possible to effectively separate the different size fractions when the injection occurs as an impulse function (Fig. 2.9).

The classification into different size fractions can be realised by gravitation (sedimentation FFF), by centrifugal fields (centrifugation FFF), by thermophoresis in temperature gradients (thermal FFF), by electric fields (electrical FFF), or by hydrodynamic fields, i.e. crossflow through the wall(s) (flow FFF). Even though the main fields of application are colloidal systems, one can also employ FFF for the classification of micrometre particles ($x > 1$ μm). In that case, diffusion can be usually neglected, yet hydrodynamic lift forces and steric effects counteract the external field and cause a reversal of the size dependency.

The particle concentration of the eluent is normally measured by means of infrared or ultraviolet photometers. Additionally, fluorescence photometer, interferometric measurements (for the refractive index), or mass-spectroscopic methods (e.g. induced coupled plasma mass spectroscopy—ICP-MS, Plathe et al. 2010) are employed. The combination of different detection systems offers an opportunity for a detailed characterisation of multi-component particle systems. Note that the classification by FFF is not ideal and the relevant material properties are not always known; moreover, the calibration of FFF is rather difficult. The attribution of particle size to residence time, thus, bears some degree of uncertainty. Recent developments of FFF instrumentation, therefore, include a particle-sizing technique additional to the flow channel and the quantity measurement (usually static and dynamic light scattering, Wyatt 1998; Cho and Hackley 2010).

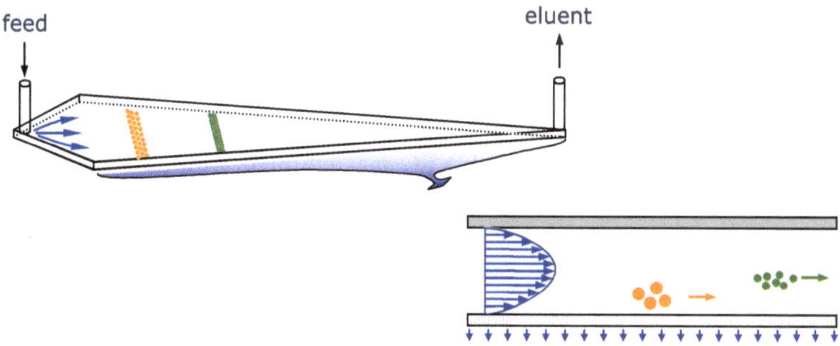

Fig. 2.10 Scheme of the asymmetric flow field-flow fractionation (AFFFF)

Field-flow fractionation was introduced as a preparation and analytic technique by Calvin Giddings (1966). In subsequent papers he and his co-workers intensively explored the FFF with regard to its different modifications, the theoretical foundation, data analysis, and possible applications (e.g. Giddings 1968; Grushka et al. 1974; Giddings et al. 1983; Yang et al. 1983). Numerous further publications by different groups since the end of the 1980s have allowed for a good estimation of the performance of FFF and the selection of optimum experimental parameters (e.g. Mori et al. 1990; Schauer 1995). A particularly promising tool for colloidal particle systems is the asymmetric flow field-flow fractionation (AFFFF, Fig. 2.10) which can be applied to a very broad range of particle sizes—from macromolecules up to micrometre particles (Cho and Hackley 2010; Laudan 2004). The classification is very effective and facilitates the analysis of protein mixtures. The decisive particle property is the hydrodynamic diameter ($x_{h,t}$). Combination with other sizing techniques, e.g. static light scattering—which delivers the diameter of gyration or the molecular weight—potentially allow insight into the particle morphology.

2.3.3.3 Size-Exclusion Chromatography (SEC)

Field-flow fractionation is closely related to chromatographic techniques, which are traditionally employed for the analysis of dissolved substances, yet can be used to a certain extent for particle characterisation as well. An example of this is the size-exclusion chromatography (SEC).

Any chromatographic technique is based on the flow through a porous medium—the stationary phase. The flow in the pores is very slow and the transport of solutes and particles is mainly diffusive. Adsorption/desorption, hydrodynamic or steric effects specifically influence the residence time of the different species and, thus, facilitate their separation. In SEC, one employs stationary media with very broad pore size distributions. Since the particles can only move into those pores that exceed their geometric dimensions, the penetrable pore volume decreases with increasing particle size. Coarse particles, therefore, pass the column more quickly than fine ones (Fedotov et al. 2011). SEC was originally developed for the separation of polymer solution, yet could be successfully applied to the characterisation of solid nanoparticles, in particular to gold nanoparticles (Wei et al. 1999; Liu 2009). Unlike FFF, the classification is related to geometric and not hydrodynamic particles properties, whereas the quantification can be conducted with the same methods. Liu (2009) cited several studies in which SEC was coupled with an optical extinction measurement. This means that, for the characterisation of metal nanoparticles, volume weighted size distributions are obtained (cf. with remarks to sedimentation).

Ideally, the particles should not interact with the stationary phase in SEC. In reality, however, electrostatic interaction and adsorption exist and have to be appropriately considered. Indeed, as these kinds of interactions are material-specific it is difficult to find a universal analytical description or an appropriate way of calibration. For that reason, it is highly recommended to supplement the classical SEC set-up with a particle sizing technique (e.g. DLS, Yamaguchi et al. 2006).

2.3.4 Static Scattering Techniques

Scattering is the dispersal of radiation at an object (particle) that differs in the relevant material properties from its environment (continuous phase). Static scattering experiments record the scattering signal as a function of the angle of observation θ or—more generally—as function of the scattering vector $\mathbf{q}$:

$$q = |\mathbf{q}| = 2k \cdot \sin(\theta/2), \tag{2.17}$$

where k is the wavenumber ($k = 2\pi/\lambda$). The resulting scattering function depends on the concentration, the size and the shape of the scatterers. Basically, scattering experiments can resolve morphological details with dimensions in the order of magnitude of $1/q$. By variation of the wavelength λ and the angle θ (Fig. 2.11) it is possible to characterise disperse systems over a broad range of size levels, e.g. with regard to interparticle distances, to particle shape, or to surface roughness).

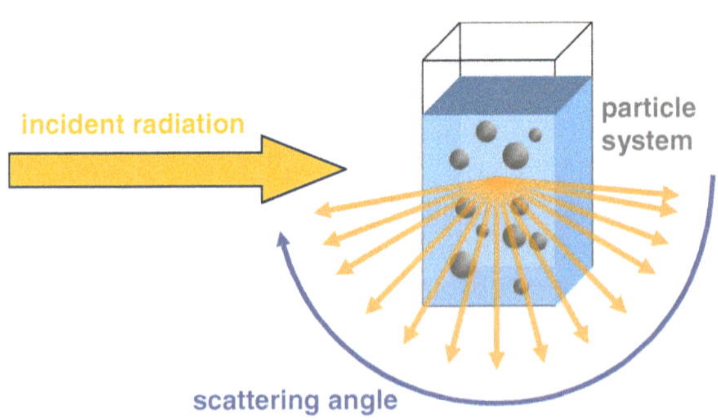

Fig. 2.11 Principal scheme for static scattering techniques

Scattering experiments can be conducted with any kind of radiation (e.g. sound, electromagnetic waves, neutron radiation). This book will be confined to the scattering of light and X-rays, as these two types are most frequently used for the characterisation of colloidal suspensions.[7] Both belong to electromagnetic radiation, yet the mechanisms of interaction with matter are completely different. This difference becomes manifest in the refractive indices, which deviate qualitatively. For this reason, both types of radiation are separately discussed.

Silica particles have a refractive index of approximately 1.45 in the domain of visible light. That is significantly different to the refractive index of water (1.33), which often serves as suspension medium. In contrast, there is only a weak contrast between the two substances in the domain of X-radiation. At a wavelength of 1 Å, one finds a complex refractive index of $0.999997–1.7 \times 10^{-8} \cdot i$ for silica and $0.999999–2.1 \times 10^{-9} \cdot i$ for water (cf. Henke et al. 1993).

Moreover, light scattering measurements have been historically realised by two different concepts of instrumentation which cover distinct size ranges. These are the *static light scattering* (SLS), which is conventionally employed for fine colloids, and the *laser diffraction*, which was originally used for micrometre particles only. Even though the measurement ranges of both techniques have actually converged in the recent past, there still remain qualitative differences in the sensor set-up and in data analysis, which justify their separate treatment.

2.3.4.1 Analysis of Static Scattering Experiments

The various static scattering techniques rely on different types of interaction between particles and radiation. Additionally, they are used for the determination of rather dissimilar measurands (e.g. average molecular weight, particle size distribution, shape parameters). This diversity of physical basis and experimental objective has led to very specific ways of data analysis, yet all of them are based on the following principal dependency of the scattering signal:

$$I_{\mathrm{sca}}(q) \propto S_{\mathrm{sus}}(q, \varphi_V) \cdot S_{\mathrm{agg}}(q, x_{\mathrm{agg}}) \cdot P(q, x_{\mathrm{p}}), \qquad (2.18)$$

wherein S_{sus} and S_{agg} denote the static structure factors of the suspension and of particle aggregates, respectively, and P is the form factor of the primary (i.e. constituent) particles. The structure factor of the suspension S_{sus} equals 1 for dilute suspensions, i.e. for very large interparticle distances. If aggregates are present, then S_{agg} reflects their size and structure as q-dependency for relatively small scattering vectors ($q < 1/x_{\mathrm{p}}$); otherwise S_{agg} is set to 1. The morphology (shape) of the primary

[7] Additionally, *small angle neutron scattering* (SANS) has some relevance for the characterisation of colloidal particle systems, in particular for dense suspensions (Romer et al. 2001; Qiu et al. 2005). With regard to particle characterisation, SANS is mainly used for disclosing the structure of particle aggregates (Hurd et al. 1987; Bugnicourt et al. 2007). A brief introduction to SANS is, for example, given by Glatter and May (2006).

particles is reflected in the form factor P and can be resolved for large q-values ($q \geq 1/x_p$).

When employed for particle sizing scattering experiments are usually conducted in dilute suspensions ($S_{sus} = 1$) and possible aggregates are typically ignored ($S_{agg} = 1$). The spectral scattering signal $I_{sca}(q)$ of the suspension is then considered as a linear superposition of all individual scattering signals:

$$I_{sca}(q) = I_{sca}(0) \cdot \int P(q, x) \mathrm{d}Q_0. \tag{2.19}$$

This is an inversion problem as defined in Eq. (2.2). Firstly, this means that its solution requires specifically adapted numerical algorithms and, secondly, that details of the size distribution can be resolved only to a certain extent.

2.3.4.2 Static Light Scattering

Static light scattering (SLS) is an established measurement technique in colloid and polymer science. The time averaged angular distribution of scattered light is commonly employed for the characterisation of macromolecules (molecular weight, radius of gyration, the second virial coefficients), but can be used to study suspensions of inorganic colloids as well (e.g. Poon et al. 1995; Heimer and Težak 2002; Wyss et al. 2004). However, the angular distribution of scattered light is insensitive to particle size for very fine particles. In that case, it cannot be exploited with regard to particle size distribution. Regardless, SLS instruments are frequently combined with a fractionating sample preparation in order to determine the concentration and mean radius of gyration of the individual size fractions (cf. Sect. 2.3.3).

The maximum scattering vector q_{max} primarily depends on the laser wavelength and the refractive index of the solvent. The minimum q-value is additionally affected by the minimum scattering angle θ_{min}. A typical commercial SLS instrument resolves an approximate angular range from 10° to 150°. Hence, for a HeNe laser (wavelength: 632.8 nm) and an aqueous solvent, a q-range from 2.3 to 26 μm^{-1} can be covered. That corresponds to a resolution of structural properties with size dimensions being not much smaller than 40 nm and not much larger than 400 nm.

The measured light intensity spectrum is usually expressed as excess Rayleigh ratio R_q (e.g. Xu 2000, p. 92):

$$R_q = \frac{I_{sus} - I_m}{I_0 \cdot f} \cdot \frac{r^2}{V_{meas}}, \tag{2.20}$$

which allows for comparison between different experimental setups ($I_{sus/m}$ = measured intensities of the suspension and the blank medium, r = distance to measurement volume V_{meas}, f = polarisation factor[8]). The excess Rayleigh ratio R_q depends on the scattering behaviour of the single particles as well as on the particle concentration. For dilute suspensions, one gets:

$$R_q = \frac{c_N}{4\pi} \frac{C_{psca}(q)}{f}, \tag{2.21}$$

where c_N denotes the particle number concentration and C_{psca} the partial scattering cross section. Alternatively, R_q can be related to the mass concentration c_m, the average molar mass $\overline{M}$, and the form factor $P(q)$ of very fine particles or macromolecules:

$$R_q = H\overline{M}c_m \cdot P(q), \tag{2.22}$$

$$H = \frac{k^4}{4\pi^2 m_F^2 N_A} \cdot \left(\frac{\partial m_{sus}}{\partial c_m}\right)^2. \tag{2.23}$$

While the expression (2.21) might be more familiar to the particle sizing community, Eq. (2.22) is customary in colloid and polymer science. The contrast factor H contains the refractive index increment $\partial m_{sus}/\partial c_m$, which is independent of size and concentration for very fine particles.

The linear relationship between scattering intensity (excess Rayleigh ratio) and particle concentration holds true only for extremely dilute suspensions. For fine, colloidal particles, a declining concentration impact can be observed, which is related to the osmotic pressure in the colloidal suspension and, therefore, depends on the virial coefficients (Einstein 1910; Zimm 1945; Debye 1947):

$$\frac{R_q}{HMP(q)c_m} = \frac{4\pi f R_q}{c_N C_{p,sca}} = \frac{1}{1 + 2A_2 M c_m + 3A_3 M c_m^2 + \ldots}. \tag{2.24}$$

Depending on the objective of SLS measurements, different methods of data processing can be utilised. According to Eq. (2.18), the dependency between R_q and q depicts the scaled structure or form factors of suspension, aggregates, or primary particles. A better distinction is possible with log-log plots. Plots of log(R_q) versus q^2 are called Guinier plots—they allow for the determination of the radius of gyration. Debye-plots show Hc_m/R_q versus c_m—they yield the average molar mass M and the second virial coefficient A_2. If measurements for different dilutions are presented in a Zimm plot, i.e. as Hc_m/R_q versus $q^2 + K \cdot c$, the average R_g can also be derived (Zimm 1948).

The calculation of the Rayleigh ratio R_q from Eq. (2.20) requires some knowledge about the instrument's optical configuration (measurement volume, detector distance, polarisation), which is difficult to obtain exactly. Besides this, the

[8] 1 for vertical polarisation, $\cos^2\theta$ for horizontal polarisation, ½(1 + $\cos^2\theta$) for unpolarised light.

conversion from detector signals to intensity values depends on the detector geometry and electronic settings. For this reason, SLS instruments are calibrated with particle-free scattering standards (e.g. toluene, benzene, and cyclohexane). And the result is calculated from $R_q = (I_{sus} - I_{med})/I_{std} \cdot R_{q,std}$ (I_{std} is measured intensity from standard). Since usually very fine, i.e. weakly scattering, particles are measured, the SLS technique is very sensitive to contamination by coarse dust particles or particle agglomerates. That requires careful sample preparation procedures (e.g. filtration of solvent and colloidal suspension, sample preparation in laminar flow box).

Like other optical characterisation techniques, SLS exhibits size-dependent concentration limits. At high concentrations, the suspension may become opaque or multiple scattering may occur which affects the angular intensity distribution. Dilution may solve the problem, but then osmotic concentration effects (virial coefficients, particle interactions) are not accessible. Conversely, very fine, weakly scattering particles may require such a high particle concentration that the mentioned concentration effects cannot be neglected and will complicate the determination of morphological structure parameters.

2.3.4.3 Laser Diffraction Spectroscopy

The term *laser diffraction* (LD) spectroscopy comprises static light scattering techniques, which are primarily designed to resolve the scattering pattern at small scattering angles. Historically, LD instruments and software were restricted to the characterisation of micrometre objects for which the scattering pattern is mainly caused by diffraction and can be explained by Fraunhofer's theory (1821). Yet, considerable enhancements in instrumentation as well as in data analyses have extended the applicability to the colloidal size range, where diffraction theory does not apply. Even though the term laser diffraction "does no longer reflect the current state of the art" (Xu 2000, p. 111) and may be replaced by *small angle light scattering* or *angular light scattering*, it is still widely accepted by the international particle sizing community.

The classical set-up of laser diffraction instruments is shown in Fig. 2.12 (left). A monochromatic light beam—typically, but not necessarily from a laser source—is passed through the particle sample and focussed on the centre of the forward scattering detector by means of a (set of) positive lens(es) (ISO 13320 2009). All light scattered in a specific direction is projected in one point of the focal plane independent of the particle's position in the measurement cell. For spherical particles, a typical diffraction pattern consists of concentric rings with outwards decreasing intensity (Fig. 2.12, right), where the radius of the rings (e.g. first minimum) depends on the sphere diameter. From a mathematical point of view the diffraction pattern can be regarded as the Fourier transform of the particle's projection image. It, therefore, reflects the particle size and shape.

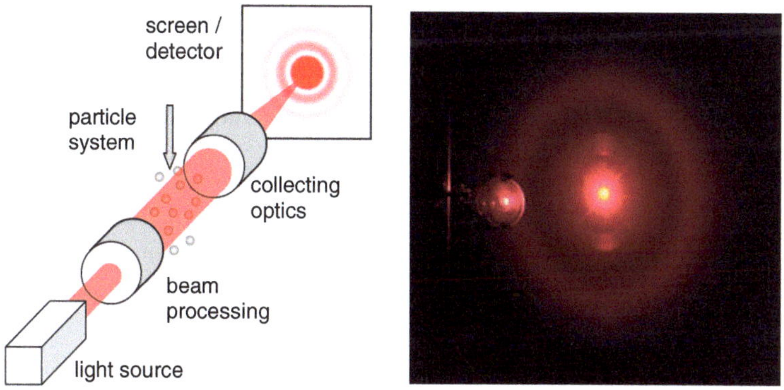

Fig. 2.12 Classical set-up of laser diffraction instruments (*left*) and demonstration of laser diffraction by illuminating a thin layer of monodisperse silica powder (© Benno Wessely 2004)

In the early stages of laser diffraction, the diffraction pattern was recorded by photographic plates (Stromgren 1975, Büchtemann et al. 1982). Today multi-element photodetectors of varying geometry are employed (Heuer and Leschonski 1985, Xu 2000, pp. 135). That means that a discrete power spectrum is measured which deviates significantly in shape from the continuous intensity spectrum (spatial intensity distribution).

For instance, a detector array of concentric half-rings with steadily increasing ring widths—as depicted in Fig. 2.13—yields a power spectrum with size-dependent, off-centre maxima and a total signal strength which is proportional to the cumulated squared particle size. Hence, for this detector configuration, the intrinsic type of quantity of measured size distribution is the particle surface. Nevertheless, size distributions from laser diffraction instruments are commonly calculated as volume weighted distributions. Note that such an interpretation changes with the detector geometry and may become rather abstract when additional detector arrays for collecting sideward and backward scattered light are installed.

The size obtained from laser diffraction is sometimes called diffraction equivalent diameter. However, in contrast to the implication of that term, it is not possible

Fig. 2.13 Intensity distribution and power spectrum for a ring-detector; cf. de Boer et al. (1987)

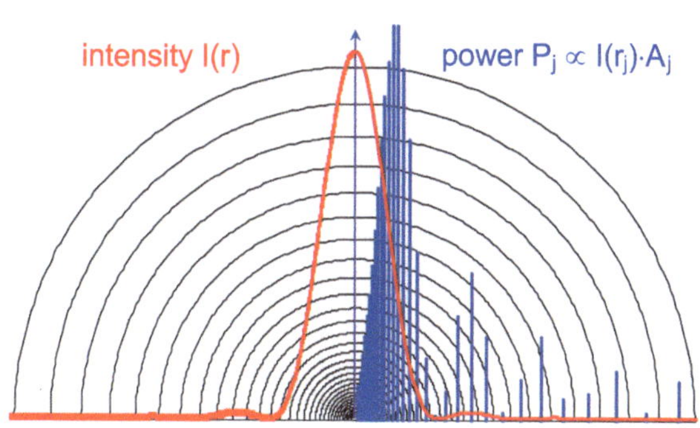

to reconstruct the diffraction pattern of a non-spherical particle by that of a sphere. Anisotropy in macro-shape (particle proportions) is clearly reflected by anisotropy in the diffraction pattern. Averaging the diffraction pattern over all spatial orientations of the particle (or obtained from an ensemble of randomly aligned monodisperse particles) broadens and smoothes the intensity spectrum similar to the way that a distribution in size does (Jones 1987; Gabas et al. 1994; Heffels et al. 1996; Stevens et al. 2007). In practice, the anisotropy of particles is commonly ignored and the analysis assumes spheres. Size distributions of nonspherical particles, therefore, typically show a pronounced mode around the minor dimension (e.g. rod diameter) and a long tail up to the size of the major dimension (e.g. rod length; cf. Matsuyama et al. 2000; Tinke et al. 2008).

The analysis of the angular spectrum (from laser diffraction instruments) traditionally refers to either Fraunhofer's diffraction theory (Fraunhofer 1821; van de Hulst 1981, p. 108, 209) or to Mie's scattering theory (Mie 1908; van de Hulst 1981, p. 114) for spherical particles. Since Fraunhofer diffraction is a limiting case of Mie's general solution, the latter is often preferred as first choice. However, Mie's theory as employed in standard instruments software firstly assumes homogenous materials and secondly requires the knowledge of the complex refractive index. Besides this, Mie calculation may run unstably for large particles if not installed numerically in the appropriate manner (Stübinger et al. 2010). Hence, for large, strongly scattering particles, Fraunhofer may perform better. However, for the characterisation of colloidal suspensions, where the measured scattering pattern is only partly due to diffraction, Mie theory should be compulsory.

First, laser diffraction instruments were developed in the 1960s (Lodi and Osmolovskaya 1975; Stromgren 1975; Thompson 1977) and were accompanied by corresponding improvements in data processing. Commercial instruments have been available since the 1970s. Today, laser diffraction instruments belong to the standard characterisation tools for particles systems. They can be applied to aerosols as well as to liquid dispersions. Measurements are quick and highly reproducible. Moreover, laser diffractometers can be run at flowing media. Hence, they offer a possibility for on-line or at-line monitoring of industrial processes. On the other hand, the particle concentration is usually restricted to values far below 1 vol % in order to avoid multiple scattering, which would affect the intensity spectrum.

The upper concentration limit depends on the particle system (size, refractive index) and the instrument (detector, optics). Some authors state transmission values above which multiple scattering can be neglected (e.g. Xu 2000, p.163: T_{min} = 85 %, Yamauchi and Ohyama 1982: 75 %, Chigier 1984: T_{min} = 40 %). Ludwig (2010) observed a weak linear correlation between sample transmission and measured size distribution, which resulted in an impact of 3 % for mean particle sizes at transmission values of 40 % (for 30 μm glass beads). If we assume an optical path of 1 mm and an aqueous solvent, then a transmission value of 40 % coincides with a solid content of 0.03 vol% for 1 μm titania spheres or 0.09 vol% of 1 μm silica spheres.

In the micrometre range, which is diffraction dominated, size distribution can be determined with high accuracy and good resolution (Knösche 2001 pp. 40–42; Mori et al. 2007; Witt et al. 2012). In order to extend the instrument applicability to colloidal particle systems, several modifications have been realised, e.g. variation of wavelength and polarisation or inclusion of wide angle scattering (Xu 2000, pp. 111–181; ISO13320). These modifications have evidently enhanced the sensitivity to colloidal particles far below 1 μm, but not in a uniform way as interlaboratory comparisons prove (Mori et al. 2008, Kuchenbecker et al. 2012). Even so, laser diffraction may serve as a useful tool for the characterisation of colloidal suspensions, in particular for monitoring dispersion procedures or product changes.

2.3.4.4 Small Angle X-ray Scattering

Shortly after its discovery, X-radiation was already being employed to study the inner structure of matter. This is possible because of the very small wavelengths of X-rays (typically in the order of 1 Å). In 1938, André Guinier proposed to evaluate X-ray scattering at small angles (i.e. in the forward direction) in order to study colloidal suspensions. He further related the intensity distribution to the radius of gyration of the scattering objects (Guinier 1938). Today, *small angle X-ray scattering* (SAXS) is an established technique for determining the size and shape of colloidal particles, analysing the conformation of macromolecules, or studying the structure of particle aggregates. Its applicability is typically limited to the range below 100 nm (Dörfler 1994, p 554, Glatter and May 2006).

SAXS instruments are usually distinguished with regard to the X-ray source. In (standard) laboratory instruments, X-ray tubes are employed which emit a broad spectrum of X-radiation with characteristic lines (e.g. at 1.54 Å for Copper anodes) and which are usually coupled with a monochromator. The emitted beam is of rather weak intensity and needs a collimation optic (e.g. pin-hole, slit, or block collimation; cf. Ilavsky et al. 2002; Anderegg et al. 1955; Kratky and Stabinger 1984, respectively). In contrast, X-radiation from synchrotron beamlines allow for a narrow adjustment of wavelengths and deliver highly intensive and well collimated beams for the SAXS measurement. Unlike laboratory instruments, the smearing of the measured spectrum is negligible and the structural parameter can be resolved in detail. However, synchrotron instrumentation is not applicable to routine analysis.

Similar to data processing in SLS, the analysis of SAXS spectra requires preliminary steps like background subtraction and—if necessary—extrapolation to zero concentration. Additionally, the turbidity of the sample may have to be considered (Kammler et al. 2004), but usually this effect is negligible. In general, the thus obtained spectrum is still a smeared version of the true scattering curve. Smearing results from the difficulty of collimating X-rays. In order to achieve sufficiently high signal-noise ratios, the detection optics collects all photons over a two-dimensional section of the scattering pattern, i.e. over a certain range of scattering angles (Fig. 2.14). In case of polychromatic irradiation there is a further smearing contribution. The knowledge of the detector geometry and the spectral distribution of the X-radiation facilitate the de-smearing, which first includes a data smoothing procedure (Glatter and May 2006; Glatter and Kratky 1982, pp. 119–165).

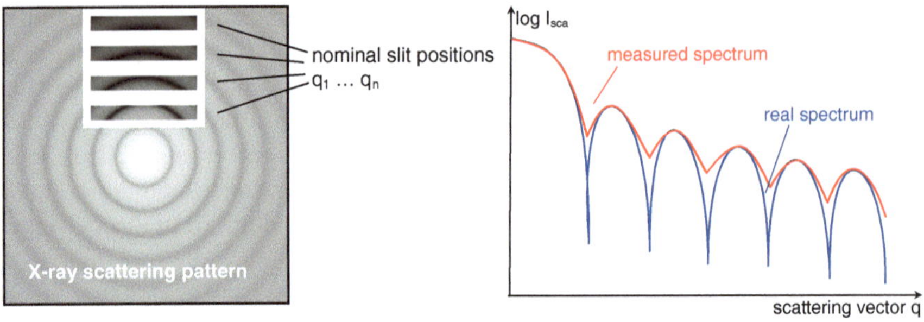

Fig. 2.14 Illustration of smearing for point source illumination and a slit detection system

Eventually, the de-smeared, "true" scattering function can be used to assess the morphological and structural properties of the colloidal suspension. Frequently, this involves some bias, e.g. on the morphology the particles (spheres, rods, core-shell, etc.) or on their size (when studying particle interactions). The algorithms of de-smearing and inversion are manifold (Glatter and May 2006; Fritz and Glatter 2006). There are, however, integral parameters that can be determined without specific assumptions on the particle shape. Based on Porod's invariant parameter Q

$$Q = \int I_{\mathrm{sca}} q^2 \mathrm{d}q \tag{2.25}$$

it is possible to calculate the total particle volume and the specific surface area (supposing the particle surface is smooth):

$$V = 2\pi^2 \frac{I(0)}{Q}, \tag{2.26}$$

$$S_V = \pi \cdot B/Q, \tag{2.27}$$

where B is derived from the asymptotic form of $I_{\mathrm{sca}}(q)$ at large scattering angles ($B = \lim_{q\to\infty} I_{\mathrm{sca}}/q^4$). The parameter B can be graphically obtained by plotting $I \times q^4$ versus q^4, which yields the following expression: $\lim_{q\to\infty} I_{\mathrm{sca}} q^4 = B + const \cdot q^4$. Here, *const* is supposed to be background and should, therefore, be zero for background corrected data. A significant deviation of *const* from zero indicates a fractal or porous particle surface (Schmidt et al. 1991).

De-smeared scattering curves of monodisperse particle systems are well structured (cf. Fig. 2.14) and allow for an accurate determination of size and shape. For polydisperse systems, smooth spectrums are obtained. The resolution of size distributions is rather low (peak distance for bidisperse distribution approx. 25 %,

Glatter and May 2006). If the particle shape is known, SAXS yields distributions of geometric particle properties (e.g. rod length), otherwise it probes essentially the radius of gyration. The distributions are intensity weighted, which typically means that they are weighted by the particle surface ($I_{sca} \propto x^2$).

When colloidal suspensions are probed by SAXS, the X-ray absorption in the solvent has to be considered because it attenuates the scattered intensities. Maximum scattering intensities are achieved for a sample transmission of 37 %, which is approximately the transmission through a 1 mm sheet of water when a typical copper anode (wavelength: 1.54 nm) is used. Because of this intrinsic background turbidity, SAXS characterisation of colloidal suspensions has to be conducted in relatively small measurement volumes.

While typical SAXS instrumentation works at scattering vector in the range of approx. 0.05–2 nm^{-1}, *ultra small angle X-ray scattering* (USAXS) instruments cover the range from 0.001 to 4 nm^{-1} and can be, therefore, applied to larger particles and particle aggregates up to the micrometre region (Ilavsky et al. 2002; Kammler et al. 2004). Since they rely on the intense radiation of synchrotrons and require special cameras they are not available as standard laboratory equipment (Glatter and May 2006).

2.3.5 *Dynamic Scattering*

The time-averaged signals from a scattering experiment provide a powerful tool to study the micro-structure of colloidal suspensions (i.e. size and shape of a particle as well as the pair correlation), as was shown in the previous section. However, scattering signals are not constant in time; they exhibit high-frequency fluctuations which are due to the dynamics of the microstructure, e.g. to the Brownian motion of the colloidal particles. Dynamic scattering techniques probe these temporal variations and, thus, add to the characterisation methods for colloidal suspensions. The most popular of these techniques is *dynamic light scattering* (DLS), which will be a focus of this section. In addition to this, diffusive wave spectroscopy, X-ray photon correlation and fluorescence correlation spectroscopy are briefly introduced.

2.3.5.1 Dynamic Light Scattering

Dynamic light scattering (DLS) techniques evaluate the fluctuations in scattered light (Fig. 2.15). These fluctuations may be caused by any changes in the microstructure of the suspension, e.g. by particle motion or vibrations of particle networks. For this reason, there are manifold applications for DLS, e.g. to particle sizing (Finsy 1994), molecular weight determination (Jena and Bohidar 1993; Siddiq and Wu 1996), studying particle aggregation (Bolle et al. 1987; Herrington and Midmore 1991; di Biasio et al. 1994), monitoring phase transition in colloidal

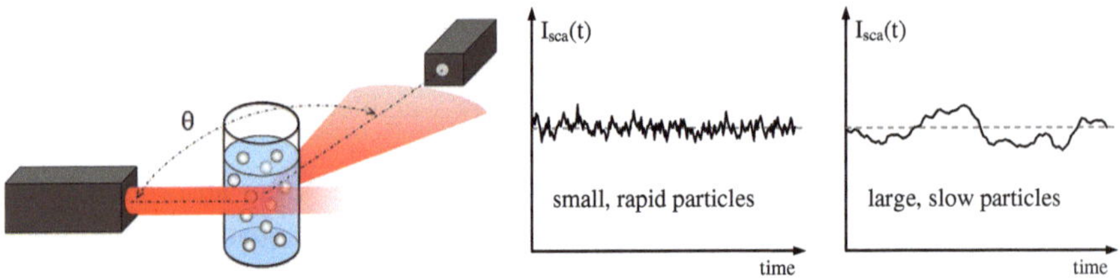

Fig. 2.15 Dynamic light scattering experiment and intensity fluctuation

suspensions (Kroon et al. 1996; Ruzicka et al. 2004; Kätzel et al. 2007), or measuring the strength of colloidal gels (Manley et al. 2005).

First, DLS measurements were conducted in the 1960s by analysing the intensity fluctuations in terms of a frequency spectrum (*frequency analysis*—FA; Cummins et al. 1964; Arrechi et al. 1967; Chu and Schones 1968; Dunning and Angus 1968). The width of the frequency spectrum is a measure of the relaxation time of the microstructural processes and can be employed for the determination of the particle diffusion coefficients (Pecora 1964). An alternative for evaluating the fluctuation of scattered light intensity is *photon correlation spectroscopy* (PCS), which has been used for the characterisation of colloidal suspensions since the end of the 1960s (Jakeman and Pike 1969; Jakeman 1970; Foord et al. 1970). PCS requires a different hardware than FA, but it can be shown that the results of both techniques are equivalent (Jakeman 1970; Xu 2000, pp. 86–89).

Today, a large variety of commercial or "self-made" DLS instruments are used. Apart from data processing (FA or PCS), they can be distinguished with regard to laser optics and signal modulation. The latter refers to the question of whether the scattered light is superposed with a reference (laser) beam. According to Fig. 2.16, three principal set-ups can be distinguished[9]:

- direct or *self-beating* measurement: no reference beam; probes interparticle distances, insensitive to mean particle velocity
- *homodyne*: reference beam with same frequency as incident beam, sensitive to the absolute position of the particles
- *heterodyne*: reference beam with frequency shift to the incident light; facilitates the measurement of particle migration velocities

The study of microstrutural processes is almost exclusively accomplished by the self-beating and homodyne technique. The light scattering experiment may be realised with conventional or fibre optics (Auwetter and Horn 1985; Bremer et al. 1993). Further differences exist with regard to scattering angle or the range of

[9]There is some confusion in the literature regarding the terms *homodyne* and *heterodyne* (cf. Xu 2000, pp. 84–86). The terminology used here agrees with the modern convention in laser technology (e.g. Paschotta 2008).

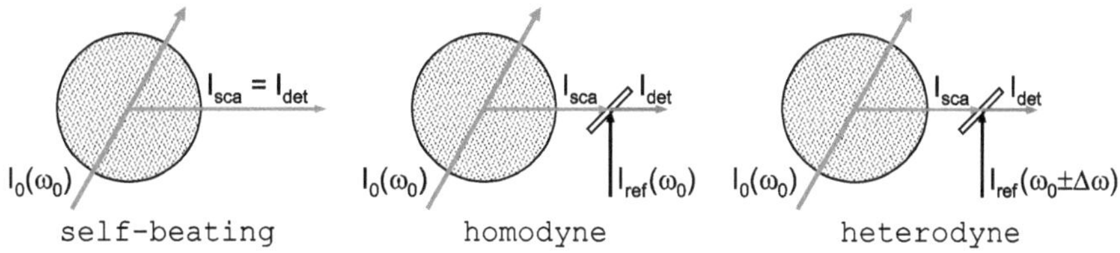

Fig. 2.16 Principal set-ups for detecting laser signals; the superposition with a reference beam allows a control of signal phase, which means for DLS measurements that the absolute particle velocity can be detected

scattering angles and with regard to cross correlation (i.e. conducting and correlating two DSL experiments in parallel at the same sample volume).

The primary result of a DLS experiment is the power spectrum of the scattering intensity which is obtained by FA (Ishii and Iwai 2008):

$$P_I(\omega) = \left| \int_0^\infty I_{\mathrm{det}}(t) \mathrm{e}^{\mathrm{i}\omega t} \mathrm{d}t \right|^2 = \int_0^\infty \langle I_{\mathrm{det}}(t) I_{\mathrm{det}}(t+\tau) \rangle \mathrm{e}^{\mathrm{i}\omega\tau} \mathrm{d}\tau \tag{2.28}$$

or the (normalised) intensity autocorrelation function as produced by PCS

$$g^{(2)}(\tau) = \langle I_{\mathrm{det}}(t) \cdot I_{\mathrm{det}}(t+\tau) \rangle \Big/ \langle I_{\mathrm{det}}(t) \rangle^2. \tag{2.29}$$

Both types of data analysis are equivalent because the spectral power function $P_I\ (\omega)$ is a Fourier transform of the corresponding autocorrelation function $g^{(2)}(\tau)$ (Wiener-Khintchine-theorem; Wiener 1930; Khintchine 1934). In the following, only PCS will be discussed.

If the light fluctuation is caused by Brownian motion, one can relate the normalised autocorrelation function of the detected light intensity $g^{(2)}$ to that of the scattered field $g^{(1)}$ by the following equations (Siegert 1943; Xu 2000, pp. 86–89):

$$\text{self-beating: } g^{(2)}(\tau) = 1 + \beta \left| g^{(1)}(\tau) \right|^2, \tag{2.30}$$

$$\text{homodyne: } g^{(2)}(\tau) = 1 + 2X\beta \cdot \left| g^{(1)}(\tau) \right| = 1 + a \cdot \left| g^{(1)}(\tau) \right|, \tag{2.31}$$

where β is a factor describing the coherence of the light received by the detector ($\beta < 1$), and X is the fraction of scattered to total light on the receiver/detector ($X \ll 1$).

The field correlation function $g^{(1)}$ directly reflects any changes in the microstructure of the suspension. In the case of purely diffusive processes, i.e. when the displacement of scattering objects follows Fick's second law, an exponential time dependency is found.

$$g^{(1)}(\tau) = \int_0^\infty \exp(-\Gamma\tau)\,\mathrm{d}Q_{\mathrm{int}}(\Gamma), \tag{2.32}$$

where Γ denotes the decay rate and $\mathrm{d}Q_{\mathrm{int}}$ the relative contribution of the corresponding diffusion process to the scattering intensity.

In quiescent, dilute suspensions, the light fluctuations result essentially from the Brownian displacement of the single particles. Thus, the decay rate Γ can be traced back to the particles' translational diffusion coefficient D_t:

$$\Gamma = 1/\tau_{\mathrm{decay}} = q^2 D_t. \tag{2.33}$$

For spherical particles, this parameter (D_t) is inversely proportional to the sphere diameter (cf. Eq. (2.4)). Thus, Eq. (2.32) expresses an integral equation for the particle size distribution (Finsy 1994). The reliable, meaningful, and robust inversion of Eq. (2.32) has been tackled by several authors (e.g. Stock and Ray 1985; Finsy et al. 1989). Most often, regularisation approaches (Provencher 1982; Maier et al. 1999) with a non-negative constraint on the weights $\mathrm{d}Q_{\mathrm{int}}$ (Lawson and Hanson 1995; Geers and Witt 2008) are employed.

The primary result of such a data analysis is the intensity weighted distribution Q_{int} of the translational hydrodynamic diameter $x_{\mathrm{h,t}}$. Instrument software usually allows for conversion in volume or number weighted distributions, but this requires a model on the relationship between $x_{\mathrm{h,t}}$ and the scattering intensity. Furthermore, distribution details that do no contribute significantly to the correlation function (e.g. very fine size fractions) and that are consequently ignored in the measured Q_{int} also cannot be revealed by numerical conversion.

Alternative routes for data analysis that avoid the inversion of Eq. (2.32) are the method of cumulants (Koppel 1972) and the Williams-Watts-analysis (Williams and Watts 1970). The former is based on a series expansion of the logarithmised autocorrelation function $g^{(1)}$, which is commonly stopped after the second term:

$$\ln\left(g^{(1)}\right) = \ln g^{(1)}(0) - \langle\Gamma\rangle \cdot \tau + \tfrac{1}{2} PDI \cdot \langle\Gamma\rangle^2 \cdot \tau^2. \tag{2.34}$$

The mean decay rate $\langle\Gamma\rangle$, which can be measured with high accuracy, corresponds to the intensity weighted harmonic mean of the size distribution (x_{cum})

$$x_{\mathrm{cum}} = \mathrm{const.} \cdot \langle\Gamma\rangle = \int x_{\mathrm{h,t}}^{-1}\,\mathrm{d}Q_{\mathrm{int}}, \tag{2.35}$$

while the polydispersity index PDI can be understood as the normalised harmonic variance:

$$PDI = x_{\mathrm{cum}}^{2} \times \int \left(x_{\mathrm{h,t}}^{-1} - x_{\mathrm{cum}}^{-1}\right)^{2} \mathrm{d}Q_{\mathrm{int}}. \tag{2.36}$$

Hence, it offers an opportunity to quickly evaluate the distribution width.

PDI values below 0.05 indicate a very narrow, quasi monodisperse distribution, while values above 0.2 usually imply a relatively broad, possibly multimodal distribution. Values above 0.5 mean that the experimental data are poorly reproduced by Eq. (2.34). For log-normal size distributions, one can easily relate the PDI to the geometric standard deviation of the distribution function (Babick et al. 2012).

The method of cumulants performs rather weakly for very broad distributions of the decay rate. In this case, the autocorrelation functions are better fitted by stretched exponentials (Williams and Watts 1970). The Williams-Watts analysis is mainly employed for phase transition in colloidal suspensions (Ruzicka et al. 2004; Kätzel et al. 2007) and for polymer suspensions.

When applied to dilute solutions of macromolecules, DLS reveals the molar mass M which is related to the diffusion coefficient D_{t} by a power-law (Mark-Hoodwink-Kuhn-Sakurada equation):

$$D_{\mathrm{t}} \propto M^{-\varepsilon}. \tag{2.37}$$

This relation reflects the conformation of the macromolecules and is, therefore, specific for each molecule-solvent combination. For random coils, the exponent is approx. 0.55, while for rigid rods and compact spheres it amounts to 0.85 and 1/3, respectively (Jena and Bohidar 1993; Harding 1995).

During the last two decades, dynamic light scattering has evolved into a major characterisation technique for colloidal suspensions. A recent interlaboratory study into the characterisation of colloidal silica (Braun et al. 2011) showed that state-of-the-art DLS instrumentation facilitate a highly reproducible and very reliable acquisition of correlation function and corresponding mean particle size x_{cum}.

The study involved 17 participants from the EU and the USA, which provided 19 independent data sets from 6 different commercial instruments covering sideward scattering (90°) and backscattering. Suspensions of colloidal silica of spherical particles with a mean diameter of approx. 19 nm and narrow distribution width were sent in several aliquots to each participant and measured by uniform SOP. There was only one outlier with regard to the mean particle size. The overall uncertainty of that parameter was expressed as 95 %-confidence interval of ±3 %. A considerably higher amount of uncertainty exists with regard to the mean values of the volume weighted distributions because of the twofold error magnification by inversion ($g^{(2)}$ in Q_{int} and conversion Q_{int} in Q_3). A parallel study with a different test material essentially confirmed the high interlaboratory comparability (Lamberty et al. 2011).

The study further confirmed high agreement with the results of electron microscopy and centrifugation analysis. However, such agreement cannot be expected for non-spherical particles. That is because the hydrodynamic diameter deviates from e.g. Stokes diameter. Additionally, there is a significant contribution of the particle rotation to the intensity fluctuation, which increases with scattering angle (Aragón and Pecora 1977; Hoffmann et al. 2009). DLS then measures an apparent hydrodynamic diameter $x_{h,app}$ which is smaller than the translational one ($x_{h,t}$). The impact of rotation on the autocorrelation function can be computed straightforwardly if the particles are very fine and their shape is known (Aragón and Pecora 1977; Xu 2000, pp. 259). If a depolarised scattering multi-angle DLS setup[10] is employed, it may even be possible to separate the impact of rotation and translation and to determine shape parameters (Nakamura et al. 2006; Hoffmann et al. 2009).

Limits for the applicability of DLS are mainly set by size and concentration of particles. First of all, the concentration should be low enough to avoid strong multiple scattering. Multiple scattering reduces the signal coherence, i.e. the signal-to-noise-ratio, affects the correlation function and contributes to depolarisation. It is of particular relevance for large particles with high contrast in the refractive index. Multiple scattering can be effectively reduced by backscattering techniques (Wiese and Horn 1991; Peters et al. 1998) and largely suppressed by cross correlation techniques (e.g. 2-colour cross-correlation: Dhont and de Kruif 1983; Drewel et al. 1990; Segrè et al. 1995; 2-dimensional cross-correlation: Phillies 1981a, b; 3-dimensional cross-correlation: Schätzel 1991; Aberle et al. 1998; Overbeck and Sinn 1999; Urban and Schurtenberger 1999), where two simultaneous scattering experiments at the same sample volume are conducted. The amplitude of the cross correlation function is proportional to the squared single scattering fraction (Aberle et al. 1998). That means that cross correlation techniques can be used as long as there is a significant portion of single scattered light in the detected signal.

When multiple scattering is discarded from the measured signal, DLS can be used to study the dynamics of concentrated suspensions, in which the Brownian motion of individual particles (self-diffusion) differs from the diffusive mass transport (gradient or collective diffusion), which causes local density fluctuations, and where the diffusion on very short time-scales ($\tau < a^2/D$) deviates from those on large time scales ($\tau \gg a^2/D$; Jones and Pusey 1991; Banchio et al. 2000). These different diffusion coefficients depend on the microstructure of the suspension, i.e. on the particle concentration and on the interparticle forces. For an unknown suspension it is not possible to state a priori which of them is probed by a DLS experiment. For this reason, a further concentration limit must be obeyed when DLS is used for basic characterisation tasks such as particle sizing. As a rule of thumb, such concentration effects vanish below concentrations of 0.01–0.1 vol%, but certainty can only be gained by experiment.

Apart from upper concentration limits, there is a further one at the lower edge, where the intensity fluctuations start to become affected by the statistic variation of

---

[10]DLS instruments typically employ vertically polarised light (I_{vv}; Xu 2000, p. 230). Depolarised scattered light (I_{vh}) results from anisometry or multiple scattering—signals are usually very weak.

particle number in the measurement zone. According to Willemse et al. (1997), a minimum of 100 particles should stay in the measurement zone. This is fulfilled for most colloidal suspensions; problems may arise for micrometre particles (i.e. ≥1 μm).

Assuming a scattering volume of 10^6 μm^3 (Willemse et al. 1997) and particles of 1 μm diameter, then a minimum concentration of 0.01 vol% is needed for ensuring 300 scattering particles. For strongly scattering particles of that size, such a concentration may already exceed the upper limit set by multiple scattering (cf. Itoh and Takahashi 1991), which means that there is no concentration range, in which the DLS measurement should be conducted. The absence of such a range is the particularly likely for particles >1 μm.

Sedimentation sets a further limit to the detection of coarse particles because the particle displacement of micrometre particles is governed by sedimentation rather than by diffusion (cf. Sect. 2.3.3.1). This is of particular importance for polydisperse particle systems, where diffusion and sedimentation are coupled (Batchelor 1982; Batchelor and Wen 1982). Moreover, sedimentation may affect the size distribution in the measurement zone. However, in the colloidal size range ($x \leq 1$ μm) there is virtually no impact of sedimentation on DLS results (e.g. Paul and Pusey 1981). A lower size limit exits only inasmuch as the scattering intensity of the particles should considerably exceed that of the fluid molecules. Xu (2000, p. 241) proposes a minimum factor of 2.5; yet, for highly reliable DLS this value should be multiplied by 10.

Occasionally, DLS is compared with the DUM (cf. Sect.2.3.1) because both methods measure the diffusion of colloidal particles. However, the two methods are not equivalent; thus, identical results cannot be expected.

Differences between the two methods exist with regard to particle property and type of quantity: DUM only evaluates the translational diffusion ($x_{h,t}$) and probes number frequencies, whereas DLS is also sensitive to the diffusive rotation ($x_{h,app}$) and yields intensity weighted distribution functions. Furthermore, the methods usually differ in sample size: Typical sample concentrations in DLS are in the range of 0.01 vol%. These are, for instance, 50,000 particles á 100 nm in a measurement volume of 10^6 μm^3, which are all observed in the order of minutes (Willemse et al. 1997), whereas with DUM, the total number of traced particles is smaller by factor 10–100, with an observation time in the order of seconds for each. Last but not least, DLS allows for a temporal resolution in the range from ns to ms, whereas DUM is subject to video processing, typically with 30 frames per second, and is, therefore, not sensitive to very fast relaxation processes (like gradient diffusion).

2.3.5.2 Diffusing Wave Spectroscopy

For the interpretation of DLS measurements, it is prerequisite that multiple scattering has been avoided by sufficiently high sample dilution or even suppressed by technical means. In contrast, *diffusing wave spectroscopy* (DWS) evaluates the light fluctuations in the presence of strong multiple scattering. The decay of the autocorrelation function is then related to the loss of coherence by multiple scattering, i.e. by the large number of scattering events a photon experiences when travelling through the suspension. In DWS the autocorrelation function reflects the Brownian motion of single particles (short-time self-diffusion) as well as the mean free path of photon transport l^*, which depends on the particle concentration and on the suspension structure (Maret and Wolf 1987; Pine et al. 1988). Consequently, DWS can be used for measuring average diffusion coefficients and mean particle size (Scheffold 2002; Rochas-Ochoa et al. 2002), for investigating aggregation processes and phase transition (Wyss et al. 2001; Alexander and Dalgleish 2007), or for probing the microrheology of gels (Mason and Weitz 1995; Ruis et al. 2008).

2.3.5.3 X-Ray Photon Correlation Spectroscopy

In principle, there is no reason why dynamic scattering techniques should be restricted to radiation in the visible domain of the electromagnetic spectrum. However, practical limits are set by the requirement of coherent radiation (at least for single particle detection) and low sample turbidity, as well as the availability of appropriate radiation sources and sensor technique. In practice, only X-radiation is employed as a further dynamic scattering technique, which is called X-ray photon correlation spectroscopy (XPCS) or X-ray intensity fluctuation spectroscopy (XIFS). In comparison to visible light X-rays allow for a greater spatial resolution of suspension microstructures (larger q-values) and virtually avoid multiple scattering because of the weak interaction of X-radiation with matter (cf. Sect. 2.3.4). On the other hand, only third generation synchrotrons provide the necessary X-radiation of high coherence and intensity. XPCS has been i. a. used to measure diffusion processes of particles in solution, to study the dynamics of polymer blends, or to characterise liquid-crystal membranes (see Sutton 2008; Nugent 2010 and references therein).

2.3.5.4 Fluorescence Correlation Spectroscopy

A further group of dynamic scattering techniques comprises such methods that analyse the fluctuation of inelastic scattering events such as Raman scattering or fluorescence. In particular, fluorescence correlation spectroscopy is widely used for studying bio-molecules (e.g. nucleic acids, proteins) in vitro or in vivo (Elson and Magde 1974; Magde et al. 1974; Krichevsky and Bonnet 2002). X-ray fluorescence correlation spectroscopy has been employed for the characterisation of colloidal suspensions, but is currently still in the stage of development and has not yet attracted wide attention in the colloidal community (Wang et al. 1998; Leupold et al. 2007).

2.3.6 *Further Spectroscopic Techniques*

Static and dynamic scattering techniques are spectroscopic characterisation methods in the sense of Sect. 2.2. These techniques evaluate the functional dependency of measurement signals on a spectral parameter, i.e. on time, space, or classically on wavelength or frequency. The major advantage of spectroscopic methods is the reduced sample preparation (no fractionation), but they involve the inversion problem. That is, the spectrum is a—most frequently incomplete and discrete—nonlinear projection of the size distribution. Beside the scattering techniques, there are further spectroscopic methods which are based on the extinction of radiation or on any other response of the particle system to an external field. This section describes optical, acoustic, and electroacoustic methods that have gained relevance for the characterisation of colloidal suspensions.

2.3.6.1 Optical Spectroscopy

The term *optical spectroscopy* refers to the evaluation of the wavelength dependency of transmitted light for the quantification of dissolved or dispersed matter (Fig. 2.17, left). Optical spectrometers deliver transmission or turbidity spectra, which commonly range from the ultraviolet (UV) to the near infrared (NIR) domain. They are standard analytical tools which are mainly used to identify solutes and to determine their concentration. The impact of particle size on the shape of the transmission spectra has been known for long and was, for instance, subject of Gustav Mie's (1908) renowned paper on the theory of light scattering. Accordingly, optical spectroscopy has been employed for the measurement of particle size distributions in many experimental studies. It is mainly applied to fine colloids, e. g to submicron polymer latices (Zollars 1980; Gulari et al. 1987, Celis et al. 2008), to crystallisation processes (Crawley et al. 1997), to fine abrasive particles in polishing slurries (Kuntzsch 2004, pp. 58–60) or to metal nanoparticles (Haiss et al. 2007).

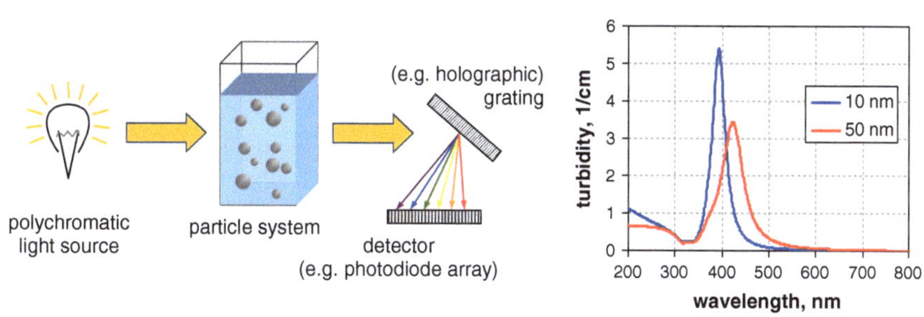

Fig. 2.17 Principle of optical spectroscopy (*left*) and turbidity spectrum for 0.0001 vol% of colloidal Ag in water (*right*)

All these applications rely on the fact that the light transmission through a colloidal suspensions depends on the morphology, the concentration and the optical properties (i.e. the complex refractive index function) of the dispersed phase. In dilute suspensions, the particles contribute independently to the extinction of light, and the transmission T obeys the Lambert-Beer law:

$$T(\lambda) = \exp\left(L \cdot c_N \cdot \int C_{ext}(x, \lambda, m(\lambda))q_0(x)\mathrm{d}x\right). \quad (2.38)$$

After conversion to turbidity τ and natural extinction E this leads to

$$\tau(\lambda) = E(\lambda)/L = c_N \cdot \int C_{ext}(x, \lambda, m(\lambda))q_0(x)\mathrm{d}x, \quad (2.39)$$

where L is the optical path, c_N the number concentration, C_{ext} the extinction cross section, and $q_0(x)$ the number weighted density function. The specific shape of an optical spectrum results from the general impact of wavelength λ on C_{ext} as well as from the material-specific wavelength dependency of the relative refractive index. The particle size distribution is obtained by inversion of the two Eqs. (2.38) and (2.39) (Elicabe and Garcia-Rubio 1989).

The optical spectroscopy is commonly applied to such colloidal suspensions that obey the Rayleigh limit or the Rayleigh-Debye-Gans limit of scattering (cf. Appendix B.2). In this case the spectra usually have a smooth and monotone shape, from which only a few details of the size distribution can be deduced. Yet, for metals with a surface plasmon resonance in the optical domain (e.g. Ag or Au), one observes a distinct, size dependent maximum in the turbidity spectra of nanoparticles (Fig. 2.17; cf. Njoki et al. 2007). The presence of such a maximum can clearly enhance the information content of the spectrum.

Size distributions from optical spectroscopy actually reflect the distribution of the extinction cross section C_{ext}, which is a function of the particles morphology and their optical properties. For very fine colloids <100 nm, i.e. for Rayleigh scatterers, the total optical behaviour is determined by the particle volume.[11] The method then yields volume equivalent diameters. In order to identify the type of quantity in which the different size fractions are weighted, one needs to determine the impact of single particles to the total signal (i.e. the integrated spectrum). The results depend on the covered frequency range and on the signal representation (transmission or extinction spectrum).

It turns out that, in the case of absorbing colloidal particles, the total signal strength correlates approximately with the third power of the particle size for both types, transmission and extinction, of spectra. They are, thus, volume weighted. For non-absorbing

[11]with C_{ext} being proportional to the particle volume for absorbing substances and to the squared volume for non-absorbing ones.

particles in the Rayleigh limit of scattering, the transmission spectra can be roughly considered as volume-weighted as well, whereas an ideal extinction spectrum is weighted by the squared particle volume.

Optical spectroscopy requires that both transmission and extinction are well above the electronic noise. This sets a lower size limit to the detection of non-absorbing colloidal particles ($C_{ext} \propto x^6$), since the maximum optical path is usually defined by the instrument and the particle concentration should be sufficiently low to avoid osmotic concentration effects (cf. Sect. 3.1.1). Such a problem does not occur for absorbing particles, where measurements can be conducted at volume concentrations in the order of ppm. However, for absorbing nanoparticles below 10 nm, the spectra are solely affected by the mass concentration and the influence of size vanishes (Njoki et al. 2007)—unless the optical properties change with size as for quantum dots (Segets et al. 2009) or for metal nanoparticles <10 nm (Römer and Fragstein 1961; Scaffardi and Tocho 2006; Santillán et al. 2013). Last but not least, the data analysis has to adequately regard the contribution of solute or adsorbed matter (ions, polymers) to the optical spectra, which is of particular relevance in the UV domain (e.g. Gaffney et al. 1992).

Although several studies have demonstrated that optical spectroscopy reliably measures the size and concentration of suspended nanoparticles, the method is only of minor relevance for particle sizing. Recent developments try to considerably increase the method's sensitivity, precision, and resolution by employing pulsed laser light with tuneable wavelength (Li et al. 2010) or to realise online-capable instrumentation, e.g. by measuring at a few selected wavelengths, which usually suffices to capture the shape of the spectrum (Melik and Fogler 1983; Gabsch et al. 2005; Steinke et al. 2009).

2.3.6.2 Ultrasonic Spectroscopy

Even though optical particle sizing techniques are very popular, most of them require sufficiently dilute suspensions with less than 1 vol% solid content. That is because dense suspensions are opaque or because light transmission and scattering are subject to unwanted (i.e. poorly understood) concentration effects like multiple scattering. In such situations, acoustic fields may serve as an alternative probe for the state of dispersion. *Ultrasonic spectroscopy* is the generic term for all particle sizing techniques that are based on the frequency dependent measurement of sound velocity or attenuation in the ultrasonic domain (mostly within 100 kHz to 200 MHz). While velocity spectroscopy is mainly used for the study of inter- and intramolecular processes, attenuation spectroscopy has found its major application in particle sizing (McClements 1996; Kachanovskaya et al. 1996). The most promising feature of acoustic characterisation techniques is their applicability to highly concentrated particle systems (up to 70 vol%) under non-equilibrium conditions (McClements 1991). That means it offers the opportunity to monitor the state of dispersion of dense product streams (Scott et al. 1998), to control the

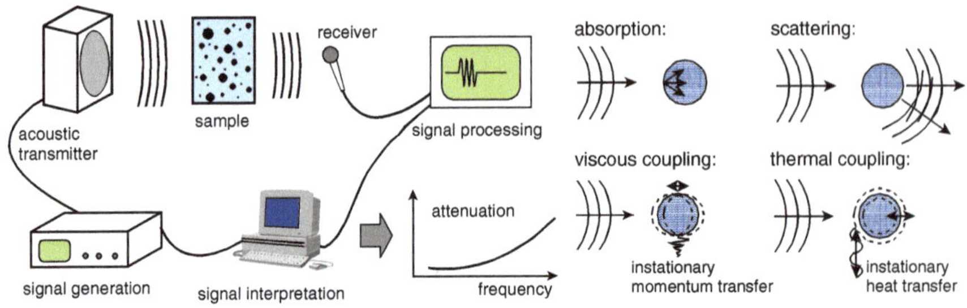

Fig. 2.18 Principle of ultrasonic attenuation spectroscopy (*left*) and kinds of particle-wave interaction in sound fields (*right*)

deagglomeration of suspension or the homogenisation of emulsions, and to study polymerisation or crystallisation processes (Hipp et al. 2000; Mougin et al. 2001).

All of these applications benefit from the size-dependency of sound-particle interactions (Fig. 2.18), which result in sound scattering and dissipation. Scattering is predominant for micrometre particles, but becomes of minor importance for colloidal particles, which are usually much smaller than the sound wavelength (in water: 15 μm at 100 MHz). In that case, sound propagation is governed by the dissipation of sound energy due to momentum and heat exchange between the oscillating and pulsating particles and the surrounding liquid. Additionally, there is always size-independent absorption of sound in the particles and in the liquid (Challis et al. 2005).

First instruments for the detection of ultrasonic velocity and attenuation were built by Pellam and Galt 1946 and Pinkerton 1947 in the middle of the 20th century. Andreae and his co-workers (Andrea et al. 1958; Andrea and Joyce 1962; Edmonts et al. 1962) eventually realised the first spectrometer for measuring ultrasonic attenuation. Today, there are several (commercial) suppliers of ultrasonic spectrometers (e.g. Hinze et al. 2000) that provide on-line instrumentation as well as laboratory instruments.

The immediate result of such instruments is the frequency spectrum of the sound velocity or the attenuation coefficient α, which is defined as the relative decrease of sound pressure per an infinitesimally small distance. The latter is often presented as ratio α/f versus the logarithm of frequency because such a spectrum has one or more size-specific peaks and its total area depends on the volume concentration only. For dilute suspensions, the data analysis reduces to a linear inversion problem (cf. Sect. 2.2):

$$\tilde{\alpha}(f,\varphi_V) = \alpha(f,\varphi_V)/f = \tilde{\alpha}_{abs}(f,\varphi_V) + \varphi_V \cdot \int k(f,x) \cdot q_3(x)\mathrm{d}x. \tag{2.40}$$

The kernel function $k(f, x)$, which quantifies the influence of size and frequency on attenuation, can be derived from the combined universal theory for the inelastic scattering of sound at spherical particles; the ECAH theory (Epstein & Carhart 1953; Allegra and Hawley 1972).[12] However, this linear theory fails for dense dispersions of colloidal particles because neighbouring particles affect each other with respect to the acoustophoretic motion and to the thermal fluctuation. The non-linearity is tackled by advanced multiple-scattering theories (Hipp et al. 2002), by effective medium approaches (Isakovich 1948; Hemar et al. 1997), or by coupled phase models (Dukhin and Goetz 1996; Babick 2005), but all of them keep restricted to the colloidal size range ($x \ll \lambda$).

In colloidal suspensions, the sound propagation is typically governed by the acoustophoretic motion of particles. For monodisperse spheroids that do not deviate too much from spherical shape (aspect ratio <10/1), the attenuation spectrum essentially reflects the volume specific surface area of the particles[13] (Babick and Richter 2006). Similar results would probably be obtained for any convex particle shape. For particle aggregates, the inner structure is decisive. Regarding the type of quantity, acoustically measured size distributions are ideally volume weighted distributions (see comments in Sect. 2.2).

It could be shown that the results of ultrasonic spectroscopy agree fairly well with those of other characterisation methods and are hardly affected by the extent of sample dilution (Dukhin and Goetz 1996; Babick et al. 1998; Babick 2005). Moreover, an interlaboratory comparison showed excellent agreement for colloidal particles (Dukhin et al. 2012). That is why ultrasonic spectroscopy is considered as a powerful tool for monitoring colloidal processes. However, the method does not allow for a very sharp resolution of size distributions in the colloidal size range (Babick and Ripperger 2002). In addition, the analytical models contain a variety of material parameters. This is of particular relevance for emulsions (14 properties), whereas for aqueous suspensions only the viscosity and sound speed of liquid and the density contrast have to be known (Babick et al. 2000). Lacking knowledge on the model parameters can be coped with by simultaneously evaluating sound speed and attenuation or by employing statistical data processing (e.g. neural networks; cf. Li et al. 2004; Babick et al. 2006).

In summary, the great potential of ultrasonic spectroscopy is the reliable characterisation of concentrated suspensions. For well-defined colloidal suspensions it may be used for fast off-line determinations of the size distribution, although the resolution is rather coarse. Its application to process monitoring is principally possible, yet this may require the development of an appropriate data analysis for the specific process. However, a sufficiently high sensitivity to relevant changes in product quality or process performance can outweigh this additional effort.

[12]The same applies to the calculation of sound dispersion, i.e. frequency impact on sound speed.

[13]Even so, the spectrum can only be reproduced as sphere-spectrum by assuming polydispersity.

2.3.6.3 Electroacoustic Mobility Spectroscopy (EMS)

The *electroacoustic mobility spectroscopy* (EMS) is a measurement technique that allows for the simultaneous determination of the particle size distribution and the effective zeta-potential of colloidal suspensions (O'Brien et al. 1995). Its primary signal is the acoustic response of a colloidal suspension to a high-frequency alternating electric field. It is called *electrokinetic sonic amplitude* (ESA) and is directly related to the dynamic electrophoretic mobility of the colloidal particles. The ESA signal results from the simultaneous electrophoretic motion of all charged particles. The corresponding oscillatory flux of momentum is transferred to the walls and subsequently transmitted as a sound wave (Fig. 2.19). Prerequisite for the existence of an ESA-signal is a surface charge and a density contrast between particles and continuous phase (O'Brien 1988).

The EMS can be applied to concentrated suspensions (>1 vol%) of particles above 50 nm and below 10 microns (Hunter 1998). Exact analytical models are available for solid concentrations below approximately 3 vol% and for thin double layers. Although there has been considerable progress in theoretical treatment of higher concentrated systems, empirical correction terms are usually used to account for the hydrodynamic and electrostatic particle-particle interactions. Depending on the quality of the data basis used for deriving these interpolating equations, very satisfactory results can be obtained (Knösche 2001; Babick et al. 2001).

Even though the particle motion originates from the polarisation of the double layer in the electric field, its magnitude and phase is decisively affected by the density contrast and the hydrodynamic drag. Hence, from the point of particle sizing, the dynamic electrophoretic mobility is closely related to the acoustophoretic mobility. The latter has been shown to be governed by the volume-specific surface when regarding spheroids with moderate aspect ratios (≤10/1, see above); acc. to Loewenberg and O'Brien (1992), this also applies to first approximation to the dynamic electrophoretic mobility.

Since the ESA signal indicates the momentum flux that is related to particle oscillation, mobility spectra are intrinsically volume weighted functions. However, size analysis is conducted at the frequencies at which a volume unit of coarse particles contributes less to the ESA signal than one of fine particles. This may lead to an underestimation of coarse size fractions. The size distribution is obtained by

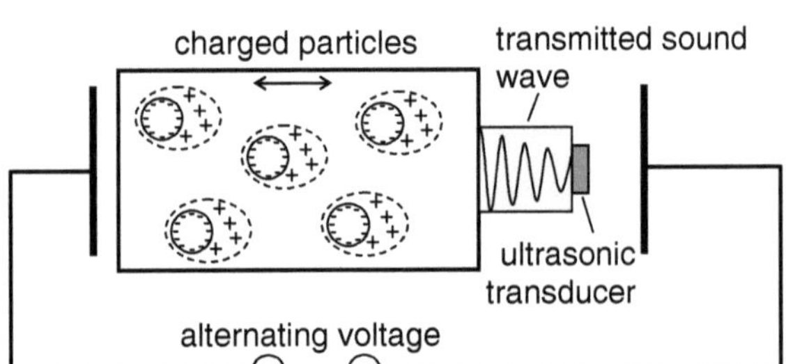

Fig. 2.19 Generation of the ESA signal in highly frequent electric fields

inversion of the mobility spectrum, yet in contrast to other spectroscopic methods, the spectral signal is a complex (i.e. two-parametric) quantity and does not only depend on the size distribution, but also on the zeta-potential. A profound analysis by Knösche (2001) revealed that, for typical measurement situations, one can extract up to 7 independent pieces of information on the size distribution. However, the ESA signal looses its sensitivity to particle size beyond the size range inidicated above (50 nm to 10 μm).

EMS has the potential advantage of being applicable in highly concentrated suspensions, which opens the way for on-line applications (Carasso et al. 1995; Hunter 1998). It delivers reliable analysis data in many circumstances, in particular for solid particles with thin diffuse layers and low polarisabilities. However, the interpretation of mobility spectra requires detailed knowledge of the particulate system, e.g. of particle shape and material properties (like dielectric constant) and sometimes of surface conductance.

2.3.7 Zeta-Potential and Interfacial Properties

When colloidal particles are dispersed in a liquid solvent, it is very likely that they acquire a surface charge (cf. Sect. 3.1.5) which has to be compensated for by a corresponding countercharge in the vicinity of the surface. Both regions, charged surface and counter-charged solvent, form the *electric double layer* (EDL). The charge separation between the two layers creates an electric potential between the particle surface and the bulk solvent, which is responsible for attractive or repulsive interactions between the colloidal particles.

Important measurands for the characterisation of the EDL are the surface charge density and the electrokinetic potential or *zeta-potential*. The zeta-potential is the electric potential at a hypothetical shear plane, which separates the mobile solvent from solvent molecules that adhere to the particle surface. The zeta-potential can be probed by imposing a relative motion between bulk solvent and particle (Delgado et al. 2007).

Such a relative motion can be induced by external electric fields or by pressure gradients or bulk forces (e.g. gravity). It is possible that particles move in a quiescent solvent or that the solvent flows through a fixed bed of particles. A detailed description on electrokinetic phenomena is e.g. given by Hunter (1988). Zeta-potential measurements on colloidal suspensions are frequently conducted via electrophoresis or by means of electroacoustics. Besides this, there are recent techniques based on non-linear optics that are sensitive to interfacial changes.

2.3.7.1 Electrophoretic Zeta-Potential Measurement

If an electric field is applied to a suspension of charged colloidal particles, the mobile ions within the double layer become spatially separated according to the

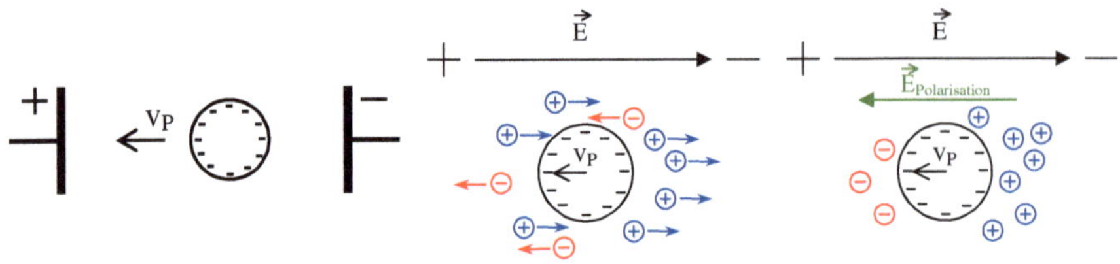

Fig. 2.20 Electrophoresis and double layer polarisation

sign of their charge (double layer polarisation), and the particles start to move along the lines of the electric field (Fig. 2.20). This electrophoretic motion is characterised by the electrophoretic mobility:

$$\mu_{\mathrm{ep}} = \frac{\mathbf{v}}{\mathbf{E}}, \tag{2.41}$$

which is a function of the zeta-potential ζ and the Debye-Hückel-parameter κ (the inverse double layer thickness $1/\kappa$):

$$\mu_{\mathrm{ep}} = \frac{2}{3} \cdot \frac{\varepsilon_{\mathrm{m}} \varepsilon_0 \zeta}{\eta} \cdot f(\kappa a, \zeta) \tag{2.42}$$

$$\text{with } \kappa^2 = \frac{\mathrm{F}^2}{\varepsilon_{\mathrm{m}} \varepsilon_0 \cdot \mathrm{R}T} \cdot \sum z_i^2 c_i^{\infty}. \tag{2.43}$$

The double layer thickness is decisive for the retarding impact of double layer polarisation on the electrophoretic motion. Equation (2.42) yields simple linear relationships only for the two limiting cases of very thin double layers ($\kappa a \to \infty$) and infinitely thick ones ($\kappa a \to 0$). The correction function $f(\kappa a)$ then simplifies to a constant value of 3/2 and 1, respectively (Fig. 2.21; von Smoluchowski 1903; Debye and Hückel 1924; O'Brien and White 1978).

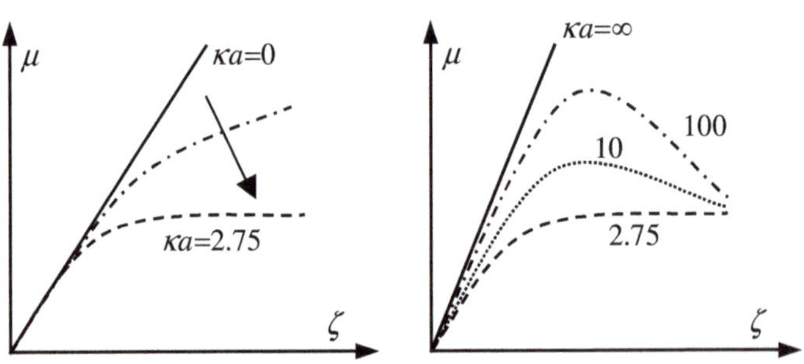

Fig. 2.21 Nonlinear relationship between zeta-potential and electrophoretic mobility acc. to O'Brien and White (1978)

The electrophoretic motion is either measured microscopically or by light scattering. The former way is called *microelectrophoresis* and usually employs ultramicroscopes when dealing with colloidal particle systems. The optical instrumentation can be identical to that of DUM, while the software has to be modified because only the displacement in the direction of the electric field is relevant. The method yields a number weighted distribution of zeta-potentials. Similar to DUM, a sufficiently large number of trajectories has to be evaluated in order to keep the statistical uncertainty within an acceptable level. Moreover, the method may be insensitive to weak scatterers within a polydisperse colloidal suspension.

An instrumental alternative to microelectrophoresis is *electrophoretic light scattering* (ELS). The light scattering at migrating particles leads to phase shift (Doppler effect), which can be detected by a heterodyne DLS set-up (i.e. reference-beating with frequency shift). The method yields an intensity weighted distribution of the zeta-potential.

The electrophoretic motion of colloidal particles is superposed by their Brownian motion. Ideally, both can be separated because of their different space-time correlations, in practice however, diffusion broadens the measured zeta-potential distribution. In microelectrophoresis, the Brownian contribution can be minimised by long observation times ($t \gg D_t/v_{ep}^2$). For ELS, diffusion is least pronounced at small scattering angles (Xu 2008).

2.3.7.2 Electroacoustic Zeta-Potential Measurement

Electroacoustic techniques are either based on the acoustic response of a suspension to an alternating electric field or on its electric response to a sound field. The former signal is the *electrokinetic sonic amplitude* (ESA), which can be employed for particle sizing (cf. Sect. 2.3.6.3) and for the determination of the zeta-potential. The latter signal is the *colloid vibration potential* (analogously the *colloid vibration current*), which is defined as the alternating electric potential between two electrodes that are separated by the half acoustic wavelength. This potential arises from the particle oscillation in the sound field (acoustophoretic motion) which coincides with a polarisation of the double layer (Fig. 2.22) (Dukhin and Goetz 2001).

The existence of the colloidal vibration potential (CVP) was predicted in the 1930s (Debye 1933; Hermans 1938a, b; Rutgers 1938) and intensively investigated—both experimentally and theoretically after the Second World War (Rutgers 1946; Enderby

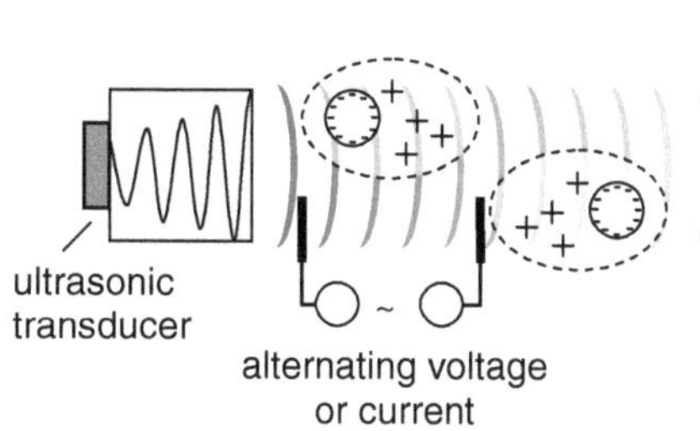

Fig. 2.22 Generation of the colloid vibration potential (CVP) or current (CVI) in ultrasonic fields

1951; Booth and Enderby 1952; Yeager et al. 1953). The discovery of the electrokinetic sonic amplitude (ESA) occurred comparatively late, in the 1980s (Oja et al. 1985), yet this effect was soon employed in commercial instruments for zeta-potential measurements. Subsequent theoretical treatments showed that both signals can be traced back to the dynamic electrophoretic mobility and that they form a reciprocal couple of the Onsager type (O'Brien 1988; O'Brien et al. 1994).

The electroacoustic techniques provide a powerful means to measure the zeta-potential in concentrated suspensions that are optically opaque (Hunter 2001, Dukhin et al. 2001, Greenwood 2003). Thus, laborious sample preparation, which aims at conserving the state of double layer during dilution, can be avoided. Yet, modelling electroacoustics for dense suspensions is not straightforward, since the hydrodynamic interaction and double layer overlap has to be adequately considered (e.g. Dukhin et al. 1999; Rasmussen 2001; O'Brien et al. 2003). Available models are based on several assumptions (e.g. on particle shape and the suspension structure) and require model parameters like particle size or surface conductance. Since the electroacoustic raw signals do not only depend on the suspension properties, but are affected by the acoustic coupling between suspension and sensor and by the instrument geometry, calibration is typically required (Hunter 1998; Dukhin et al. 2001). Last but not least, it should be noted that electroacoustic signals are essentially volume weighted. That means they provide a better sensitivity to very fine colloidal particles than an optical detection system and may be more appropriated for the electrokinetic characterisation of multi-component suspensions (containing weak and strong scatterers).

2.3.7.3 Surface Characterisation with Second Harmonic Generation

The zeta-potential, which is traditionally measured with electrokinetic and electroacoustic techniques, is an important parameter for the computation of particle-particle interactions and the corresponding electroviscous effects. However, it does not reveal the complete structure of the electric double layer, in particular its immobile part (Stern layer). New arising techniques that are sensitive to the real interface without affecting the double layer equilibrium may facilitate a complementary characterisation.

One such technique is the second harmonic generation (SHG), which relies on the effect that a minute portion of light is scattered with twice the frequency (i.e. energy) of the incident radiation (Fig. 2.23). This can be understood as the spontaneous re-emission of one double-energy photon after the simultaneous absorption of two normal photons (Hollis 1988). In contrast to fluorescence, there is no phase lag. Hence, the SHG is a coherent scattering event, which—like Rayleigh scattering—would be completely cancelled by destructive interference in a homogenous medium with randomly aligned molecules for any direction besides forwards. At interfaces with symmetry breaks and molecule orientation, the second harmonics superpose constructively and produce a corresponding scattering field (prerequisite that the interface curvature is not much smaller than the wavelength).

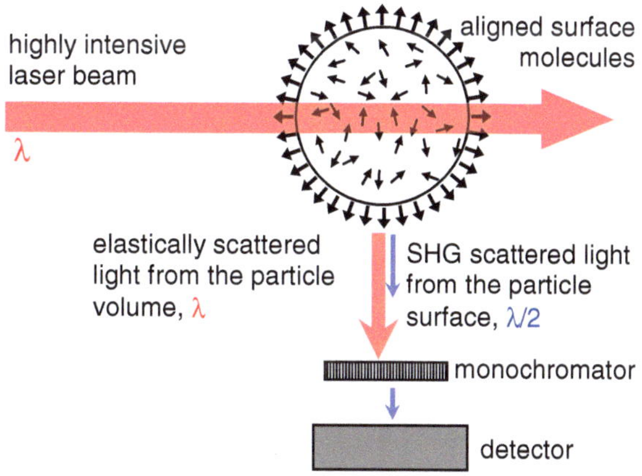

Fig. 2.23 Second harmonic generation at particle interface

In order attain measurable SHG signals, pulsed femtosecond lasers with large intensities are usually employed (Yan et al. 1998; Schneider et al. 2007). It was possible to show that the SHG scales with surface potential and the independently measured zeta-potentials can be reproduced by adopting appropriate models for the electric double layer (Yan et al. 1998). More generally, SHG is directly related to the surface excess of adsorbate as shown for malachite green on polystyrene (Eckenrode et al. 2005). This technique offers the opportunity for online and in situ characterisation of colloidal suspensions with particle sizes considerably larger than 5 nm (Schneider and Peukert 2007; Schürer and Peukert 2010).

SHG is far from being a mature characterisation technique of colloidal interfaces. But with regard to the on-going developments in instrumental set-up and the growing number of application experiences its potential benefits seems very promising.

2.3.8 Final Remarks on Particle Sizing

This chapter introduced import characterisation techniques for colloidal suspensions. The focus was on particle sizing methods, which are specified by the particle property employed to distinguish the different size fractions as well as by the type of quantity intrinsically used to quantify the size fractions. Additionally, they differ in the degree to which details of size distributions can be resolved and in their sensitivity to certain parts of the size distribution (e.g. sensitivity towards coarse, contaminant particles). Further differences concern the ranges of particle size and concentration in which sizing is possible. Table 2.2 summarises the discussed sizing methods with regard to the mentioned aspects.

The differences between the various sizing techniques mean that a measured size distribution is inevitably affected by the employed characterisation method. That frequently causes confusion on the correctness of measurements when deviations

Table 2.2 Characteristics of selected particle sizing methods

Method	Classification	Particle property	Equiv. diameter or related geometric parameter	Type of quantity (index, index for nanoparticles)	Size and concentration range (conservative)	Comments
Dynam. ultramicroscopy (DUM)	Counting	Diffusion coeff.	$x_{\mathrm{h,t}}$	Number (0)	10 nm–1 μm ≪ 1 vol%	Requires inversion, statistics
Image analysis (IA)	Counting	Any length/area of projected image	e.g. perimeter, area	Number (0)	1 nm to ≪ 1 vol%	Various imaging methods
Optical centrifugation (OCA)	Fractionating	Settling velocity	$X_{\mathrm{Stokes}} = X_V^3/X_{h,t}$	Extinction (ext, NP: 3 or 6)	10 nm–1 μm < 1 vol%	Different realisations
Field-flow fractionation (FFF)	Fractionating	e.g. diffusion coeff., settling velocity	e.g. $X_{h,t}$, X_{Stokes}	e.g. scattering intensity (int, NP: 6)	10 nm–1 μm ≪ 1 vol%	Different fields, different detection systems
Static light scattering (SLS)	Spectroscopic	Volume, diameter of gyration molecular weight	X_V, X_g	Scattering intensity (int, NP: 6)	10 nm–1 μm ≪ 1 vol%	Impact of colloid concentration
Laser diffraction spectroscopy (LD)	Spectroscopic	Main proportions of projected image	(Orientation averaged) Feret diameters	e.g. projection area (2)	500 nm–1 mm ≪ 1 vol%	Not only diffraction, impact of detection optics
Small angle X-ray scattering (SAXS)	Spectroscopic	Volume, diameter of gyration	$X_V X_g$	Scattering intensity (int, NP: 2)	1 nm–50 nm 0.1–10 vol%	Best quality with synchrotron X-ray sources

(continued)

Table 2.2 (continued)

Method	Classification	Particle property	Equiv. diameter or related geometric parameter	Type of quantity (index, index for nanoparticles)	Size and concentration range (conservative)	Comments
Dynamic light scattering (DLS)	Spectroscopic	(Apparent) diffusion coefficient	$X_{h,t}$ or $X_{h,app}$	Scattering intensity (int, NP: 6)	1 nm–1 nm < 1 vol%	Correlation spectroscopy or frequency analysis
Optical spectroscopy (OS)	Spectroscopic	Volume, diameter of gyration	$X_V X g$	Extinction (ext, NP: 3 or 6)	10 nm–1 μm < 0.1 vol%	Plasmon resonance for metals
Ultrasonic spectroscopy (US)	Spectroscopic	Scattering, acoustoph. mobility	Approx.: volume, specific surface	Volume (3)	10 nm–100 μm 1–70 vol%	Evaluation of sound speed or attenuation

between differently determined size distributions are observed. Again, apart from uncertainties due to sample preparation and measurement, there are intrinsic differences between particle sizing techniques. It is, therefore, necessary to ask for the most appropriate characterisation technique for a specific characterisation task. For that purpose, one has to clarify the objective(s) of measurement in advance (cf. Polke et al. 2003):

- which property defines the quality of the suspension or its process behaviour (particle volume, particle diffusivity, …)?
- which parameters of the size distribution are required (mean size, modal values, amount of nanoparticles, …)?
- where will the measurement be conducted (at process line, in laboratories, …)?
- what level of accuracy and precision is required?

The answers to those questions define the requirements on the measurement instrument and reduce the list of applicable characterisation techniques. It might be even possible that the characterisation aims at the quantification of two or more independent or weakly correlating parameters of the size distribution (e.g. maximum size and average size). In that case, it is quite likely that at least two methods have to be employed.

Ruscitti et al. (2008) reported on the performance of different sizing techniques with regard to the qualification of product quality after comminution. The product quality was described by a set of three macroscopic suspension properties, two of which correlate with "average" particle size (below 1 μm) and one with the maximum size (above 1 μm). The outcome of this study was twofold: On the one hand, it was not possible to determine average and maximum particle size with one inline-capable sizing technique. On the other hand, the authors significantly improved their understanding of the process, in particular with regard to the interdependency between the quality parameters.

In general, by combining different characterisation techniques, one can benefit from their specific characteristics, e.g. probed particle property, and gain (more) comprehensive insight into the microstructure of colloidal suspensions.

References

General Matter

R. Polke, M. Schäfer, N. Scholz, Charakterisierung disperser Systeme, in *Handbuch der Mechanischen Verfahrenstechnik, Bd. 1*; Chapter 2, ed. by H. Schubert (Wiley-VCH, Weinheim, 2003), pp. 7–100. ISBN 3-527-30577-7

R. Xu, *Particle Characterization: Light Scattering Methods* (Kluwer Academic Publishers, Dortrecht, 2000). ISBN 0-7923-6300-0

Sizing

F. Babick, S. Ripperger, Information content of acoustic attenuation spectra. Part. Part. Syst. Charact. **19**(3), 176–185 (2002). doi:10.1002/1521-4117(200207)19:3<176:AID-PPSC176>3.0.CO;2-8

R. Finsy, N. De Jaeger, R. Sneyers, E. Geladé, Particle sizing by photon correlation spectroscopy. Part III: Mono and Bimodal Distributions and Data Analysis. Part. Part. Sys. Charact. **9**(1–4), 125–137 (1992). doi:10.1002/ppsc.19920090117

ISO 9276-1:1998, *Representation of Results of Particle Size Analysis—Part 1: Graphical Representation* (Beuth-Verlag, Berlin, 2004)

M. Kandlikar, G. Ramachandran, Inverse methods for analysing aerosol spectrometer measurements: a critical review. J. Aerosol Sci. **30**(4), 413–437 (1999). doi:10.1016/0021-8502(98)00066-4

C. Knösche, *Möglichkeiten und Grenzen der elektroakustischen Spektroskopie zur Gewinnung von Partikelgrößeninformationen*. PhD thesis, Technische Universität Dresden, 2001

J.H. Koo, E.D. Hirleman, Synthesis of integral transform solutions for the reconstruction of particle-size distributions from forward-scattered light. Appl. Opt. **31**(12), 2130–2140 (1992). doi:10.1364/AO.31.00213

K. Leschonski, Representation and evaluation of particle size analysis data. Part. Part. Syst. Char. **1**(1–4), 89–95 (1984). doi:10.1002/ppsc.19840010115

U. Riebel, F. Löffler, The fundamentals of particle size analysis by means of ultrasonic spectrometry. Part. Part. Syst. Charact. **6**(1–4), 135–143 (1989). doi:10.1002/ppsc.19890060124

H. Rumpf, K.F. Ebert, Darstellung von Kornverteilungen und Berechnung der spezifischen Oberfläche. Chem. Ing. Tech. **36**(5), 523–537 (1964). doi:10.1002/cite.330360516

M. Stintz, *Technologie-relevante Charakterisierung von Partikeln und Partikelsystemen* Habilitation thesis, Technische Universität Dresden, 2005

M. Stintz, F. Babick, G. Roebben, *Metrology for nanoparticle characterisation: instruments, standard methods and reference materials* (Workshop report, Nuremberg, 28./29.04.2010). Co-nanomet Consortium, c/o euspen, Cranfield, 2010. ISBN 978-0-9566809-3-8

R.S. Stock, W.H. Ray, Interpretation of photon correlation spectroscopy data: A comparison of analysis methods. J. Polym. Sci. Part B: Polym. Phys. **23**(7), 1393–1447 (1985). doi:10.1002/pol.1985.180230707

S. Twomey, *Introduction to the mathematics of inversion in remote sensing and indirect measurements* (Elsevier Scientific Publishing Company, Amsterdam, 1977). ISBN 0-444-41547-5

W. Witt, U. Köhler, J. List, Current limits of particle size and shape analysis with high speed image analysis. On the CD-ROM: *PARTEC 2007, International Congress for Particle Technology, Nuremberg*, 27–29 Mar 2007. (available from NürnbergMesse GmbH, Messezentrum, 90471 Nürnberg, Germany), paper S35_2

Ultramicroscopy

C. Bigg, Evident atoms: visuality in Jean Perrin's Brownian motion research. Stud. Hist. Phil. Sci. **39**(3), 312–322 (2008). doi:10.1016/j.shpsa.2008.06.003

R.F. Domingos, M.A. Baalousha, Y. Ju-Nam, M.M. Reid, N. Tufenkji, J.R. Lead, G.G. Leppard, K.J. Wilkinson, Characterizing manufactured nanoparticles in the environment: multimethod determination of particle sizes. Environ. Sci. Technol. **43**(19), 7277–7284 (2009). doi:10.1021/es900249m

A. Einstein, Über die von der molekulartheoretischen Theorie der Wärme geforderten Bewegung von in ruhenden Flüssigkeiten suspendierten Teilchen. Ann. Phys. IV **17**(8), 549–560 (1905). doi:10.1002/andp.19053220806

J. Perrin, L'agitation molèculaire et le mouvement brownien. CR Hebd. Seance Acad. Sci. **146** (19), 967–970 (1908)

J. Perrin, Mouvement brownien et realité moléculaire. Ann. Chim. Phys. **8**(18), 1–114 (1909)

J. Reissig, Ultramikroskopische Beobachtungen. Ann. Phys. IV **27**(11), 186–212 (1908). doi:10.1002/andp.19083321110

H. Saveyn, B. De Baets, O. Thas, P. Hole, J. Smith, P. Van der Meeren, Accurate particle size distribution determination by nanoparticle tracking analysis based on 2-D Brownian dynamics simulation. J. Colloid Interface Sci. **352**(2), 593–600 (2010). doi:10.1016/j.jcis.2010.09.006

H. Siedentopf, R. Zsigmondy, Über Sichtbarmachung und Größenbestimmung ultramikroskopischer Teilchen, mit besonderer Anwendung auf Goldrubingläser. Ann. Phys. IV **10**(1), 1–39 (1903). doi 10.1002/andp.19023150102

R. Zsigmondy, Über ein neues Ultramikroskop. Phys. Z. **14**, 975–979 (1913)

R. Zsigmondy, W. Bachmann, Handhabung des Immersionsultramikroskops. Kolloid Z. **14**(6), 281–295 (1914). doi:10.1007/BF01423340

Imaging

M. V. Ardenne, Das Elektronen-Rastermikroskop. Theoretische Grundlagen. Z. Phys. **109**(9–10), 553–572 (1938). doi:10.1007/BF01341584

G. Binnig, C.F. Quate, C. Gerber, Atomic force microscope. Phys. Rev. Lett. **56**(9), 930–933 (1986). doi:10.1103/PhysRevLett.56.930

H.-J. Butt, R. Berger, E. Bonaccurso, Y. Chen, J. Wang, Impact of atomic force microscopy on interface and colloid science. Adv. Colloid Interface Sci. **133**(2), 91–104 (2007). doi:10.1016/j.cis.2007.06.001

G.D. Danilatos, Bibliography of environmental scanning electron microscopy. Microsc. Res. Tech. **25**(5–6), 529–534 (1993). doi:10.1002/jemt.1070250526

H.-U. Danzebrink, L. Koenders, G. Wilkening, A. Yacoot, H. Kunzmann, Advances in scanning force microscopy for dimensional metrology *Ann.* CIRP **55**(2), 841–878 (2006). doi:10.1016/j.cirp.2006.10.010

P. Fiala, D. Göhler, E. Buhr, T. Dziomba, T. Klein, *Realisierung und Optimierung von Präparationsmethoden für zuverlässige Größenmessungen mit AFM und TSEM*. DIN-INS report,-Technische Universität Dresden, Physikalisch-Technische Bundesanstalt Braunschweig, 2011

F.J. Giessibl, Advances in atomic force microscopy. Rev. Mod. Phys. **75**(3), 949–983 (2003). doi:10.1103/RevModPhys.75.949

L.-O. Heim, J. Blum, M. Preuss, H.-J. Butt, Adhesion and friction forces between spherical micrometre-sized particles. Phys. Rev. Lett. **83**(16), 3328–3331 (1999). doi:10.1103/PhysRevLett.83.3328

S.W. Hell, Far-field optical nanoscopy. Science **316**(5828), 1153–1158 (2007). doi:10.1126/science.1137395

M. Knoll, E. Ruska, Das Elektronenmikroskop. Z. Phys. **78**(5–6), 318–339 (1932). doi:10.1007/BF01342199

P. F. Schmidt et al., *Praxis der Rasterelektronenmikroskopie und Mikrobereichsanalyse*. In series: *Kontakt & Studium*, ed. by W. J. Bartz, vol. 444 (Messtechnik. expert Verlag, Renningen-Malmsheim, 1994). ISBN 3-8169-1038-6

Sedimentation

M. Beiser, *Sedimentationsverhalten submikroner Partiklen in Abhängigkeit physikalisch-chemischer Einflüsse und ihr Separationverhalten in Dekantierzentrifugen*. PhD thesis, Universität Fridericiana Karlsruhe (TH), 2005

C. Bernhardt, Sedimentation in particle size analysis, in *Encyclopedia of analytical chemistry*, ed. by R. A. Meyers (Wiley, 2010). doi:10.1002/9780470027318.a1513

G. Bickert, *Sedimentation feinster suspendierter Partikeln im Zentrifugalfeld*. PhD thesis, Universität Fridericiana Karlsruhe (TH), 1997

S.G. Conlin, W.J. Levene, W.F. Volume, An instrument for size analysis of fine powders by X-ray absorption. J. Sci. Instrum. **44**(8), 606–610 (1967). doi:10.1088/0950-7671/44/8/306

T. Detloff, T. Sobisch, D. Lerche, Particle size distribution by space or time dependent extinction profiles obtained by analytical centrifugation. Part. Part. Syst. Charact. **23**(2), 184–187 (2006). doi:10.1002/ppsc.200601028

K. Edelmann, *Lehrbuch der Kolloidchemie*, vol. I (Verlag der Wissenschaften, Berlin, 1962)

S. Enomoto, T. Hiroshi, N. Tachikawa, M. Senoo, Particle size determination by use of 55Fe X-ray absorption. Int. J. Appl. Radiat. Isot. **30**(1), 51–54 (1979). doi:10.1016/0020-708X(79)90097-8

B. L. Henke, E. M. Gullikson, J. C. Davis, X-Ray Interactions: Photoabsorption, scattering, transmission, and reflection at E = 50-30,000 eV, Z = 1-92. At. Data Nucl. Data Tables **54**(2), 181–342 (1993). doi:10.1006/adnd.1993.1013

H.J. Kamack, A note on centrifugal particle size analysis. J. Phys. D Appl. Phys. **5**(10), 1962–1968 (1972). doi:10.1088/0022-3727/5/10/330

K. Leschonski, *Hochschulkurs Grundlagen und Moderne Verfahren der Partikelmesstechnik*. Technische Universität Clausthal, Institut für Mechanische Verfahrenstechnik und Umweltverfahrenstechnik, 1982

S. Odén, Eine neue Methode zur Bestimmung der Körnerverteilung in Suspensionen. Kolloid Z. **18**(2), 33–48 (1916). doi:10.1007/BF01432659

W. Ostwald, F.-V. V. Hahn, Ueber kinetische Flockungsmesser, I. Kolloid-Z. **30**(1), 62–70, 1922. doi:10.1007/BF01430381

K. M. Paciejewska, *Untersuchung des Stabilitätsverhaltens von binären kolloidalen Suspensionen*. PhD thesis, Technische Universität Dresden, 2010. http://nbn-resolving.de/urn:nbn:de:bsz:14-qucosa-65050

G. E. Salinas Salas, *Sedimentationsverhalten von Submikrometerpartikeln in wässrigen Suspensionen*. PhD thesis, Technische Universität Dresden, 2007

K. Schilling, *Charakterisierung mizellarer Systeme mit der Analytischen Ultrazentrifuge—Synthese von Titandioxid in mizellaren Reaktionsräumen*, PhD thesis, Universität Potsdam, 1999

T. Svedberg, J.B. Nichols, Determination of size and distribution of size of particle by centrifugal methods. J. Am. Chem. Soc. **45**(12), 2910–2917 (1923). doi:10.1021/ja01665a016

T. Svedberg, H. Rinde, The ultra-centrifuge, a new instrument for the determination of size and distribution of size of particle in amicroscopic colloids. J. Am. Chem. Soc. **46**(2), 2677–2693 (1924). doi:10.1021/ja01677a011

H. C. van de Hulst, *Light Scattering by Small Particles* (Dover Publications, New York, 1981). ISBN 0-486-64228-3

Field-Flow Fractionation (FFF)

T.J. Cho, V.A. Hackley, Fractionation and characterization of gold nanoparticles in aqueous solution: asymmetric-flow field flow fractionation with MALS, DLS, and UV-Vis detection. Anal. Bioanal. Chem. **398**(5), 2003–2018 (2010). doi:10.1007/s00216-010-4133-6

J.C. Giddings, A new separation concept based on a coupling of concentration and flow nonuniformities. Separ. Sci. **1**(1), 123–125 (1966). doi:10.1080/01496396608049439
J.C. Giddings, Nonequilibrium theory of field-flow fractionation. J. Chem. Phys. **49**(1), 81–85 (1968). doi:10.1063/1.1669863
J.C. Giddings, G. Karaiskakis, K.D. Caldwell, M.N. Myers, Colloid characterization by sedimentation field-flow fractionation. I. Monodisperse population. J. Colloid Interface Sci. **92**(1), 66–80 (1983). doi:10.1016/0021-9797(83)90117-0
E. Grushka, K.D. Caldwell, M.N. Myers, Marcus, J. C. Giddings, Field flow fractionation. Sep. Purif. Rev. **2**(1), 127–151 (1974). doi:10.1080/03602547408068793
J. Laudan, *Untersuchung von ultrahochmolekularen Polymeren mittels asymmetrischer Fluss Feld-Fluss-Fraktionierung und Lichtstreu-/Konzentrationsdetektor-Kombination*. PhD thesis, Universität Hamburg, 2004
Y. Mori, B. Scarlett, H.G. Merkus, Effects of ionic strength of eluent on size analysis of submicrometre particles by sedimentation field-flow fractionation. J. Chromatogr. A **515**, 21–35 (1990). doi:10.1016/S0021-9673(01)89298-0
K.L. Plathe, F. von der Kammer, M. Hassellov, J. Moore, M. Murayama, T. Hofmann, M.F. Hochella Jr, Using FlFFF and aTEM to determine trace metal-nanoparticle associations in riverbed sediment. Environ. Chem. **7**(1), 82–93 (2010). doi:10.1071/EN09111
T. Schauer, Symmetrical and asymmetrical flow field-flow fractionation for particle size determination. Part. Part. Syst. Charact. **12**(6), 284–288 (1995). doi:10.1002/ppsc.19950120606
P.J. Wyatt, Submicrometre particle sizing by multiangle light scattering following fractionation. J. Colloid Interface Sci. **197**(1), 9–20 (1998). doi:10.1006/jcis.1997.5215
F.-S. Yang, K.D. Caldwell, J.C. Giddings, Colloid characterization by sedimentation field-flow fractionation. II. Particle size distribution. J. Colloid Interface Sci. **92**(1), 81–91 (1983). doi:10.1016/0021-9797(83)90118-2

Size-Exclusive Chromatography (SEC)

P.S. Fedotov, N.G. Vanifatova, V.M. Shkinev, B.Y. Spivakov, Fractionation and characterization of nano- and microparticles in liquid media. Anal. Bioanal. Chem. **400**(6), 1787–1804 (2011). doi:10.1007/s00216-011-4704-1
F.-K. Liu, Analysis and applications of nanoparticles in the separation sciences: a case of gold nanoparticles. J. Chromatogr. A **1216**(52), 9034–9047 (2009). doi:10.1016/j.chroma.2009.07.026
G.T. Wei, F.-K. Liu, C.R.C. Wang, Shape separation of nanometre cold particles by size-exclusion chromotography. Anal. Chem. **71**(11), 2085–2091 (1999). doi:10.1021/ac990044u
T. Yamaguchi, Y. Azuma, K. Okuyama, Development of a photon correlation spectroscopy instrument to measure size distributions of nanoparticles. Part. Part. Syst. Charact. **23**(2), 188–192 (2006). doi:10.1002/ppsc.200601029

Static Light Scattering (SLS)

P. Debye, Molecular-weight determination by light scattering. J. Phys. Colloid Chem. **51**(1), 18–32 (1947). doi:10.1021/j150451a002
A. Einstein, Theorie der Opaleszenz von homogenen Flüssigkeiten und Flüssigkeitsgemischen in der Nähe des kritischen Zustandes (Theory of opalescence of homogenous liquids and liquid mixtures near critical conditions). Ann. Phys. IV **33**(16), 1275–1298 (1910). doi:10.1002/andp.19103381612

S. Heimer, D. Težak, Structure of polydispersed colloids characterised by light scattering and electron microscopy. Adv. Colloid Interface Sci. **98**(1), 1–23 (2002). doi:10.1016/S0001-8686 (01)00090-2

W.C.K. Poon, A.D. Pirie, P.N. Pusey, Gelation in colloid-polymer mixtures. Faraday Discuss. **101**, 65–76 (1995). doi:10.1039/fd9950100065

H.M. Wyss, J. Innerlohinger, L.P. Meier, L.J. Gauckler, O. Glatter, Small-angle static light scattering of concentrated silica suspensions during in situ destabilization. J. Colloid Interface Sci. **271**(2), 388–399 (2004). doi:10.1016/j.jcis.2003.09.051

R. Xu, *Particle Characterization: Light Scattering Methods* (Kluwer Academic Publishers, Dortrecht, 2000). ISBN 0-7923-6300-0

B.H. Zimm, Molecular theory of the scattering of light in fluids. J. Chem. Phys. **13**(4), 141–145 (1945). doi:10.1063/1.1724013

B.H. Zimm, Apparatus and methods for measurement and interpretation of the angular variation of light scattering; Preliminary results on polystyrene solutions. J. Chem. Phys. **16**(12), 1099–1116 (1948). doi:10.1063/1.1746740

Laser Diffraction

A. Büchtemann, H. Dautzenberg, I. Müller, Ermittlung der Größenverteilung sphärischer Teilchen mittels Laserdiffraktion an elektronenmikroskopischen Aufnahmen. Acta Polym. **33**(10), 605–612 (1982). doi:10.1002/actp.1982.010331008

N. Chigier, Comparative measurements using different particle size instruments, in *Liquid particle size measurement techniques, ASTM STP 848*, eds. by J. M. Tishkoff, R. D. Ingelbo, J. B. Kennedy (American Society for Testing and Materials, Philadelphia, 1984), pp. 169–189. ISBN 0-8031-0227-5

G.B.J. de Boer, C. de Weerd, D. Thoenes, H.W.J. Goossens, Laser diffraction spectrometry: Fraunhofer diffraction versus Mie scattering. Part. Charact. **4**(1), 14–19 (1987). doi:10.1002/ppsc.19870040104

N. Gabas, N. Hiquily, C. Laguérie, Response of laser diffraction particle sizer to anisometric particles. Part. Part. Syst. Charact. **11**(2), 121–126 (1994). doi:10.1002/ppsc.19940110203

C.M.G. Heffels, P.J.T. Verheijen, D. Heitzmann, B. Scarlett, Correction of the effect of particle shape on the size distribution measured with a laser diffraction instrument. Part. Part. Syst. Charact. **13**(5), 271–279 (1996). doi:10.1002/ppsc.19960130504

M. Heuer, K. Leschonski, Results obtained with a new instrument for the measurement of particle size distributions from diffraction patterns. Part. Charact. **2**(1-4), 7-13 (1985). doi: 10.1002/ppsc.1985002010210.1002/ppsc.19850020102

ISO 13320:2009, *Particle Size Analysis—Laser Diffraction Methods* (Beuth-Verlag Berlin, 2009)

A.R. Jones, Fraunhofer diffraction by random irregular particles. Part. Part. Syst. Charact. **4**(4), 123–127 (1987). doi:10.1002/ppsc.19870040126

C. Knösche, *Möglichkeiten und Grenzen der elektroakustischen Spektroskopie zur Gewinnung von Partikelgrößeninformationen*. PhD thesis, Technische Universität Dresden, 2001

P. Kuchenbecker, M. Gemeinert, T. Rabe, Interlaboratory study of particle size distribution measurements by laser diffraction. Part. Part. Syst. Charact. **29**(4), 304–310 (2012). doi:10.1002/ppsc.201000026

M.N. Lodi, E.P. Osmolovskaya, Laser diffraction systems for measuring small objects. Meas. Techn. **18**(7), 1006–1008 (1975). doi:10.1007/BF00818603

M. Ludwig, *Methodeneruierung zur Charakterisierung von Sprühtröpfchen verschiedener Sprühlösungen*, public report, Technische Universität Dresden, Institut für Verfahrenstechnik und Umwelttechnik, 2010

T. Matsuyama, H. Yamamoto, B. Scarlett, Transformation of diffraction pattern due to ellipsoids into equivalent diameter distribution for spheres. Part. Part. Syst. Charact. **17**(2), 41–46 (2000). doi:10.1002/1521-4117(200006)17:2<41:AID-PPSC41>3.0.CO;2-W

G. Mie, Beiträge zur Optik trüber Medien, speziell kolloidaler Goldlösungen. Ann. Phys. IV **25**(3), 377–445 (1908). doi:10.1002/andp.19083300302

Y. Mori, H. Yoshida, H. Masuda, Characterization of reference particles of transparent glass by laser diffraction method. Part. Part. Syst. Charact. **24**(2), 91–96 (2007). doi:10.1002/ppsc.200601048

Y. Mori, H. Yoshida, H. Masuda, Round robin test results of reference particle candidates of submicrometer size range. On the CD-ROM: *Particulate Systems Analysis 2008, Stratford-Upon-Avon,* 02–04 Sept 2008. (available from PCIG, C/o Particle Technology Ltd, Station Yard Industrial Estate, Hatton, Derbyshire, DE65 5DU, United Kingdom), paper 38

N. Stevens, J. Shrimpton, M. Palmer, D. Prime, B. Johal, Accuracy assessments for laser diffraction measurements of pharmaceutical lactose. Meas. Sci. Technol. **18**(12), 3697–3706 (2007). doi:10.1088/0957-0233/18/12/004

T. Stromgren, Linear measurements of growth of shells using laser diffraction. Limnol. Oceanogr. **20**(5), 845–848 (1975). doi:10.4319/lo.1975.20.5.0845

T. Stübinger, U. Köhler, W. Witt, Verification of Mie scattering algorithms by extreme precision calculations. On the CD-ROM: *WCPT6 2010, World Congress on Particle Technology, Nuremberg,* 2–29 Apr 2010. ISBN 978-3-00-030570-2. (available from NürnbergMesse GmbH, Messezentrum, 90471 Nürnberg, Germany), paper 00272

B.J. Thompson, Hybrid processing systems—assessment. Proc. IEEE **65**(1), 62–76 (1977). doi:10.1109/PROC.1977.10431

A.P. Tinke, A. Carnicer, R. Govoreanu, G. Scheltjens, L. Lauwerysen, N. Mertens, Particle shape and orientation in laser diffraction and static image analysis. Powder Technol. **186**(2), 154–167 (2008). doi:10.1016/j.powtec.2007.11.017

H. C. van de Hulst, *Light Scattering by Small Particles* (Dover Publications, New York, 1981). ISBN 0-486-64228-3

J. von Fraunhofer, Neue Modifikation des Lichtes durch gegenseitige Einwirkung und Beugung der Strahlen, und Gesetze derselben. Denkschr. K. Akad. Wiss. München **8**, 3–76 (1821)

W. Witt, T. Stübinger, U. Köhler, J. List, J. Jordan, Partikelgrößenanalyse mit absoluter Genauigkeit. Chem. Ing. Tech. **84**(3), 211–222 (2012). doi:10.1002/cite.201100172

R. Xu, *Particle Characterization: Light Scattering Methods* (Kluwer Academic Publishers, Dortrecht, 2000). ISBN 0-7923-6300-0

T. Yamauchi, Y. Ohyama, A study on the measurement of particle size distribution with laser diffraction systems. Bull. Jpn. Soc. Mech. Eng. **25**(210), 1931–1937 (1982). doi:10.1299/jsme1958.25.1931

Small Angle Neutron and X-ray Scattering (SANS and SAXS)

J.W. Anderegg, W.W. Beeman, S. Shulman, P. Kaesberg, An investigation of the size, shape and hydration of serum albumin by small-angle X-ray scattering. J. Am. Chem. Soc. **77**(11), 2927–2937 (1955). doi:10.1021/ja01616a002

E. Bugnicourt, J. Galy, J.-F. Gérarda, F. Bouéb, H. Barthel, Structural investigations of pyrogenic silica–epoxy composites: combining small-angle neutron scattering and transmission electron microscopy. Polymer **48**(4), 949–958 (2007). doi:10.1016/j.polymer.2006.12.012

H.-D. Dörfler, *Grenzflächen- und Kolloidchemie* (Wiley-VCH, Weinheim, 1994). ISBN 3-527-29072-9

G. Fritz, O. Glatter, Structure and interaction in dense colloidal systems: evaluation of scattering data by the generalized indirect Fourier transformation method. J. Phys. Condens. Matter **18** (36), S2403–S2419 (2006). doi:10.1088/0953-8984/18/36/S14

O. Glatter, R. May, Small-angle techniques, in *International tables for crystallography volume C: Mathematical, physical and chemical tables*, Chapter 2.6, *IUCr Series. International Tables for Crystallography*, ed. by E. Prince (Springer, Luxembourg, 2006), pp. 89–112. ISBN 978-1-4020-1900-5; doi:10.1107/97809553602060000581
O. Glatter, O. Kratky (eds.), *Small angle X-ray scattering* (Academic Press, London, 1982). ISBN 0-12-286280-5
A. Guinier, La diffusion des rayons X sous les très faibles angles appliquèe á l'etude de fines particules et de suspensions colloidales (The diffusion of X-rays under the extremely weak angles applied to the study of fine particles and colloidal suspension). CR Hebd. Seance Acad. Sci. **206**(19), 1374–1376 (1938)
B. L. Henke, E. M. Gullikson, J. C. Davis, X-Ray Interactions: Photoabsorption, scattering, transmission, and reflection at E = 50-30,000 eV, Z = 1-92. At. Data Nucl. Data Tables **54**(2), 181–342 (1993). doi:10.1006/adnd.1993.1013
A.J. Hurd, D.W. Schaefer, J.E. Martin, Surface and mass fractals in vapor-phase aggregates. Phys. Rev. A **35**(5), 2361–2364 (1987). doi:10.1103/PhysRevA.35.2361
J. Ilavsky, A.J. Allen, G.G. Long, P.R. Jemian, Effective pinhole-collimated ultrasmall-angle X-ray scattering instrument for measuring anisotropic microstructures. Rev. Sci. Instrum. **73** (3), 1660–1662 (2002). doi:10.1063/1.1425387
H.K. Kammler, G. Beaucage, R. Mueller, S.E. Pratsinis, Structure of flame-made silica nanoparticles by ultra-small-angle X-ray scattering. Langmuir **20**(5), 1915–1921 (2004). doi:10.1021/la030155v
O. Kratky, H. Stabinger, X-ray small angle camera with block-collimation system an instrument of colloid research. Colloid Polym. Sci. **262**(5), 345–360 (1984). doi:10.1007/BF01410252
D. Qiu, C.A. Dreiss, T. Cosgrove, A.M. Howe, Small-angle neutron scattering study of concentrated colloidal dispersions: The interparticle interactions between sterically stabilized particles. Langmuir **21**(22), 9964–9969 (2005). doi:10.1021/la050322m
S. Romer, C. Urban, H. Bissig, A. Stradner, F. Scheffold, P. Schurtenberger, Dynamics of concentrated colloidal suspensions: diffusion, aggregation and gelation. Phil. Trans. R. Soc. Lond. A **359**(1782), 977–984 (2001). doi:10.1098/rsta.2000.0812
P.W. Schmidt, D. Avnir, D. Levy, A. Höhr, M. Steiner, A. Röll, Small angle X-ray scattering from the surfaces of reversed phase silicas: power-law scattering exponents of magnitudes greater than four. J. Chem. Phys. **94**(2), 1474–1479 (1991). doi:10.1063/1.460006

Dynamic Light Scattering (DLS)

L.B. Aberle, P. Hülstede, S. Wiegand, W. Schröer, W. Staude, Effective suppression of multiply scattered light in static and dynamic light scattering. Appl. Opt. **37**(27), 6511–6524 (1998). doi:10.1364/AO.37.006511
S.R. Aragón, R. Pecora, Theory of dynamic light scattering from large anisotropic particles. J. Chem. Phys. **66**(6), 2506–2516 (1977). doi:10.1063/1.434246
F.T. Arecchi, M. Giglio, U. Tartari, Scattering of coherent light by a statistical medium. Phys. Rev. **163**(1), 186–197 (1967). doi:10.1103/PhysRev.163.186
H. Auweter, D. Horn, Fiber-optical quasi-elastic light scattering of concentrated dispersions. J. Colloid Interface Sci. **105**(2), 399–409 (1985). doi:10.1016/0021-9797(85)90313-3
F. Babick, M. Vorbau, M. Stintz, Characterization of pyrogenic powders with conventional particle sizing technique: II. Experimental data. Part. Part. Syst. Charact. **29**(2), 116–127 (2012). doi:10.1002/ppsc.201000025
A. Banchio, G. Nägele, J. Bergenholtz, Collective diffusion, self-diffusion and freezing criteria of colloidal suspensions. J. Chem. Phys. **113**(8), 3381–3396 (2000). doi:10.1063/1.1286964
G.K. Batchelor, Sedimentation in a dilute polydisperse system of interacting spheres. Part 1. General theory. J. Fluid Mech. **119**, 379–408 (1982). doi:10.1017/S0022112082001402

G.K. Batchelor, C.-S. Wen, Sedimentation in a dilute polydisperse system of interacting spheres. Part 2. Numerical results. J. Fluid Mech. **124**, 495–528 (1982). doi:10.1017/S0022112082002602

G. Bolle, C. Cametti, P. Codastefano, P. Tartaglia, Kinetics of salt-induced aggregation in polystyrene lattices studied by quasielastic light scattering. Phys. Rev. A **35**(2), 837–841 (1987). doi:10.1103/PhysRevA.35.837

A. Braun, K. Franks, V. Kestens, G. Roebben, A. Lamberty, T. Linsinger, *Certification Report. Certification of Equivalent Spherical Diameters of Silica Nanoparticles in Water. Certified Reference Material ERM®-FD100* (Publications Office of the European Union, Luxembourg, 2011). ISSN 1018-5593, doi:10.2787/33725

L.G.B. Bremer, L. Deriemaeker, R. Finsy, E. Gelad, J.G.H. Joosten, Fiber optic dynamic light scattering, neither homodyne nor heterodyne. Langmuir **9**(8), 2008–2014 (1993). doi:10.1021/la00032a019

B. Chu, F.J. Schoenes, The laser homodyne "self-beating" technique in light scattering. J. Colloid Interface Sci. **27**(3), 424–431 (1968). doi:10.1016/0021-9797(68)90180-X

H.Z. Cummins, Y. Yeh, N. Knable, Observation of diffusion broadening of Rayleigh scattered light. Phys. Rev. Lett. **12**(6), 150–153 (1964). doi:10.1103/PhysRevLett.12.150

J.K.G. Dhont, C.G. de Kruif, Scattered light intensity cross correlation. I. Theory. J. Chem. Phys. **79**(4), 1658–1683 (1983). doi:10.1063/1.446009

A. di Biasio, G. Bolle, C. Cametti, P. Codastefano, F. Sciortino, P. Tartaglia, Crossover region in the aggregation of colloids. Phys. Rev. E **50**(2), 1649–1652 (1994). doi:10.1103/PhysRevE.50.1649

M. Drewel, J. Ahrens, U. Podschus, Decorrelation of multiple scattering for an arbitrary scattering angle. J. Opt. Soc. Am. A: **7**(2), 206–210 (1990). doi:10.1364/JOSAA.7.000206

J.W. Dunning Jr, J.C. Angus, Particle-size measurement by doppler-shifted laser light, a test of the Stokes-Einstein relation. J. Appl. Phys. **39**(5), 2479–2480 (1968). doi:10.1063/1.1656588

R. Finsy, P. de Groen, L. Deriemaeker, M. van Laethem, Singular value analysis and reconstruction of photon correlation data equidistant in time. J. Chem. Phys. **91**(12), 7374–7383 (1989). doi:10.1063/1.457260

R. Finsy, Particle sizing by quasi-elastic light scattering. Adv. Colloid Interface Sci. **52**, 79–143 (1994). doi:10.1016/0001-8686(94)80041-3

R. Foord, E.R. Pike, E. Jakeman, R.J. Blagove, E. Wood, A.R. Peacocke, Determination of diffusion coefficients of haemocyanin at low concentration by intensity fluctuation spectroscopy of scattered laser light. Nature **227**(5255), 242–245 (1970). doi:10.1038/227242a0

H. Geers, W. Witt, Direct calculation of the volume based particle size distribution from PCS or PCCS measurements. On the CD-ROM: *Particulate Systems Analysis 2008, Stratford-Upon-Avon,* 02–04 Sept 2008. (available from PCIG, C/o Particle Technology Ltd, Station Yard Industrial Estate, Hatton, Derbyshire, DE65 5DU, United Kingdom), paper 28

S.E. Harding, On the hydrodynamic analysis of macromolecular conformation. Biophys. Chem. **55**(1–2), 69–93 (1995). doi:10.1016/0301-4622(94)00143-8

T.M. Herrington, B.R. Midmore, The rapid aggregation of dilute suspensions—an experimental investigation of Smoluchowski theorem using photon-correlation spectroscopy. Powder Technol. **65**(1–3), 251–256 (1991). doi:10.1016/0032-5910(91)80188-O

M. Hoffmann, C.S. Wagner, L. Harnau, A. Wittemann, 3D Brownian diffusion of submicron-sized particle clusters. ACS Nano **3**(10), 3326–3334 (2009). doi:10.1021/nn900902b

K. Ishii, T. Iwai, Theoretical analysis of path-length-resolved power spectrum measurement using low-coherence dynamic light scattering. Jpn. J. Appl. Phys. **47**(11), 8397–8401 (2008). doi:10.1143/JJAP.47.8397

M. Itoh, K. Takahashi, Measurement of aerosol particles by dynamic light scattering—I. Effects of non-Gaussian concentration fluctuation in real time photon correlation spectroscopy. J. Aerosol Sci. **22**(7), 815–822 (1991). doi:10.1016/0021-8502(91)90076-T

E. Jakeman, E.R. Pike, Spectrum of clipped photon-counting fluctuations of Gaussian light. Phys. A **2**(3), 411–412 (1969). doi:10.1088/0305-4470/2/3/021

E. Jakeman, Theory of optical spectroscopy by digital autocorrelation of photon-counting fluctuations. J. Phys. A **3**(2), 201–215 (1970). doi:10.1088/0305-4470/3/2/012

S.S. Jena, H.B. Bohidar, Determination of absolute polydispersity and molecular-weight distribution of high-molecular-weight polymers from dynamic light-scattering. J. Chem. Phys. **99**(1), 673–681 (1993). doi:10.1063/1.465739

R.B. Jones, P.N. Pusey, Dynamics of suspended colloidal spheres. Ann. Rev. Phys. Chem. **42**, 137–169 (1991). doi:10.1146/annurev.physchem.42.1.137

U. Kätzel, T. Richter, M. Stintz, H. Barthel, T. Gottschalk-Gaudig, Phase transitions of pyrogenic silica suspensions: a comparison to model laponite. Phys. Rev. E **76**(3), 031402 (2007). doi:10.1103/PhysRevE.76.031402

A. Khintchine, Korrelationstheorie der stationären stochastischen Prozesse. Math. Ann. **109**(1), 604–615 (1934). doi:10.1007/BF01449156

D.E. Koppel, Analysis of macromolecular polydispersity in intensity correlation spectroscopy: the method of cumulants. J. Chem. Phys. **57**(11), 4814–4820 (1972). doi:10.1063/1.1678153

M. Kroon, G.H. Wegdam, R. Sprik, Dynamic light scattering studies on the sol-gel transition of a suspension of anisotropic colloidal particles. Phys. Rev. E **54**(6), 6541–6550 (1996). doi:10.1103/PhysRevE.54.6541

A. Lamberty, K. Franks, A. Braun, V. Kestens, G. Roebben, T. Linsinger, Interlaboratory comparison for the measurement of particle size and zeta potential of silica nanoparticles in an aqueous suspension. J. Nanopart. Res. **13**(12), 7317–7329 (2011). doi:10.1007/s11051-011-0624-4

C. L. Lawson, R. J. Hanson, *Solving least squares problems*. In series: *Classics in applied mathematics*, vol. 15 (Society for Industrial Mathematics, Philadelphia, 1995), pp. 158–165. ISBN 0-898-71356-0

D. Maier, M. Marth, J. Honerkamp, J. Weese, Influence of correlated errors on the estimation of the relaxation time spectrum in dynamic light scattering. J. Appl. Opt. **38**(21), 4671–4680 (1999). doi:10.1364/AO.38.004671

S. Manley, B. Davidovitch, N.R. Davies, L. Cipelletti, A.E. Bailey et al., Time-dependent strength of colloidal gels. Phys. Rev. Lett. **95**(4), 048302 (2005). doi:10.1103/PhysRevLett.95.048302

M. Nakamura, M. Suzuki, H. Takano, M. Itoh, Simultaneous rotational and translational diffusion measurement of ellipsoidal fine particles by dual-polarization photon correlation spectroscopy. *On the CD-ROM: 10th International Conference on Liquid Atomization and Spray Systems ICLASS-2006*, Kyoto, 27 Aug–01 Sept 2006. (available from Institute for Liquid Atomization and Spray Systems—Japan, 2-14-9, Kasugadenaka, Konohana-ku, Osaka-shi, 554-0022, JAPAN), paper 217

E. Overbeck, C. Sinn, Three-dimensional dynamic light scattering. J. Mod. Opt. **46**(2), 303–326 (1999). doi:10.1080/09500349908231273

R. Paschotta, *Encyclopedia of Laser Physics and Technology,* vol. A—M (Wiley-VCH, Berlin, 2008). ISBN 978-3-527-40828-3

G.L. Paul, P.N. Pusey, Observation of a long-time tail in Brownian motion. J. Phys. A: Math. Gen. **14**(12), 3301–3327 (1981). doi:10.1088/0305-4470/14/12/025

R. Pecora, Doppler shifts in light scattering from pure liquids and polymer solutions. J. Chem. Phys. **40**(6), 1604–1614 (1964). doi:10.1063/1.1725368

R. Peters, Y. Georgalis, W. Sänger, Accessing lysozyme nucleation with a novel dynamic light scattering detector. Acta Cryst. D **54**(5), 873–877 (1998). doi:10.1107/S0907444998002455

G.D.J. Phillies, Suppression of multiple scattering effects in quasielastic light scattering by homodyne cross-correlation techniques. J. Chem. Phys. **74**(1), 260–262 (1981a). doi:10.1063/1.440884

G.D.J. Phillies, Experimental demonstration of multiple-scattering suppression in quasielastic-light-scattering spectroscopy by homodyne coincidence techniques. Phys. Rev. A **24**(4), 1939–1943 (1981b). doi:10.1103/PhysRevA.24.1939

S.W. Provencher, CONTIN. A general purpose constrained regularization program for inverting noisy linear algebraic and integral equations. Comput. Phys. Commun. **27**(3), 229–242 (1982). doi:10.1016/0010-4655(82)90174-6

V. Ruzicka, L. Zulian, G. Ruocco, Routes to gelation in a clay suspension. Phys. Rev. Lett. **93**(25), 258301 (2004). doi:10.1103/PhysRevLett.93.258301

K. Schätzel, Suppression of multiple scattering by photon cross-correlation techniques. J. Mod. Opt. **38**(9), 1849–1865 (1991). doi:10.1080/09500349114551951

P.N. Segré, W. van Megen, P.N. Pusey, K. Schätzel, W. Peters, Two-colour dynamic light scattering. J. Mod. Opt. **42**(9), 1929–1952 (1995). doi:10.1080/09500349514551681

M. Siddiq, C. Wu, B. Li, Dynamic light-scattering characterization of the molecular weight distribution of a broadly distributed phenolphthalein poly(ary1 ether ketone). J. Appl. Polymer Sci. **60**(11), 1995–1999 (1996). doi:10.1002/(SICI)1097-4628(19960613)60:11<1995:AID-APP24>3.0.CO;2-Z

A. J. F. Siegert, *On the fluctuations in signals returned by many independently moving scatterers.* MIT Rad. Lab. Rep. 465, Massachusetts Institute of Technology, 1943

R.S. Stock, W.H. Ray, Interpretation of photon correlation spectroscopy data: a comparison of analysis methods. J. Polym. Sci. Part B Polym. Phys. **23**(7), 1393–1447 (1985). doi:10.1002/pol.1985.180230707

C. Urban, P. Schurtenberger, Application of a new light scattering technique to avoid the influence of dilution in light scattering experiments with milk. Phys. Chem. Chem. Phys. **1**(17), 3911–3915 (1999). doi:10.1039/a903906f

N. Wiener, Generalized harmonic analysis. Acta Math. **55**(1), 117–258 (1930). doi:10.1007/BF02546511

H. Wiese, D. Horn, Single-mode fibers in fiber-optic quasielastic light scattering: a study of the dynamics of concentrated latex dispersions. J. Chem. Phys. **94**(10), 6429–6443 (1991). doi:10.1063/1.460272

A.W. Willemse, J.C.M. Marijnissen, A.L. van Wuyckhuyse, R. Roos, H.G. Merkus, B. Scarlett, Low-concentration photon correlation spectroscopy. Part. Part. Syst. Charact. **14**(4), 157–162 (1997)

G. Williams, D.C. Watts, Non-symmetrical dielectric relaxation behaviour arising from a simple empirical decay function. Trans. Faraday Soc. **66**, 80–85 (1970). doi:10.1039/tf9706600080

R. Xu, *Particle Characterization: Light Scattering Methods* (Kluwer Academic Publishers, Dortrecht, 2000). ISBN 0-7923-6300-0

Diffusive Wave Spectroscopy (DWS)

M. Alexander, D.G. Dalgleish, Diffusing wave spectroscopy of aggregating and gelling systems. Curr. Opin. Colloid Interface Sci. **12**(4–5), 179–186 (2007). doi:10.1016/j.cocis.2007.07.008

G. Maret, P.E. Wolf, Multiple light scattering from disordered media. The effect of Brownian motion of scatterers. Z. Phys. B **65**(4), 409–413 (1987). doi:10.1007/BF01303762

T.G. Mason, D.A. Weitz, Optical measurement of frequency-dependent linear viscoelastic moduli of complex fluid. Phys. Rev. Lett. **74**(7), 1250–1253 (1995). doi:10.1103/PhysRevLett.74.1250

D.J. Pine, D.A. Weitz, P.M. Chaikin, E. Herbolzheimer, Diffusing wave spectroscopy. Phys. Rev. Lett. **60**(12), 1134–1137 (1988). doi:10.1103/PhysRevLett.60.1134

L.F. Rojas-Ochoa, S. Romer, F. Scheffold, P. Schurtenberger, Diffusing wave spectroscopy and small-angle neutron scattering from concentrated colloidal suspensions. Phys. Rev. E **65**(5), 051403 (2002). doi:10.1103/PhysRevE.65.051403

H.G.M. Ruis, P. Venema, E. van der Linden, Diffusing wave spectroscopy used to study the influence of shear on aggregation. Langmuir **24**(14), 7117–7123 (2008). doi:10.1021/la800517h

F. Scheffold, Particle sizing with diffusing wave spectroscopy. J. Disp. Sci. Technol. **23**(5), 591–599 (2002). doi:10.1081/DIS-120015365

H.M. Wyss, S. Romer, F. Scheffold, P. Schurtenberger, L.J. Gauckler, Diffusing-wave spectroscopy of concentrated alumina suspensions during gelation. J. Colloid Interface Sci. **241**(1), 89–97 (2001). doi:10.1006/jcis.2001.7668

Fluorescence and X-ray Spectroscopy

E.L. Elson, D. Magde, Fluorescence correlation spectroscopy. I. Conceptual basis and theory. Biopolymers **13**(1), 1–27 (1974). doi:10.1002/bip.1974.360130102
O. Krichevsky, G. Bonnet, Fluorescence correlation spectroscopy: the technique and its applications. Rep. Prog. Phys. **65**(2), 251–297 (2002). doi:10.1088/0034-4885/65/2/203
O. Leupold, G. Grübel, S.V. Roth, C. Schroer, W. Roseker, M. Sikorski, A. Robert, X-ray fluorescence correlation spectroscopy—a tool to study element-specific dynamics. J. Appl. Cryst. **40**(Suppl. 1), s283–s285 (2007). doi:10.1107/S0021889807017852
D. Magde, E.L. Elson, W.W. Webb, Fluorescence correlation spectroscopy. II. An experimental realization. Biopolymers **13**(1), 29–61 (1974). doi:10.1002/bip.1974.360130103
K.A. Nugent, Coherent methods in the X-ray sciences. Adv. Phys. **59**(1), 1–99 (2010). doi:10.1080/00018730903270926
M. Sutton, A review of X-ray intensity fluctuation spectroscopy. C. R. Physique **9**(5–6), 657–667 (2008). doi:10.1016/j.crhy.2007.04.008
J. Wang, A.K. Sood, P.V. Satyam, Y. Feng, X.-Z. Wu, Z. Cai, W. Yun, S.K. Sinha, X-ray fluorescence correlation spectroscopy: A method for studying particle dynamics in condensed matter. Phys. Rev. Lett. **80**(5), 1110–1113 (1998). doi:10.1103/PhysRevLett.80.1110

Optical Spectroscopy

M.-T. Celis, A. Forgiarini, M.-I. Briceno, L.H. García-Rubio, Spectroscopy measurements for determination of polymer particle size distribution. Colloids Surf. A **331**(1–2), 91–96 (2008). doi:10.1016/j.colsurfa.2008.07.024
G. Crawley, M. Cournil, D.D. Benedetto, Size analysis of fine suspensions by spectral turbidimetry. Powder Technol. **91**(3), 197–208 (1997). doi:10.1016/S0032-5910(96)03252-4
G.E. Elicabe, L.H. Garcia-Rubio, Latex particle size distribution from turbidimetry using inversion techniques. J. Coll. Int. Sci. **129**(1), 192–200 (1989). doi:10.1016/0021-9797(89)90430-X
S. Gabsch, B. Wessely, S. Ripperger, L. Steinke, Dynamic extinction spectroscopy as a new method to investigate precipitation processes in micro scale, in *VDI Berichte, Band 1901*, ed. by J. Ulrich (VDI Verlag GmbH, Düsseldorf, 2005), pp. 1189–1193. ISBN 3-18-091901-9
J.S. Gaffney, N.A. Marley, M.M. Cunningham, Measurement of the absorption constants for nitrate in water between 270 and 335 nm. Environ. Sci. Technol. **26**(1), 207–209 (1992). doi:10.1021/es00025a027
E. Gulari, G. Bazzi, E. Gulari, A. Annapragada, Latex particle size distributions from multiwavelength turbidity spectra. Part. Part. Syst. Charact. **4**(1–4), 96–100 (1987). doi:10.1002/ppsc.19870040120
W. Haiss, N. T. K. Thanh J. Aveyard, D. G. Fernig, Determination of size and concentration of gold nanoparticles from uv-vis spectra. Anal. Chem. **79**(11), 4215–4221 (2007). doi:10.1021/ac0702084
T. Kuntzsch, *Erfassung und Beeinflussung des Zustandes von Poliersuspensionen für das chemisch-mechanische Polieren (CMP) in der Halbleiterbauelementfertigung*. PhD thesis, Technische Universität Dresden, 2004
F. Li, R. Schafer, C.-T. Hwang, C.E. Tanner, S.T. Ruggiero, High-precision sizing of nanoparticles by laser transmission spectroscopy. Appl. Opt. **49**(34), 6602–6611 (2010). doi:10.1364/AO.49.006602
D.H. Melik, H.S. Fogler, Turbidimetric determination of particle size distributions of colloidal systems. J. Coll. Int. Sci. **92**(1), 161–180 (1983). doi:10.1016/0021-9797(83)90125-X
G. Mie, Beiträge zur Optik trüber Medien, speziell kolloidaler Goldlösungen. Ann. Phys. IV **25**(3), 377–445 (1908). doi:10.1002/andp.19083300302

P.N. Njoki, I.S. Lim, D. Mott, H.-Y. Park, B. Khan, S. Mishra, R. Sujakumar, J. Luo, C.-J. Zhong, Size correlation of optical and spectroscopic properties for gold nanoparticles. J. Phys. Chem. C **111**(40), 14664–14669 (2007). doi:10.1021/jp074902z

H. Römer, C. V. Fragstein, Bestimmung des Absorptionskoeffizienten und des Brechungsquotienten von kolloidalem Gold. Ein Beitrag zur Anomalie der optischen Konstanten. Z. Phys. **163**(1), 27-43 (1961). doi:10.1007/BF01328913

J.M.J. Santillán, F.A. Videla, M.B. Fernández van Raap, D. Muraca, L.B. Scaffardi, D.C. Schinca, Influence of size-corrected bound-electron contribution on nanometric silver dielectric function. Sizing through optical extinction spectroscopy. J. Phys. D Appl. Phys. **46**(43), 435301 (2013). doi:10.1088/0022-3727/46/43/435301

L.B. Scaffardi, J.O. Tocho, Size dependence of refractive index of gold nanoparticles. Nanotechnol **17**(5), 1309–1315 (2006). doi:10.1088/0957-4484/17/5/024

D. Segets, J. Gradl, R. Klupp Taylor, V. Vassilev, W. Peukert, Analysis of optical absorbance spectra for the determination of zno nanoparticle size distribution, solubility, and surface energy. ACS Nano **3**(7), 1703–1710 (2009). doi:10.1021/nn900223b

L. Steinke, B. Wessely, S. Ripperger, Optische Extinktionsmessverfahren zur Inline-Kontrolle disperser Stoffsysteme. (Optical extinction measurement procedures to control Inline substance dispersed systems). Chem. Ing. Tech. **81**(6), 735–747 (2009). doi:10.1002/cite.200800129

R.L. Zollars, Turbidimetric method for on-line determination of latex particle number and particle size distribution. J. Colloid Interface Sci. **74**(1), 163–172 (1980). doi:10.1016/0021-9797(80) 90179-4

Ultrasonic Spectroscopy

J.R. Allegra, S.A. Hawley, Attenuation of sound in suspensions and emulsions: theory and experiments. J. Acoust. Soc. Am. **51**(5), 1545–1564 (1972). doi:10.1121/1.1912999

J. Andreae, R. Bass, E. Heasell, J. Lamb, Pulse technique for measuring ultrasonic absorption in liquids. Acustica **8**, 131–142 (1958)

P. Andreae, J.H. Joyce, 30 to 230 megacycle puls technique for ultrasonic absorption measurements in liquids. Brit. J. Appl. Phys. **13**(9), 462–467 (1962). doi:10.1088/0508-3443/13/9/307

F. Babick, F. Hinze, M. Stintz, S. Ripperger, Ultrasonic spectrometry for particle size analysis in dense submicron suspensions. Part. Part. Syst. Charact. **15**(5), 230–236 (1998). doi:10.1002/(SICI)1521-4117(199810)15:5<230:AID-PPSC230>3.0.CO;2-D

F. Babick, F. Hinze, S. Ripperger, Dependence of ultrasonic attenuation on the material properties. Colloids Surf. A **172**(1–3), 33–46 (2000). doi:10.1016/S0927-7757(00)00571-9

F. Babick, S. Ripperger, Information content of acoustic attenuation spectra. Part. Part. Syst. Charact. **19**(3), 176–185 (2002). doi:10.1002/1521-4117(200207)19:3<176:AID-PPSC176>3.0.CO;2-8

F. Babick, *Schallspektroskopische Charakterisierung von submikronen Emulsionen*. PhD thesis, Technische Universität Dresden, 2005. http://nbn-resolving.de/urn:nbn:de:swb:14-1112809488055-07176

F. Babick, M. Stintz, A. Richter, Ultrasonic particle sizing of disperse systems with partly unknown properties. Part. Part. Syst. Charact. **23**(2), 175–183 (2006). doi:10.1002/ppsc.200601027

F. Babick, A. Richter, Sound attenuation by small spheroidal particles due to visco-inertial coupling. J. Acoust. Soc. Am. **119**(3), 1441–1448 (2006). doi:10.1121/1.2168427

R.E. Challis, M.J.W. Povey, M.L. Mather, A.K. Holmes, Ultrasound techniques for characterizing colloidal dispersions. Rep. Prog. Phys. **68**(7), 1541–1637 (2005). doi:10.1088/0034-4885/68/7/R01

A.S. Dukhin, P.J. Goetz, Acoustic spectroscopy for concentrated polydisperse colloids with high density contrast. Langmuir **12**(21), 4987–4997 (1996). doi:10.1021/la951085y

A. Dukhin, S. Parlia, D. Klank, M. Lesti, Particle sizing and zeta potential of silica Koestrosol (basis for certified reference material ERM-FD100 for nanoparticles) by acoustics and electroacoustics. Part. Part. Syst. Charact. **27**(5–6), 165–171 (2012). doi:10.1002/ppsc.201100038

J. Edmonds, P.D. Pearce, V.F. Andreae, 1.5 to 28.5 Mc/s pulse apparatus for automatic measurement of sound absorption in liquids and some results for aqueous and other solutions. *Brit*. J. Appl. Phys. **13**(11), 551–560 (1962). doi:10.1088/0508-3443/13/11/310

P.S. Epstein, R.R. Carhart, The absorption of sound in suspensions and emulsions. I. Water fog in air. J. Acoust. Soc. Am. **25**(3), 553–565 (1953). doi:10.1121/1.1907107

Y. Hemar, N. Herrmann, P. Lemaréchal, R. Hocquart, F. Lequeux, Effective medium model for ultrasonic attenuation due to the thermo-elastic effect in concentrated emulsions. J. Phys. II France **7**(4), 637–647 (1997). doi:10.1051/jp2:1997148

F. Hinze, S. Ripperger, M. Stintz, Charakterisierung von Suspensionen nanoskaliger Partikel mittels Ultraschallspektroskopie und elektroakustischer Methoden. Chem. Ing. Tech. **72**(4), 322–332 (2000). doi:10.1002/1522-2640(200004)72:4<322:AID-CITE322>3.0.CO;2-2

A. K. Hipp, B. Walker, M. Mazzotti, M. Morbidelli, In-situ monitoring of batch crystallization by ultrasound spectroscopy. Ind. Eng. Chem. Res. **393**, 783–789 (2000). doi:10.1021/ie990448c

A.K. Hipp, G. Storti, M. Morbidelli, Acoustic characterization of concentrated suspensions and emulsions. I. Model analysis. Langmuir **18**(2), 391–404 (2002). doi:10.1021/la015538c

M.A. Isakovich, On the propagation of sound in emulsions (in Russian: O rasprostranenii zvuka v emulsiyakh). Zh. Eksperim. i Teor. Fiz. **18**(10), 907–912 (1948)

L.D. Kachanovskaya, E.V. Datskevich, V.S. Sperkach, Y.D. Usenko, Acoustic studies of aqueous solutions of biomacromolecules. Colloids Surf. A **106**(2–3), 103–107 (1996). doi:10.1016/0927-7757(95)03361-0

M. Li, D. Wilkinson, K. Patchigolla, P. Mougin, K.J. Roberts, R. Tweedie, On-line crystallization process parameter measurements using ultrasonic attenuation spectroscopy. Cryst. Growth Des. **4**(5), 955–963 (2004). doi:10.1021/cg030041h

D.J. McClements, Ultrasonic characterisation of emulsions and suspensions. Adv. Colloid Interface Sci. **37**(1–2), 34–72 (1991). doi:10.1016/0001-8686(91)80038-L

D.J. McClements, Principles of ultrasonic droplet size determination in emulsion. Langmuir **12**(2), 3454–3461 (1996). doi:10.1021/la960083q

P. Mougin, D. Wilkinson, K.J. Roberts, R. Tweedie, Characterization of particle size and its distribution during the crystallization of organic fine chemical products as measured in situ using ultrasonic attenuation spectroscopy. J. Acoust. Soc. Am. **109**(1), 274–282 (2001). doi:10.1121/1.1331113

J. Pellam, J.R. Galt, Ultrasonic Propagation in Liquids: I. Application of pulse technique to velocity and absorption measurements at 15 megacycles. J. Chem. Phys. **14**(10), 608–614 (1946). doi:10.1063/1.1724072

J.M.M. Pinkerton, A pulse method for measurement of ultrasonic absorption in liquids. Nature **160** (4056), 128–129 (1947). doi:10.1038/160128b0

D.M. Scott, A. Boxman, C.E. Jochen, In-line particle characterization. Part. Part. Syst. Charact. **15** (1), 47–50 (1998). doi:10.1002/(SICI)1521-4117(199802)15:1<47:AID-PPSC47>3.0.CO;2-Q

Electroacoustic Mobility Spectroscoppy

F. Babick, C. Knösche, M. Schäfer, F. Stenger, Characterisation of emulsions with the ultrasonic spectroscopy. On the CD-ROM: *PARTEC 2001, International Congress for Particle Technology, Nuremberg,* 27–29 Mar 2001. (available from NürnbergMesse GmbH, Messezentrum, 90471 Nürnberg, Germany), paper 10/1

M.L. Carasso, W.N. Rowlands, R.A. Kennedy, Electroacoustic determination of droplet size and zeta potential in concentrated intravenous fat emulsions. J. Colloid Interface Sci. **174**(2), 405–413 (1995). doi:10.1006/jcis.1995.1408

R. J. Hunter, Recent developments in the electroacoustic characterisation of colloidal suspensions and emulsions. Colloids Surf. A **141**(1), 37–65 (1998). doi: 10.1016/S0927-7757(98)00202-7
C. Knösche, *Möglichkeiten und Grenzen der elektroakustischen Spektroskopie zur Gewinnung von Partikelgrößeninformationen*. PhD thesis, Technische Universität Dresden, 2001
M. Loewenberg, R.W. O'Brien, The dynamic mobility of nonspherical particles. J. Colloid Interface Sci. **150**(1), 158–168 (1992). doi:10.1016/0021-9797(92)90276-R
R.W. O'Brien, Electro-acoustic effects in a dilute suspension of spherical particles. J. Fluid Mech. **190**, 71–86 (1988). doi:10.1017/S0022112088001211
R.W. O'Brien, D.W. Cannon, W.N. Rowlands, Electroacoustic determination of particle size and zeta potential. J. Colloid Interface Sci. **173**(2), 406–418 (1995). doi:10.1006/jcis.1995.1341

Zeta-Potential and Electrophoresis

P. Debye, E. Hückel, Bemerkungen zu einem Satze über die kataphoretische Wanderungsgeschwindigkeit suspendierter Teilchen (Remarks on a rate concerning cataphoretic appearances in suspended parts). Phys. Z. **25**(3), 49–52 (1924)
A.V. Delgado, F. González-Caballero, R.J. Hunter, L.K. Koopal, J. Lyklema, Measurement and interpretation of electrokinetic phenomena. J. Colloid Interface Sci. **309**(2), 194–224 (2007). doi:10.1016/j.jcis.2006.12.075
R. J. Hunter, *Zeta Potential in Colloid Science: Principles and Applications*. In Series: *Colloid science*, vol. 2; 3rd edn. (Academic Press, London, 1988). ISBN 0-12-361961-0
R.W. O'Brien, L.R. White, Electrophoretic mobility of a spherical colloidal particle. J. Chem. Soc. Faraday Trans. 2 **74**(9), 1607–1626 (1978). doi:10.1039/F29787401607
M. von Smoluchowski, Przyczynek do teorji endosmozy elektrycznej i kilku pokrewnych zjawisk. *Rozprawy Wydziału matematyczno-przyrodiczego Akademji Umiejętnosci w Krakowie*, T. XLIII, Serja A, 110–127 (1903). (reprint in French: Contribution à la théorie de l'endosmose électrique et de quelques phénomènes corrélatifs. Bull. Int. Acad. Sci. Cracovie, Cl. Sci. Math. Nat. **8**, 182–200 (1903))
R. Xu, Progress in nanoparticles characterization: Sizing and zeta potential measurement. Particuology **6**(2), 112–115 (2008). doi:10.1016/j.partic.2007.12.002

Electroacoustic Zeta-Potential Measurement

F. Booth, J.A. Enderby, On electrical effects due to sound waves in colloidal suspensions. Proc. Phys. Soc. A **65**(5), 321–324 (1952). doi:10.1088/0370-1298/65/5/303
P. Debye, A method for the determination of the mass of electrolyte ions. J. Chem. Phys. **1**(1), 13–16 (1933). doi:10.1063/1.1749213
A.S. Dukhin, V.N. Shilov, H. Ohshima, P.J. Goetz, Electroacoustic phenomena in concentrated dispersions: new theory and CVI experiment. Langmuir **15**(20), 6692–6706 (1999). doi:10.1021/la990317g
A.S. Dukhin, P.J. Goetz, Acoustic and electroacoustic spectroscopy for characterizing concentrated dispersions and emulsions. Adv. Colloid Interface Sci. **92**, 73–132 (2001). doi:10.1016/S0001-8686(00)00035-X
A.S. Dukhin, P.J. Goetz, S. Truesdail, Titration of concentrated dispersions using electroacoustic zeta-potential probe. Langmuir **17**(4), 964–968 (2001). doi:10.1021/la001024m
J.A. Enderby, On electrical effects due to sound waves in colloidal suspensions. Proc. Roy. Soc. A **207**(1090), 329–342 (1951). doi:10.1098/rspa.1951.0121
R. Greenwood, Review of the measurement of zeta potentials in concentrated aqeous suspensions using electroacoustics. Adv. Colloid Interface Sci. **106**(1–3), 55–81 (2003). doi:10.1016/S0001-8686(03)00105-2

J.J. Hermans, Charged colloidal particles in an ultrasonic field. Philos. Mag. **25**(168), 426–438 (1938a)

J.J. Hermans, Charged colloid particles in an ultrasonic field. II. Particles surrounded by a thin double layer. Philos. Mag. **26**(177), 674–683 (1938b)

R. J. Hunter, Recent developments in the electroacoustic characterisation of colloidal suspensions and emulsions. Colloids Surf. A **141**(1), 37–65 (1998). doi: 10.1016/S0927-7757(98)00202-7

R.J. Hunter, Measuring zeta potential in concentrated industrial slurries. Colloids Surf. A **195**(1–3), 205–214 (2001). doi:10.1016/S0927-7757(01)00844-5

R.W. O'Brien, Electro-acoustic effects in a dilute suspension of spherical particles. J. Fluid Mech. **190**, 71–86 (1988). doi:10.1017/S0022112088001211

R.W. O'Brien, P. Garside, R.J. Hunter, The electroacoustic reciprocal relation. Langmuir **10**(3), 931–935 (1994). doi:10.1021/la00015a053

R.W. O'Brien, A. Jones, W.N. Rowlands, A new formula for the dynamic mobility in a concentrated colloid. Colloids Surf. A **218**(1–3), 89–101 (2003). doi:10.1016/S0927-7757(02)00593-9

T. Oja, G. L.Petersen, D. Cannon, Measurement of electro-kinetic properties of a solution. *US Patent* 4 497 208, 1985

M. Rasmusson, Volume fraction effects in electroacoustic measurements. J. Colloid Interface Sci. **240**(2), 432–447 (2001). doi:10.1006/jcis.2001.7559

A.J. Rutgers, Comments on the difference of potential caused by ultrasonic waves in solutions. Physica **5**(1), 46 (1938). doi:10.1016/S0031-8914(38)80106-3

A.J. Rutgers, Supersonic vibration potentials and centrifugation potentials. Nature **157**(3977), 74–76 (1946). doi:10.1038/157074b0

E. Yeager, H. Dietrick, F. Hovorka, Ultrasonic waves and electrochemistry. II. Colloidal and ionic vibration potentials. J. Acoust. Soc. Am. **25**(3), 456–460 (1953). doi:10.1121/1.1907063

Second Harmonic Generation (SHG)

H.M. Eckenrode, S.-H. Jen, J. Han, A.-G. Yeh, H.-L. Dai, Adsorption of a cationic dye molecule on polystyrene microspheres in colloids: Effect of surface charge and composition probed by second harmonic generation. J. Phys. Chem. B **109**(10), 4646–4653 (2005). doi:10.1021/jp045610q

D.B. Hollis, Review of hyper-Rayleigh and second-harmonic scattering in minerals and other inorganic solids. Am. Mineral. **73**(7–8), 701–706 (1988)

L. Schneider, W. Peukert, Review: Second harmonic generation spectroscopy as a method for in situ and online characterization of particle surface properties. Part. Part. Syst. Charact. **23**(5), 351–359 (2007). doi:10.1002/ppsc.200601084

L. Schneider, H.J. Schmid, W. Peukert, Influence of particle size and concentration on the second-harmonic signal generated at colloidal surfaces. Appl. Phys. B **87**(2), 333–339 (2007). doi:10.1007/s00340-007-2597-7

B. Schürer, W. Peukert, In situ surface characterization of polydisperse colloidal particles by second harmonic generation. Part. Sci. Technol. **28**(5), 458–471 (2010). doi:10.1080/02726351.2010.504131

E.C.Y. Yan, Y. Liu, K.B. Eisenthal, New method for determination of surface potential of microscopic particles by second harmonic generation. J. Phys. Chem. B **102**(33), 6331–6336 (1998). doi:10.1021/jp981335u

Final Remarks

O. Ruscitti, R. Franke, H. Hahn, F. Babick, T. Richter, M. Stintz, Application of particle measurement technology in process intensification. Chem. Eng. Technol. **31**(2), 270–277 (2008). doi:10.1002/ceat.200700465

Chapter 3
Fundamentals in Colloid Science

Preparation and handling of colloidal suspension or even their characterisation is not possible without a basic understanding of the physical effects that prevail on the microscopic level. Some of these effects are directly related to the fineness of the particles because it coincides with a large specific surface area (which corresponds, e.g., to a high adsorption capacity), a considerably curvature of the particle surface (which i. a. promotes dissolution), a high number concentration (i.e. significant osmotic pressure and large collision frequencies), small interparticle distances (i.e. strong interactions between scattered waves), low particle mass (i.e. low weight and inertia), or with the fact that the wavelength of light is larger than particle size (i.e. weak optical scattering). A further effect of the small size of colloidal particles is that their diffusive motion is much faster than for micrometre particles. Moreover, the interactions between colloidal particles are decisive for the macroscopic behaviour of the whole suspension (e.g. with regard to rheology or coagulation). These interactions may be attractive and/or repulsive and depend on the chemical nature and the material properties of the particles. Most of all, they are affected by the interfacial properties which reflect the physico-chemical interaction between the particulate phase, the solvent, and the solutes (e.g. hydrophobicity, adsorption, surface charging).

This chapter provides a brief introduction to the fundamentals in colloid science. It addresses the physico-chemical properties of single colloidal particles as well as the processes at the interface and the structure of the interfacial layer. It further examines the non-viscous interactions that occur among colloidal particles. For detailed explanation, the reader is referred to the textbooks of Adamson and Gast (1997), Hunter (1988, 1993), Lyklema (1991, 1995, 2000), or Israelachvili (1992).

F. Babick, *Suspensions of Colloidal Particles and Aggregates*,
Particle Technology Series 20, DOI 10.1007/978-3-319-30663-6_3

3.1 Physico-Chemical Properties of Colloids and Interfaces

Many macroscopic phenomena of colloidal suspensions are related to the light scattering and the Brownian motion of the single particles. Both properties depend largely on particle size; they are, therefore, frequently employed for the characterisation of particle systems (cf. previous chapter). Additionally, the small size of colloids enhances the significance of the interface to the particles physical behaviour. The interfacial properties additionally affect the interaction bet-ween particles and are, thus, crucial for the macroscopic suspension behaviour (e.g. stability). A particular characteristic of interfaces is the electric double layer (EDL), which cannot be ignored in most situations. Its formation and structure is closely related to dissolved ionic species and their interaction with the particle surface (e.g. adsorption, precipitation). Last but not least, the interfacial properties can be affected by the solubility behaviour of the particle phase.

3.1.1 Scattering in Colloidal Suspensions

Since the mid of the 19th century, the phenomenon of light scattering at sky or in colloidal sols had attracted physicists' interest. It was soon associated with the presence of fine particles, even though the particles themselves were "invisible" for contemporary measurement instrumentation. Scattering phenomena on colloidal systems were reported by Faraday (1857) for gold sols and were extensively studied by Tyndall (1869) for aerosol particles ("precipitated clouds"). In 1871, J.W. Strutt (Lord Rayleigh), who referred to Tyndall's experiments, explained light scattering with transversal vibrations of small particles and derived a functional relationship between the scattered intensity I_{sca}, the particle volume V_p, its relative refractive index m_r and the wavelength λ (Rayleigh 1871a, b, c):

$$\frac{I_{sca}}{I_0} \propto \left|m_r^2 - 1\right|^2 \cdot \frac{V_p^2}{r^2 \lambda^4} \cdot \sin^2 \gamma, \tag{3.1}$$

where r is the distance from the scattering volume and γ the angle between the scattering direction and the direction of the vibration, i.e. the electric field vector. For the vibrational component that is perpendicular to the scattering plane, γ remains unchanged (90°), while for the parallel component, γ is complementary to the scattering angle θ ($\gamma = 90° - \theta$). Hence, the scattering field of a *Rayleigh scatterer* forms a monopole field for vertical polarisation and a dipole field for parallel polarisation. Rayleigh's theory did not only explain the blue colour of the sky, but accurately described the polarisation of scattered light as well. This is remarkable, since at that time the ether was a widely accepted concept and the electromagnetic nature of light was unknown. Moreover, it was not even clear which kind of particles was responsible for scattering.

Rayleigh considered light scattering on objects that are much finer than the wavelength ($\lambda \gg x$ or $ka \ll 1$) and that move because of their thermal energy. This motion is independent and causes a phase shift between the scattered waves of two neighbouring particles. Rayleigh argued that, due to the smallness of the light wavelength, this phase shift comprises several periods of oscillations and that, hence, the scattered waves of neighbouring particles are completely incoherent.[1] Therefore, the total intensity of the scattered light is obtained by summation over all individual scattering intensities. However, Rayleigh's assumption applies to air molecules in the stratosphere, but not to the atmosphere on the ground or to most colloidal suspensions (Mandelstam 1907; Sobel'man 2002). In fact, in dense colloidal suspensions, the interparticle distances are usually much smaller than the wavelength of visible light and the scattered waves of neighbouring particles are partially coherent. If the particles in such a suspension were homogenously and regularly distributed in space, the interference of scattered waves would lead to a complete extinction of sidewards and backwards scattered light. von Smoluchowski (1908) concluded that macroscopic scattering phenomena (like the Tyndall effect or the opalescence of liquids near the critical point) are caused by local density fluctuations, which are equivalent to an optical inhomogeneity of the system. In colloidal suspensions, these density variations are due to the Brownian motion of the particles. The total scattering intensity is proportional to the variance $\sigma^2(c_N)$ of the local particle number concentration, which is just the average particle concentration $\bar{c}_N$ for dilute suspensions. In that case, both approaches—incoherent single particle scattering and density fluctuation scattering—agree. The mathematical framework of the latter approach was developed by Einstein (1910). It reveals that the light scattering of suspensions does not grow linearly with the particle concentration because particle interactions (e.g. steric effects, hydrodynamic interactions) reduce the particle diffusivity—similarly as they affect the osmotic pressure of colloidal suspensions (Debye 1947). For that reason, it is possible to measure the virial coefficients of colloidal suspensions by light scattering experiments (cf. Sect. 2.3.4.2). Since the fluctuation approach supposes continuous local density gradients on length scales of the wavelength and Rayleigh scattering of the particles ($I_{\mathrm{sca}} \propto V_{\mathrm{p}}^2$), it is restricted to macromolecules and nanoparticles.

For large and medium-sized particles, the Rayleigh approach (3.1) cannot be employed because volume elements from opposite sides of the particle scatter with significantly different phases. In contrast to Rayleigh scattering, the particle shape matters. Lorenz (1898) and Mie (1908) independently developed a mathematical formalism to calculate the light scattering on spherical particles of arbitrary size. This solution is based on a multipole expansion of the electromagnetic waves, as was previously developed by Lord Rayleigh (1880) for sound waves. The light scattering in suspensions of large particles is typically incoherent, i.e. the phases of the scattered waves are completely uncorrelated. Non-linear concentration effects

[1]With the exception of forward-scattered light ($\theta = 0$), which is in phase for all particles.

are then mainly related to multiple scattering. The significance of multiple scattering depends i. a. on the material system and the instrumentation optic. It is usually described as a function of particle concentration, even though the influence of interparticle forces via the suspension structure cannot be ignored (e.g. Foldy 1945; Lax 1952; Waterman and Truell 1961). In experimental practice, multiple scattering is frequently avoided in order to simplify the data analysis. However, by proper experimental design, one can separate the multiply scattered light (e.g. Ishimaru 1994; Bryant et al. 1998) and even employ this signal for the characterisation of colloidal suspensions (cf. Sect. 2.3.5.2).

3.1.2 *Particle Diffusion and the Sedimentation-Diffusion Equilibrium*

The origin and the consequences of the Brownian motion of colloidal particles have already been discussed in the context of particle characterisation (Sect. 2.3). It was shown that this phenomenon is exploited by several sizing techniques (e.g. ultramicroscopy or dynamic light scattering) because the particle diffusion coefficient is inversely proportional to the particle size (Eq. (2.4)). On the other hand, it was further shown that particle diffusion may interfere with the particle characterisation when the measuring principle is based on particle migration, like in electrophoresis or sedimentation analysis. For very fine particles, the diffusive transport ($-D_t\nabla c$) may even outweigh the migration transport (e.g. $v_{\text{Stokes}}\cdot c$), which result in a characteristic, stationary distribution of the local particle concentration. For sedimentation, such a state is called sedimentation-diffusion equilibrium; which yields a "barometric" concentration profile for dilute suspensions (Perrin 1908):

$$c(h_2) = c(h_1) \cdot \exp(-\Delta h / l_g), \tag{3.2}$$

which is described by the characteristic or gravitational length l_g:

$$l_g = \frac{k_B T}{\Delta\rho V g}. \tag{3.3}$$

For high particle concentrations and long-ranging interparticle forces, the real profile considerably deviates from that in Eq. (3.3) and may even be used to study the particle interaction (Philipse and Koenderink 2003; Belloni 2005).

The sedimentation-diffusion equilibrium is not just a balance between diffusive transport and settling, but can also be regarded as equilibrium between chemical potential and gravity potential ($RT \cdot d\ln c = -g\Delta\rho V \cdot N_A \cdot dh$). Based on this interpretation, Perrin (1909) employed this colloidal phenomenon to measure Avogradro's constant N_A (and, thereby, to confirm the molecular origin of Brownian motion). It is rather fascinating that he obtained values between 6.5×10^{23} mol^{-1} and 7.5×10^{23} mol^{-1}, which are close to the real value.

The high diffusivity of colloidal particles has further implications for the processing of suspensions. That concerns, e.g., the purification of liquids by depth filtration or cross flow filtration—in particular for particles below 100 nm: Diffusion enhances the efficiency of depth filters (Hinds 1999, pp. 196–200) and counteracts particle deposition on the membrane in cross flow filtration, it thus warrants stable permeate flux (Ripperger and Altmann 2002). In general, the diffusive mass transport of colloidal particles (e.g. to walls) cannot be ignored.

3.1.3 Interfacial Energy

The physical behaviour of colloidal particles is largely affected by the properties of the interface between the particle and the continuous phase. The term *interface* does not describe a conceptual 2-dimensional boundary, but refers to a structured region around the geometric surface of the particle where neither the bulk properties of the particulate phase nor those of the continuous phase prevail. It thus includes the surface atoms of the particle as well as layers of adsorbed ions or molecules or even the ion cloud that surrounds a charged suspended particle.

Since interfaces are binary or multi-component systems, their thermodynamics can be described by a *Gibbs-Duhem-equation*:

$$S^{\text{if}}\,dT - A_{\text{if}}\,\mathrm{d}\pi + A_{\text{if}} \sum \Gamma_i \mathrm{d}\mu_i^{\text{if}} = 0, \tag{3.4}$$

where A^{if}, S^{if}, and μ^{if} are the interfacial area, the interfacial entropy, and the interfacial chemical potential of component *i*, respectively. The interfacial pressure π is frequently replaced by the interfacial tension γ ($\mathrm{d}\pi = -\mathrm{d}\gamma$), which can be considered as Gibbs free energy per surface area at a constant temperature and pressure ($\gamma = (\partial G/\partial A)_{T,p,n}$). The concentration Γ_i of component *i* at the interface can principally not be measured (cf. Lyklema 2000, pp. 4.2–4.6). Accessible by experiment is solely the excess concentration, i.e. the excess of component *i* at the interface when compared to its molar fraction x_i in the bulk phase ($\Gamma_i^{\mathrm{E}} = \Gamma_i - x_i \cdot \sum \Gamma_i$). However, for surface active components which strongly adhere on the particle surface, there is no significant difference between both values (because $x_i \ll 1$).

In the case of equilibrium ($\mathrm{d}T = 0$), Eq. (3.4) can be transformed to the Gibbs adsorption isotherm:

$$\Gamma_i = -\frac{1}{RT} \cdot \frac{\mathrm{d}\gamma}{\mathrm{d}\ln a_i}, \tag{3.5}$$

where a_i is the activity of component *i* in the bulk phase of the solvent. The Gibbs isotherm applies to the adsorption of ionic surfactants in the presence of excess counter-ions and to the adsorption of non-ionic components.

The Gibbs isotherm states that the adsorption of surfactants leads to a decrease of the interfacial tension γ. However, the interfacial tension γ will not completely vanish as long as the particulate and fluid phase can be distinguished.

In general, interfacial energy is a material-specific property influenced by temperature and pressure. Its significance results from the fact that its value rises with any increase in the interfacial area. From an energetic point of view it is, therefore, favourable when, e.g., the droplets of a colloidal emulsion coalesce and, thus, reduce the interfacial area, or when the coarse particles of a colloidal phase grow at the expense of the fine particles (*Ostwald ripening*).

3.1.4 *Influence of the Interfacial Energy on the Bulk Properties of Fine Particles*

A further consequence of the interfacial energy is the capillary pressure, i.e. the pressure excess in the inner part of a droplet or particle resulting from the interfacial tension:

$$\Delta p = \frac{2\gamma}{R}. \tag{3.6}$$

Equation (3.6) is called *Young-Laplace equation*, in which R is the harmonic mean of the principal radii of curvature. The capillary pressure promotes the release of atoms or molecules from the particle surface. This leads to a decrease of the equilibrium vapour pressure p^{eq} with increasing droplet size (*Kelvin equation*):

$$\ln \frac{p^{\text{eq}}(r)}{p^{\text{eq}}(\infty)} = \frac{2\gamma \cdot V_{\text{m}}}{R \cdot \text{R}T}, \tag{3.7}$$

to an enhanced solubility c_{sat} of very fine particles compared to plane surfaces (*Ostwald-Freundlich equation*):

$$\ln \frac{c_{\text{sat}}(r)}{c_{\text{sat}}(\infty)} = \frac{2\gamma \cdot V_{\text{m}}}{R \cdot \text{R}T}, \tag{3.8}$$

and to lowered melting points T_{m} in the case of nano-sized particles:

$$\ln \frac{T_{\text{m}}(r)}{T_{\text{m},\infty}} = -\frac{2\gamma}{R \cdot \rho_{\text{s}} q_{\text{lat}}}, \tag{3.9}$$

where V_{m}, ρ_{s}, q_{lat} are the molar volume, the solid density, and the latent heat, respectively.

The relevance of these effects is illustrated in Fig. 3.1 at the solubility of amorphous silica and the melting temperature of aluminium. A notable change in

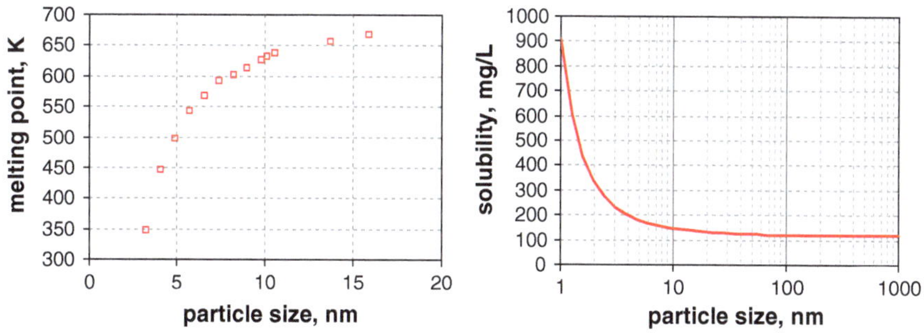

Fig. 3.1 Impact of particle size on the melting temperature of free aluminium acc. to Kreth (2001, p. 73), and on solubility of amorphous SiO_2 acc. to Iler (1979, pp. 47–56)

these properties is visible only for very fine, nanosized particles ($x < 10$ nm). Nevertheless, the dependency of bulk properties on the particle size is of major technical relevance, e.g. for crystallisation and emulsion preparation, since it affects microprocesses like nucleation and Ostwald ripening.

3.1.5 *Electric Double Layer*

Electric charges on the surface of colloidal particles are of elementary importance for the behaviour of these particles in their environment. For aqueous systems, the effect of such surface charges is limited to the close proximity of the particle's surface. This is because of the high permittivity (i.e. polarity) of water and the high activity of the dissolved ions. As a result, the surface charge is neutralised by an excess of counter-ions in the vicinity of the surface. This region and the surface can be regarded more simply as two layers of opposite charge, and are captured by the term *electric double layer* (EDL).

The origins for the electric charges at the interface between surface and solvent are manifold. They result from dissociation (e.g. of ionic crystals), from acid/base-reactions of functional groups (e.g. hydroxyl groups) or from the adsorption/desorption of ionic species (e.g. ionic surfactants/polyelectrolytes).

An important class of colloidal particles are (metal) oxides like SiO_2 or TiO_2. When these materials are suspended in aqueous media, amphoteric hydroxyl-groups are formed which can be protonated or deprotonated in dependence on the pH value (Hunter 1993, p. 379).

$$\mathrm{M-OH} \overset{\mathrm{OH^-},\,K_{a1}}{\longleftrightarrow} \mathrm{M-O^-} + \mathrm{H_2O}, \tag{3.10}$$

$$\mathrm{M-OH_2^+} \overset{\mathrm{H^+},\,K_{a2}}{\longleftrightarrow} \mathrm{M-OH} + \mathrm{H^+}. \tag{3.11}$$

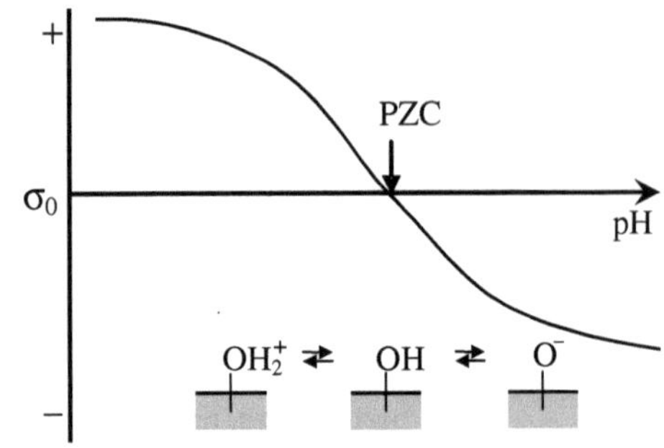

Fig. 3.2 Dependency of the surface charge on the pH for oxide particles in aqueous solutions

The processes of protonation and deprotonation are described by the acid constants K_{a1} and K_{a2}. The surface charge is quantified by the surface charge density σ_0:

$$\sigma_0 = \sum z_i F \Gamma_i, \tag{3.12}$$

where z_i is the valency of the charged surface groups, F is the Faraday constant, and Γ_i is the surface concentration of positive and negatives sites.

The pH-value at which the positive and negative surface charges balance each other is named *point of zero charge* (PZC). It is a characteristic property of the particle phase. It can be calculated from the acid constants $Ka,_i$ of the protonation reactions (Hunter 1993, p. 374):

$$pH_{PZC} = -\frac{1}{2}\lg(K_{a1}K_{a2}) = \frac{1}{2}(pK_{a1} + pK_{a2}). \tag{3.13}$$

The surface charge is positive when the pH is below the PZC and negative when the pH exceeds the PZC (Fig. 3.2). The concept that the surface charge is caused by the amphoteric properties of a functional group is called 2-pK-model.

Ionic crystals, like silver iodid (AgI), form a further material class with regard to surface charging. The responsible charging mechanism is the crystal dissociation, e.g. AgI $\leftrightarrow$ $Ag^+ + I^-$ ($pK_S = 16$, i.e. weakly soluble), together with a different affinity of the anions and cations to the crystal. For instance, iodid ions are more strongly bound to the AgI surface than silver ions. That leads to a negative surface charge which can be neutralised or even positively recharged by an excess of the silver ions (e.g. by adding $AgNO_3$ to the continuous phase). Hence, the surface charge is solely determined by silver ions and the PZC is a characteristic concentration of the silver ions ($pAg_{PZC} = 5.6$).

3.1.5.1 Structure of the Electric Double Layer (EDL)

The electric double layer is formed by the charged surface and the surface-close region, which exhibits an excess of counter-ions for charge neutralisation. It is characterised by the spatial distribution of the co-ions and counter-ions and by the presence of an electric potential.

The structure of the EDL is neither simple nor universal; it depends to a great extent on the physico-chemical properties of particles and dispersion medium. In general, it is assumed that some ions from the solvent adhere on the particle surface and partially neutralise the surface charge. This layer of immobile ions is called *Stern layer*. The other ions spread in the solvent by thermal motion yet are subject to the electric field generated from the charged surface. With growing surface distance the concentrations of the ionic species tend to their equilibrium values of the free solvent. The region adjacent to the Stern layer with excess of counter-ions is called the *diffuse layer*. In this part of the EDL, the ion distribution results from the balance of electrostatic and osmotic forces.

This general view on the EDL ignores the mechanism of surface charging and ion adsorption in the Stern layer. It cannot, therefore, reveal the fine details of real EDLs. For instance, ions may adsorb in the Stern layer with or without keeping their hydration shells and may accordingly be located on different planes parallel to the surface (*inner* and *outer Helmholtz plane*—IHP and OHP); co-ions may be covalently bound to the surface and, thus, virtually increase the surface charge; and last but not least, the structure of the EDL may be affected by surface morphology (Hunter 1988, Chap. 2).

The charge separation between surface and the ions in the Stern and diffuse layer generates an electric field which is described by an electric potential ψ. The potential ψ_0 on the surface is finite and has the same sign as the surface charge. Within the diffuse layer, the electric potential decays exponentially from the diffuse layer potential ψ_d to zero in the free solvent ($\psi(\infty) = 0$, Fig. 3.3).

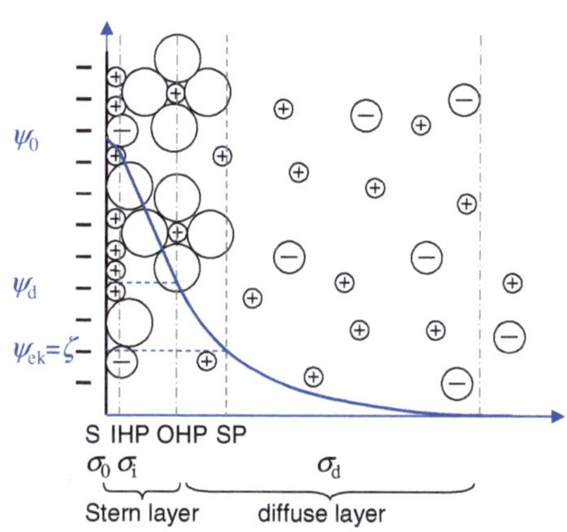

Fig. 3.3 Schematic representation of the electric double layer with immobilised ions in the *Stern layer* and mobile ions in the *diffuse layer*; a further subdivision of the Stern layer is indicated by specifically adsorbed ions located at the *inner Helmholtz plane* (IHP) and immobilised hydrated counter-ions at the *outer Helmholtz plane* (OHP); the OHP marks the beginning of the diffuse layer, where the electric potential exponentially decays with surface distance; the regions between the surface, the IHP and the OHP are free of charge; the *shear plane* or *slip plane* (SP) marks the transition from stagnant to mobile solvent

The extension of the double layer, or more precisely of the diffuse layer, is indicated by the Debye length $1/\kappa$, where κ is called Debye-Hückel parameter, which can be calculated from the ionic strength I of the solution:

$$\kappa = \sqrt{\frac{2\mathrm{F}^2 I}{\varepsilon_\mathrm{m}\varepsilon_0 \mathrm{R}T}} = \sqrt{\frac{\mathrm{F}^2 \sum z_i^2 c_{n,i}}{\varepsilon_\mathrm{m}\varepsilon_0 \mathrm{R}T}}. \tag{3.14}$$

Here, ε_m is the relative permittivity of the solvent, ε_0 is the electric constant, R the universal gas constant, T the temperature, I the ionic strength, and z_i and $c_{n,i}$ are the valency and molar concentration of the ionic species. For a purely exponential decay, the Debye length equals the distance from the Stern layer, at which the potential ψ is decreased by the factor e from ψ_d (i.e. $0.37 \times \psi_\mathrm{d}$). Equation (3.14) has to be modified for extremely low electrolyte concentrations in the solvent in order to account for the counter ions that compensate for the surface charge (Thies-Weesie et al. 1995; Téllez and Biben 2000; Belloni 2005).

3.1.5.2 Potential Distribution in the Diffuse Layer

There are different mathematical models of the EDL which reflect the possible variety of its inner structure at different levels of sophistication (Hunter 1993, pp. 379). All of them refer to the approach of Gouy (1909) und Chapman (1913) which neglects the existence of the Stern layer and solely accounts for a diffuse layer in immediate contact with the surface. The potential distribution $\psi(r)$ inside the diffuse layer is defined by the Poisson-Boltzmann-equation (PBE):

$$\nabla(\varepsilon_\mathrm{m}\varepsilon_0 \nabla\psi) = -\rho_\mathrm{e,solvent} - \sum \mathrm{F} z_i c_{n,i}^{\infty} \exp\left(-\frac{z_i \psi \mathrm{F}}{\mathrm{R}T}\right) \tag{3.15}$$

or in its simplified form:

$$\Delta\psi = -\frac{1}{\varepsilon_\mathrm{m}\varepsilon_0} \sum \mathrm{F} z_i c_{n,i}^{\infty} \exp\left(-\frac{z_i \psi \mathrm{F}}{\mathrm{R}T}\right). \tag{3.16}$$

Here, F, R, and T denote the Faraday constant, the molar gas constant, and the absolute temperature, respectively. The parameter ε_m is the static permittivity of the liquid, $c_{n,i}^{\infty}$ the bulk concentration of the ionic species i, and z_i its valency. Note that $\varepsilon_\mathrm{m}\varepsilon_0$ is replaced by $\varepsilon_\mathrm{m}/4\pi$ when an unrationalised notation of electric quantities, e.g., in electrostatic CGS units, is employed (cf. Hunter 1988, pp. 357).

The PBE describes a continuous distribution of charge inside the solvent and on the surface—any effect related to the discreteness of charge and the ion volume is ignored. In general, the PBE has to be solved numerically. Analytical solutions exist only for simple geometries and/or symmetrical electrolytes (e.g. Lyklema 1995, Chap. 3.5). Even for an isolated sphere there is no exact analytical solution.

For this reason, the PBE is frequently reduced to its linearised formulation (Debye-Hückel approximation):

$$\Delta\psi = -\kappa^2\psi, \tag{3.17}$$

where κ is the Debye-Hückel parameter as defined in Eq. (3.14). By means of the linearised PBE (3.17), the potential field ψ inside the diffuse layer can be analytically solved for several geometric situations. For an isolated sphere with the surface potential ψ_0^{iso}, one obtains (Hunter 1993, pp. 388; Lyklema 1995, pp. 3.35):

$$\psi^{\mathrm{iso}} = \psi_0^{\mathrm{iso}}\frac{a}{r}\cdot \mathrm{e}^{-\kappa(r-a)} \tag{3.18}$$

The potential distribution is coupled with the spatial distribution of the ionic species in the diffuse layer. Their total net charge is just the opposite of the surface charge within the Gouy-Chapman model. Therefore, the surface charge σ_0^{iso} can be related to the surface potential ψ_0^{iso}:

$$\sigma_0^{\mathrm{iso}} = \frac{\varepsilon_{\mathrm{m}}\varepsilon_0\psi_0^{\mathrm{iso}}}{a}(1+\kappa a). \tag{3.19}$$

The Guy-Chapman model can be applied to EDLs with Stern layers when the surface potential ψ_0 is replaced by the diffuse layer potential ψ_{d} and when the thickness of the Stern layer is properly accounted for. Additionally, Eq. (3.19) would refer to the total net charge of surface and Stern layer. If, however, a quantitative description of the complete EDL is required, the Gouy-Chapman model of the diffuse layer has to be supplemented with according models for the ion adsorption in the Stern layer and the charging of the particle surface. Note that such models inherently depend on material properties which considerably increases the effort for model parametrisation (e.g. Sonnefeld et al. 2001).

3.1.5.3 Zeta-potential

Both the surface potential ψ_0 and the diffuse layer potential ψ_{d} defy a direct experimental determination. What can be measured are the surface charge density σ_0 (via titration) and the *electrokinetic potential* or *zeta-potential* ζ, which corresponds to the conceptual hydrodynamic *slip plane* (or *shear plane*, SP in Fig. 3.3) between the Stern layer and the diffuse layer. Any measurement of the zeta-potential is, therefore, based on relative motion between the particle surface and the diffuse layer (cf. Sect. 2.3.7, Delgado et al. 2007).

The zeta-potential is affected by

- the charge determining ions, i.e. those ions that control the surface charge (e.g. H^+ and OH^- for metal oxides, Fig. 3.2),

- specifically adsorbed ions which are located in the Stern layer,
- the concentration and valency of all ionic species present in the solvent.

The pH at which the zeta-potential is zero is called *isoelectric point* (IEP). Because of specific ion adsorption, the IEP may deviate from the PZC, which is a purely material property. For this reason, the IEP is less a parameter of the particle phase but rather a characteristic parameter of the suspension. In a couple of papers, Kosmulski reviewed experimentally determined PZC and IEP-values for numerous materials in varying environments (Kosmulski 2002, 2004, 2006, 2009).

The zeta-potential is frequently used to predict the stability of a suspension or the adhesion of suspended particles on macroscopic surfaces (e.g. cellulose fibres, tubing). This is because double layer interaction between particles or between particles and surfaces is governed by the ion distribution in the diffuse layer, which primarily depends on the Debye-Hückel parameter κ and the diffuse layer potential ψ_d. The latter, however, is commonly approximated by the zeta potential ζ (Lyklema 2010, cf. Fig. 3.3). It is quite obvious that repulsion requires high zeta-potential values of equal sign, whereas adhesion occurs in the absence of surface charge or for oppositely charged surfaces.

In practice, zeta-potential measurements are, therefore, frequently used to optimise the dose of dispersion or flocculation agents (Greenwood 2003), or to find optimum surface modifications for membranes and filter fabrics (Afonso et al. 2001; Bellman et al. 2002; Schnitzer and Ripperger 2008) or cellulose fibres (Carrasco et al. 1998) that warrant maximum retention of particles and pigments, respectively. A further purpose of zeta-potential measurements is to monitor changes of interfacial properties. Such monitoring may concern the coating of pigments (van Lent and Klinksiek 1997; Wilhelm and Stephan 2006), the mechanical activation of particle surfaces during grinding processes (Mende et al. 2004; Skwarek et al. 2011), or the fouling of walls and membranes (Park et al. 2003). Sometimes the zeta-potential is employed as an indication for the interfacial properties (e.g. oxygene content at the surface of ceramic particle, Nitzsche et al. 1991). In any case, the zeta-potential serves as an important characteristic of the EDL in colloidal suspensions.

3.1.5.4 Electroviscous Effects

The presence of the EDL affects the hydrodynamic drag of charged particles as if the viscosity of the liquid phase were slightly increased. Likewise, the flow through a porous material is apparently retarded when an EDL is formed in the pores. This observation can be related to different phenomena that are called *electroviscous effects* (Hunter 1988, Chap. 5).

The *primary electroviscous* effect is the retardation of the particle velocity (or the fluid flow in pores) due the polarisation of the EDL, which is a consequence of the relative motion between diffuse layer and the particle surface (Fig. 3.4). The polarisation, in turn, causes a flux of ions and solvent that opposes and decelerates the relative motion. This kind of deceleration affects particle sedimentation (von Smoluchowski 1903; Ohshima et al. 1984; Keh and Ding 2000); capillary flow

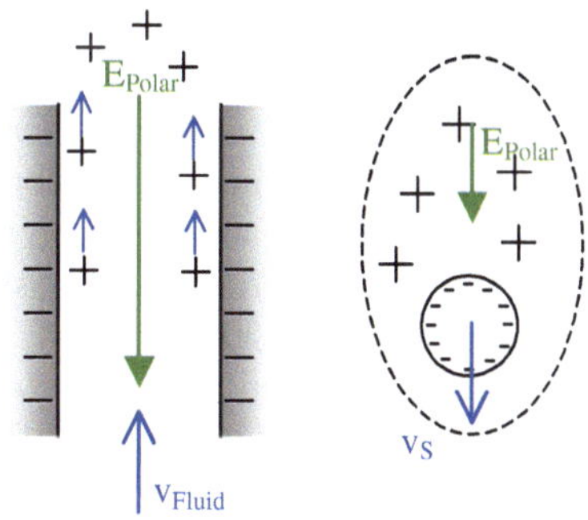

Fig. 3.4 Double layer polarisation due to capillary flow (*left*) and sedimentation (*right*); the corresponding electric field has a decelerating effect on the relative motion between surface and bulk liquid

(Bull 1932, Levine et al. 1975), the hydrodynamic drag on single particles (Booth 1954) or the viscosity of a suspension (Booth 1950; Ruiz-Reina et al. 2003). Note that the primary electroviscous effect is related to individual particles and surfaces; any double layer interaction is excluded. It is essentially a second-order effect which—to first approximation—obeys the following equation:

$$\frac{\Delta p_{\mathrm{cap}}}{\Delta p_{\mathrm{cap},\zeta=0}}, \frac{v_{\mathrm{sett},\zeta=0}}{v_{\mathrm{sett}}}, \frac{F_{\mathrm{d}}}{F_{\mathrm{d},\zeta=0}}, \frac{\eta_{\mathrm{sp}}}{\eta_{\mathrm{sp},\zeta=0}} = 1 + \mathrm{const.} \times \left(\frac{\zeta}{\kappa a}\right)^2. \tag{3.20}$$

For thin double layers ($\kappa a \gg 1$) and small potentials ($|\zeta| \leq 25$ mV), it can usually be neglected (Rubio-Hernández et al. 2004).

In concentrated suspensions, the motion of particles is crucially affected by hydrodynamic interaction between neighbouring particles, which strongly depends on the interparticle distances, i.e. on the suspension structure (cf. Overbeck et al. 1999; Watzlawek and Nägele 1997, 1999). This structure is clearly influenced by the interparticle forces, in particular by the forces that occur when the EDL of two particles overlap (e.g. Russel 1978; Quemada and Berli 2002). When a suspension contains only a single particulate component, such a double layer overlap leads to repulsions and, thus, decreases the particle mobility and increases the suspension viscosity (Fig. 3.5). This effect is called *secondary electroviscous effect*. Its

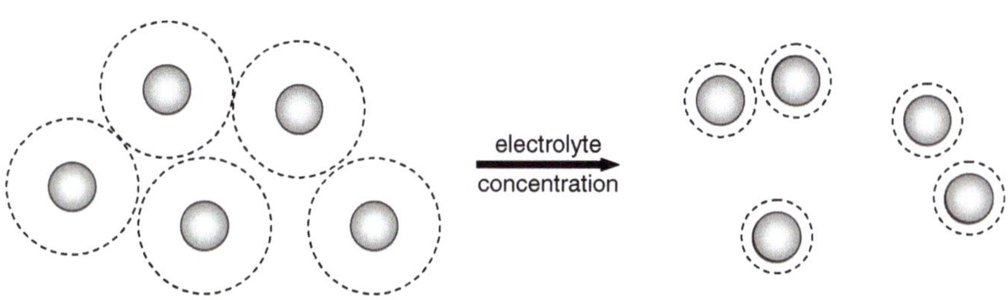

Fig. 3.5 Secondary electroviscous effect: strong repulsion between similarly charged particles reduces their mobility since the overlap of EDL (indicated by dotted lines) would require additional energy and, as a result, the suspension is not randomly ordered; the decisive parameter is the double layer thickness, which can be controlled via the electrolyte concentration

relevance grows with the absolute value of the zeta-potential and with the ratio of double layer thickness to mean interparticle distance.

Additionally, a *tertiary electroviscous* effect exists, which is related to the dependence of the interfacial geometry on the parameters of the EDL. This can be, e.g., observed for adsorbed polyelectrolytes, whose conformation on the particle surface (stretched, random etc.) is determined by surface charge and ionic strength (Hunter 1988, pp. 200–204).

Some authors report a *visco-electrical effect*, which is defined as a real change of the liquid viscosity within the EDL and which is explained with an orientation of (polar) solvent molecules in vicinity of the charged surface. This effect is considered to be relevant to a thin region of a few layers of molecules immediately adjacent to the particle surface (Lyklema and Overbeek 1961; Lyklema 2010).

3.1.6 Interaction of Charged Surfaces with Ions and Molecules

The Coulombic attraction between the charged surface and the counter-ions in the solvent gives rise to ion adsorption on the surface. Additionally, ion adsorption can be promoted by physical forces and chemical bonds and, thus, may even apply to co-ions. In that case, the ions are called *specifically adsorbing*. High charge densities and high electrolyte concentration cause strong ion adsorption that coincides with further ion specific effects. For instance, hydrolysed metal ions can precipitate on the surface as hydroxide phase. A particular case is the adsorption of surfactants and polymers because they form associated structures at particle surfaces. This section focuses on oxide surfaces whose charge is determined by the acid-base reaction of the amphoteric hydroxyl groups (Eqs. (3.10) and (3.11)).

3.1.6.1 Adsorption of Ions at Oxidic Surfaces

Although the formation of the EDL is governed by the surface charge and, thus, by the charge determining ions, e.g., H^+ and OH^- for oxides, the interaction with other EDLs is mainly determined by the charge of the diffuse layer. Hence, the adsorption of ions in the Stern layer is crucial for the effective behaviour of charged colloidal particles. Traditionally, one distinguishes between non-specifically adsorbing (or indifferent) and specifically adsorbing ions (Hunter 1988, pp. 223).

Unspecific adsorption refers to the electrostatic attraction of counter-ions and is, therefore, independent from ion size or chemical nature. Such an adsorption reduces the effective surface charge, but can never reverse its sign. That means that the isoelectric point (IEP), which refers to the zeta potential, agrees with the point of zero charge (PZC), which refers to the surface charge (Fig. 3.6). Since the attraction is purely Coulombic, the ions retain their hydration shells; hence, they are located at the OHP.

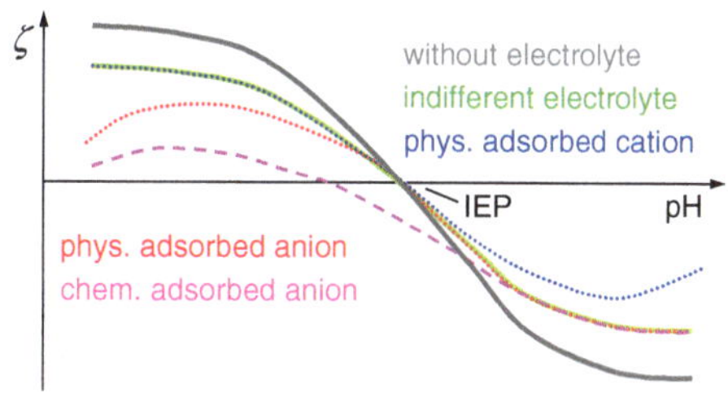

Fig. 3.6 Principal impact of indifferent and specifically adsorbing ions on the pH dependency of zeta-potential, acc. to Fuerstenau (as cited in Hunter 1988, p. 234)

When the adsorption results not only from electrostatic, but (additionally) from physical adhesion (e.g. van-der Waals forces, hydrogen bonding) and/or chemical bonds (covalent or ionic) too, it is inevitably affected by the ion type (e.g. by mass and electronegativity) and solid phase (e.g. by atomic lattice). Such a specific adsorption can increase the value of the effective surface charge or may lead to a charge reversal. Specifically-adsorbed ions are partially or completely dehydrated and directly associated to the surface. They are, consequently, located at the IHP (cf. Fig. 3.3).

According to Fuerstenau and Hunter (Hunter 1988, p. 233), the physical adsorption is relevant for counter-ions; it does not affect the IEP, but can result in charge reversal at sufficiently high ion concentrations. In contrast, chemisorption can occur even when the surface charge has the same sign as the ion. Consequently, chemically adsorbing ions can affect the IEP.

This traditional view on specific adsorption, which relates the EDL formation to the interaction between surface and single ions, is far from being complete. In fact, further ion-specific effects that are related to the discrete nature of ions have to be considered. That means that there is no continuous distribution of charge, that ion size limits the charge density in the diffuse layer and in the Stern layer, as well as that the electrostatic and steric interactions between ions result in a certain spatial correlation (cf. Levine et al. 1962; Wiese at al. 1971; Quesada-Pérez et al. 2003). It was shown that the discreteness in charge and size can provide a good explanation of the charge reversal at high ion concentration without accounting for specific adsorption (Quesada-Pérez et al. 2005). Furthermore, it has been shown by several authors that the hydration of ions and surfaces also influences the ion adsorption in the Stern layer. Systematic studies for monovalent cations on weakly and strongly hydrated surfaces (e.g. mica and silica, respectively) revealed ion-specific effects which correlate with the thickness of the hydration layer (Colic et al. 1998; Johnson et al. 1999; Franks 2002). The origin of this hydration effect is not fully understood yet. It is, however, related to the discreteness of charge and size. Ninham and Yaminsky (1997) gave attention to the fact that the mobile ions in the diffuse layer interact with the surface not only by Coulombic forces, but also by van-der-Waals forces. Hence, the PBE (3.16) has to be modified in order to account for the corresponding potential energy additional to the electrostatic energy potential $z_i\psi F/RT$. Boström et al. (2006) showed that the non-electrostatic ion-surface interaction can explain the systematic differences observed for alkali cations, which were previously attributed to hydration.

Yet, although ion-ion correlation can explain ion-specificity to a great extent, there are some indications that the chemical interactions between ions and surface groups cannot be neglected (Lyklema 2009). In general, ion-specific effects are observed for high ionic strengths (≥0.1 M) and the higher the ion's valency is, the more likely their occurrence is.

Whether an ionic species interacts specifically with a particle surface is a matter of experimental investigation. There are, however, some rules of thumb. For example, halide ions (Cl^-, I^-) and nitrate ions (NO_3^-) react indifferently to most oxides, whereas the oxyanions of the halogenes (BrO_3^-, ClO_4^-, SO_4^{2-}) show distinct specific adsorption. Besides this, ion-specific effects have to be considered for multivalent ions (e.g. SO_4^{2-}, Ba^{2+}, La^{3+}). Yet, such rules are not universal. Medium sized and large alkali ions (Na^+, K^+, Cs^+) behave non-specifically towards titania and alumina surfaces but clearly affect the IEP of silica suspensions. In contrast, the IEP of silica is hardly affected by the strongly hydrated Li^+-ions, but that of alumina suspension does (references for Al_2O_3: Modi and Furstenau 1957; Franks et al. 1999; Johnson et al. 1999; Szczepaniak and Koœcielna 2002; for TiO_2: Kosmulski et al. 1999b; for SiO_2: Wiese et al. 1971; Colic et al. 1998; Franks 2002; Paciejewska 2010, cf. Fig. 3.7).

The adsorption of ionic species, or surface ionisation, is frequently portrayed as *surface complexation* (e.g. Osseo-Asare 1996; Kosmulski et al. 1999a). However, this terminology "does not imply a particular bonding structure" (Davis et al. 1978), as the ion bonding may be caused by physical or chemical interactions, yet it offers an opportunity to quantify the ionisation of surfaces as reversible chemical reactions in equilibrium. Appropriate calculation schemes for negatively charged surfaces and positive (hydrated) counter-ions are presented by Davis et al. (1978), Osseo-Asare (1996), Kosmulski et al. (1999a), and Jabłoński et al. (2000).

In summary, the adsorption of ionic species on surfaces depends to a large extent on the surface charge, thus on the concentration of the charge determining ions, and is affected by the non-electrostatic ion affinity to the surface and by ion–ion interactions. Additionally, there is a significant influence of ion hydration—in

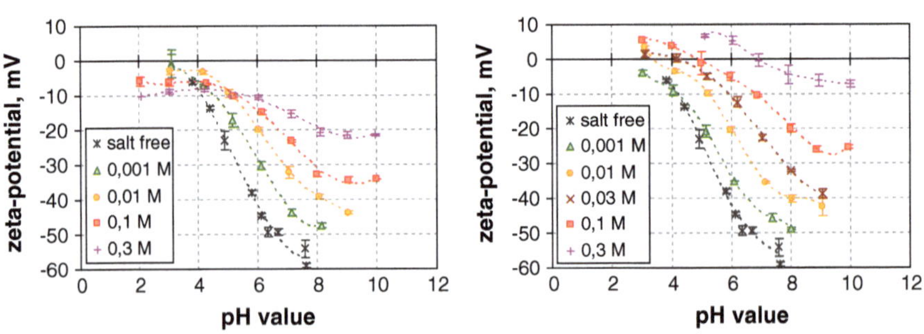

Fig. 3.7 Zeta-potential curves of amorphous silica (5 wt%, 150 m^2/g) in aqueous solutions of $LiNO_3$ (*left*) and KNO_3 (*right*); the former is an example of an indifferent electrolyte, while the latter leads to a clear shift of the IEP with increasing salt concentration (specific effect); reproduced from Paciejewska (2010), copyright 2010, with permission from K. Paciejewska

particular for metal cations—which essentially depends on the pH-value. Hydration shells hydrolyse with increasing pH, which may result in weakly soluble hydroxides and negatively charged hydroxide complex-ions (e.g. Jiang and Graham 1998; Leong 2005). Hence, the adsorption of ionic species and its effect on the net charge of surface and Stern layer is a function of pH.

3.1.6.2 Surface Precipitation

The concept of surface complexation of individual (hydrated) ions works well for low amounts of adsorbed ions. If, however, the surface concentration of hydrated multivalent cations or the pH exceed a critical value, surface precipitation of the respective hydroxide is induced. This qualitatively changes the surface properties and frequently coincides with charge reversal.

Figure 3.8 depicts the typical pH-dependency of the zeta-potential for colloidal suspensions in the presence hydrolysable metal ions. There are three *points of charge reversal* (PCR), which correspond to the PZC of the surface (PCR 1), to the neutralisation of the negative surface charge due to positively charged surface precipitate (PCR 2) and to the charge reversal of the hydroxide phase (James and Helay 1972; Leong 2005). The second and third PCR can only be observed for a sufficiently high concentration of the hydrolysable metal ions, i.e. for a significantly high coverage of the surface with precipitate. In the case of a complete coverage, the PCR 3 agrees with the PZC of the hydroxide phase. The second PCR depends on the ion concentration and lies above the critical pH at which the precipitation starts. According to Kuan et al. (2004), this pH is approximately one pH unit lower than the critical pH for bulk precipitation. Note that the zeta-potential curves of hydrolysable metal ions do not only result from surface precipitation but also reflect the adsorption of the different hydrolysed ionic species (e.g. $Al(H_2O)_6^{3+}$, $Al(H_2O)_5OH^-$ etc.). Moreover, for surfaces with significant solubility (e.g. amorphous SiO_2), the precipitate may consist of hydroxide and mixed oxides (e.g. alumino-silicate).

Surface precipitation increases the complexity of surface charging and can qualitatively change the behaviour of colloidal particles in electrolyte solutions. Models that allow for a quantification of surface precipitation are, e.g., discussed by

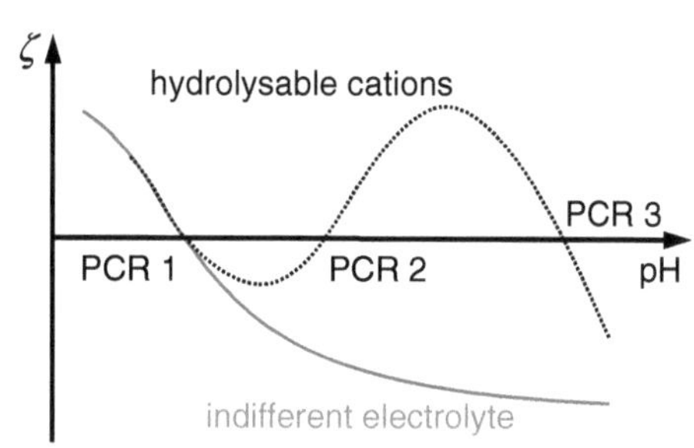

Fig. 3.8 Effect of the surface precipitation of hydrolysable cations (e.g. Al(III)-ions) on the zeta-potential curves of oxides (cf. James and Healy 1972; Kuan et al. 2004); up to 3 points of charge reversal (PCR) can be observed

Farley et al. (1985) and Kuan et al. (2004). From a technical point of view, the surface precipitation offers an interesting opportunity to create core-shell particles (e.g. Firment et al. 1989).

3.1.6.3 Adsorption of Surfactants on Charged Surfaces

Surfactants are molecules that consist of a non-polar hydrophobic (e.g. alkyl-chains) and a polar hydrophilic part, which can be cationic, anionic, zwitter-ionic, or non-ionic. The relevance of the polar part for the molecule behaviour is quantified in form of the hydrophilic-lipophilic balance (HLB), for which values above 7 indicate dominance of the hydrophilic groups (Myers 1999, pp. 270–273). The heterogeneous composition leads to amphiphilic properties and interfacial activity. That means surfactants accumulate on interfaces and, thus, reduce the interfacial energy (cf. Sect. 3.1.3). Surfactant molecules associate to nanosized aggregates, so-called *micelles*, above a critical concentration (the critical micelle concentration—CMC). Such molecule associations do not only exist in the bulk solution but can also be formed by adsorbed surfactant molecules (Holmberg et al. 2002, pp. 39–66).

The adsorption of surfactants on solids is affected by the surface properties (hydrophilicity/hydrophobicity, surface charge) and by the surfactant properties (ionic/non-ionic, CMC, HLB). Figure 3.9 illustrates the possible configurations of non-ionic surfactants on hydrophobic (A) and on hydrophilic (B) solid surfaces as well as those of cationic surfactants on negatively charged oxide layers (C). The surface concentrations increases successively from row I (ideal gas) to row IV or V (saturated surface). The picture reveals a broad variety including monolayers,

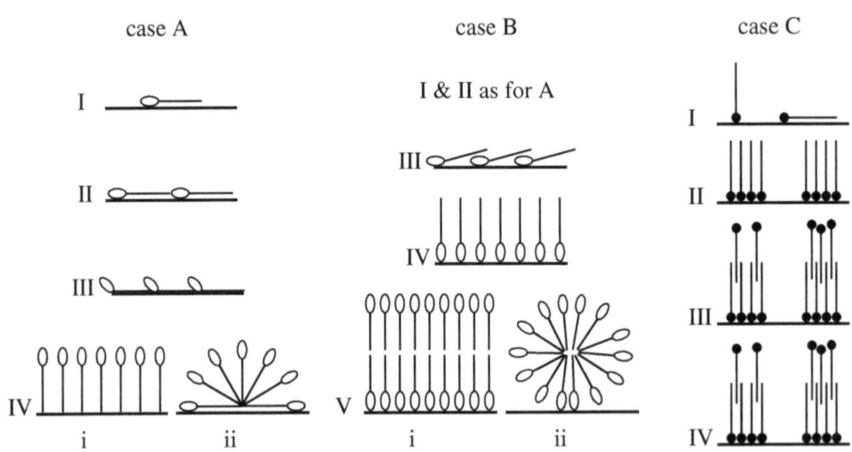

Fig. 3.9 Structure of the adsorbate layer for non-ionic surfactants on hydrophobic **A** and hydrophilic **B** surfaces and for cationic surfactants on negatively charged oxides **C**; the surfactant concentration increases from I to V (acc. to Fan et al. 1997; Atkin et al. 2003; Paria and Khilar 2004); the hydrophobic parts of the surfactant molecules are indicated as *straight lines*

bilayers, hemi-micelles, and surface micelles (according to Fan et al. 1997; Atkin et al. 2003; Paria and Khilar 2004).

For very low surfactant concentrations, the adsorption is solely determined by the hydrophobic and electrostatic interaction between single surfactant molecules and the surface; there is no interaction between the adsorbed molecules, and the molecular surface area is rather large. With increasing surface concentration, the surfactant molecules start to interact (attraction of hydrophobic parts, repulsion of identically charged groups) and tend to a preferred vertical orientation. As a result, the effective surface properties change, e.g. the hydrophobicity or the charge density. The interaction of adsorbed molecules (via hydrophobic parts and charge) eventually results in bi-layers or molecule aggregates at very high surface concentrations.

The signs of surface and surfactant charge and the total electrolyte concentration are crucial factors for the adsorbate structure: Opposite signs of charge promote the adsorption of surfactants and lead to a neutralisation or even reversal of surface charge. The background electrolyte determines the EDLs around the particle and around the surfactant molecules. High salt concentrations screen the Coulombic interactions and, thus, allow for a dense packing of ionic surfactants. A further important factor is temperature because it affects the configuration of the molecules, the HLB, and structure of micelles.

In summary, the adsorption of surfactants on colloidal particles is influenced by the EDL and it affects the EDL. Surfactant molecules compete with simple ions from the background electrolyte for adsorption in the Stern layer (Dimov et al. 2002), yet the interaction between adsorbed surfactant molecules introduces additional complexity to the problem of EDL formation. Detailed introduction into the surfactant adsorption on solid surfaces is, e.g., given by Myers (1999, Chap. 9) and Holmberg et al. (2002, Chap. 17).

3.1.7 Surface Dissolution

The interfacial properties of colloidal suspensions are determined by the chemical reactions (e.g. protonation) and adsorption of solutes. Additionally, the interface can be affected by the dissolution-precipitation equilibrium of the particle phase. This is because precipitation changes the surface morphology (Vigil et al. 1994) or leads to phase transition (Lefèvre et al. 2002; Carrier et al. 2007). In addition, dissolution means degradation of the particles and may result in the loss of the finest particle fractions (i.e. Ostwald ripening). For this reason, it is necessary to understand the factors governing the dissolution of solid particles.

3.1.7.1 Dissolution of Oxides

Solubility is a material property; only at the lower nanoscale does particle size matter (cf. Sect. 3.1.4). The dissolution of oxides strongly depends on the reducing element. Even so, the dissolution of metal oxides can be described by the following general equation:

$$MO_{x/2(s)} + \frac{x}{2} H_2O \Leftrightarrow M^{x+} + x\,OH^-. \tag{3.21}$$

The released metal ions M^{x+} are covered by a hydration shell, the size of which depends on the valency of the ion and the size of the nucleus. The hydrate complexes $[M(H_2O)_n]^{x+}$ of multi-valent ions can deprotonate (i.e. they act as acids), which affects the chemical equilibrium of the reaction (3.21). As a result, the apparent solubility is a function of the pH-value. Deprotonation leads to a transformation of the hydrate complexes into hydroxide complexes $[M(OH)_n]^{x-n}$. The neutral hydroxide complexes are hardly soluble. The total amount of dissolved material and its composition can be calculated from the acid-base-equilibria of the complex ions and the pH. In general, the lower the main group of the metal is, the higher the oxide solubility.

For γ-alumina, the dissolution can be described as $0.5\ Al_2O_{3(\gamma)} + 3\ H^+ \leftrightarrow Al^{3+} + 1.5\ H_2O$ (with $pK_{diss} = -8.5$, Lindsay 1979). The hydrated Al^{3+} ion exists only in acid environment. For pH > 2, hydrolysis leads to a variety of mononuclear and even polynuclear aluminum complexes (e.g. $Al_2(OH)_2^{4+}$, $Al_3(OH)_4^{5+}$ and $Al_{13}O_4(OH)_{24}^{7+}$; Fig. 3.10). All solute species of the Al^{3+} ion can be expressed by the general equation $xAl^{3+} + yH_2O \leftrightarrow Al_x(OH)_y^{3x-y} + yH^+$. The corresponding equilibrium constants can be taken from the literature (e.g. Baes and Mesmer 1976; Jiang and Graham 1998).

Equation (3.21) cannot be applied to the oxides of metalloids and nonmetals. In that case, the dissolution reaction produces acid molecules with oxyanions. Deprotonisation of these acids eventually causes a pH-dependency of the solubility, similar to metal oxides.

For instance, the dissolution reaction of amorphous silica can be written as a synthesis of orthosilicic acid $SiO_{2(s)} + 2\ H_2O \leftrightarrow H_4SiO_4$ ($pK_{diss} = 2.7$, Gunnarsson and Arnórsson 2000). At pH-values above 8, the H_4SiO_4 molecules dissociate into $H_3SiO_4^-$ and $H_2SiO_4^{2-}$ and, thus, affect the dissolution equilibrium. Hence, the solubility of silica in a basic environment exceeds that of the solubility in neutral milieu considerably (approx. by factor 2 at pH 10 and by factor 50 at pH 11, cf. Fig. 3.11; Iler 1979, pp. 47 and 123).

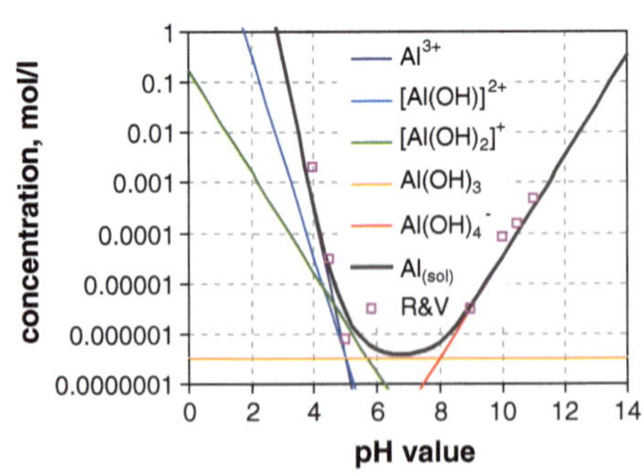

Fig. 3.10 Equilibrium concentrations of the hydrolysed aqua-complexes of Al(III)-ions versus pH calculated according to Baes and Mesmer (1976) for 25 °C; the *thick solid line* is the total amount of dissolved aluminium; squares refer to the experiments of Roelofs and Vogelsberger (2006)

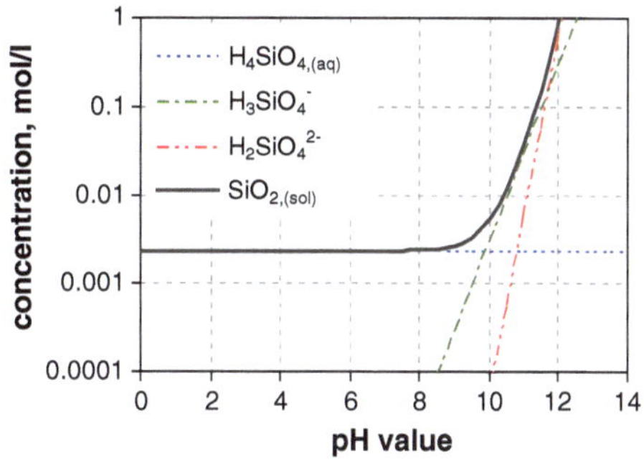

Fig. 3.11 Equilibrium concentration of orthosilicic acid (H_4SiO_4) and of its dissociated ionic species as a function of the pH (25 °C); the *thick solid line* marks the total amount of dissolved silica based on data from Iler (1979, pp. 47 and 123)

The relevance of surface dissolution is not only determined by solubility, but also by the dissolution kinetics, which are influenced by the degree of undersaturation of the solvent phase and by the area of the interface. For dense suspensions the rate of dissolution can be estimated with (Vogelsberger et al. 1999):

$$\frac{\mathrm{d}c_n}{\mathrm{d}t} = k_t \cdot \frac{\varphi_V}{1-\varphi_V} \cdot S_V \cdot \left(1 - \frac{c_n}{c_{n,\mathrm{sat}}}\right), \tag{3.22}$$

where c_n denotes the molar concentration of the dissolved species, φ_V the solid volume fraction, S_V the volume specific surface area of the particle system, and $c_{n,\mathrm{sat}}$ the saturation concentration. The kinetic constant k_t depends on temperature, pressure and, for oxidic materials, the pH and ionic strength (Fig. 3.12).

Dissolution is an equilibrium reaction. That is, the dissolution of the surface is accompanied by precipitation of the solute *oxide species. Some authors argue that this* results in transformation of the surface morphology. For instance, Vigil et al. (1994) proposed the formation of a gel-like layer on a silica surface, which results from re-precipitation of dissolved silica and that consists of polymeric –Si(OH)$_2$–O–Si(OH)$_2$–OH chains. The thickness of this layer is estimated with 10 Å. A similar concept was stated by Adamczyk et al. (2004).

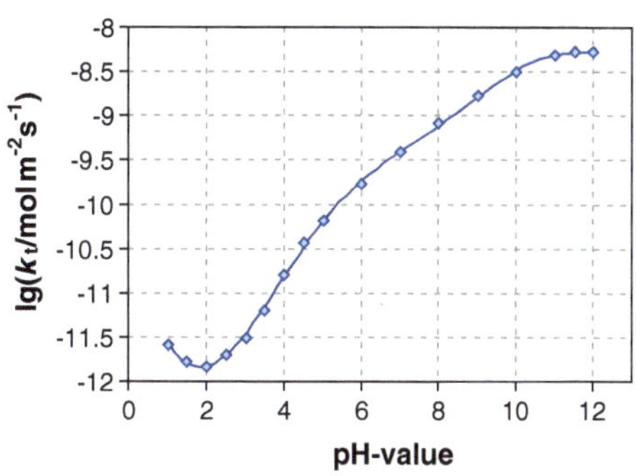

Fig. 3.12 Dissolution rate k_t of amorphous (pyrogenic) silica versus pH for background electrolyte 0.01 M NaCl, acc. to Vogelsberger et al. (1999)

Moreover, solutes may correspond to different solid phases with different solubility. Surface dissolution can, therefore, be accompanied by phase transition on the particle surface. For instance, surfaces of fused quartz swell under water to form gel layers of amorphous silica (Iler 1979, pp. 34–40; Yaminsky et al. 1998). Similarly, it could be shown that the dissolution of γ-alumina (γ-Al_2O_3) is coupled with the gradual formation of a hydroxide phase on the particle surface (e.g. Roelofs and Vogelsberger 2006; Carrier et al. 2007; Bonelli et al. 2010).

3.1.7.2 Impact of Multi-valent Ions on the Dissolution of Oxides

A few studies have shown that the solubility of silica is negatively affected by surface-coordinated ions like Cu(II) and Al(III) (Stumm and Wollast 1990; Kuan et al. 2004; Rickert 2000). This is first explained with a blocking of the oxide surface groups by such ions. In the case of Al(III), additional effects are the formation of alumino-silicate complexes between Al(III)-ions and the silanol (Si–OH) groups and the precipitation of hydroxide phase at high Al(III)-concentrations (see above). Moreover, Kuan et al. (2004) proposed a scenario for moderate Al(III) concentrations, according to which dissolved orthosilicic acid eventually precipitates on such alumina-silicate layers, which yields surface properties similar to the original one. The relevance of such dissolution and precipitation processes strongly depends on ion concentration and total particle surface area. Paciejewska (2010) proved its relevance for multi-component suspensions where dissolved species from the particulate components compete in adsorption and precipitation on the interfaces with dramatic effects on the suspensions' stability. The complexity of this topic is even increased by the fact that the dissolution kinetics depend on the concentration of all ionic species. Dissolution is accelerated with increasing ionic strength and, in the case of oxides, additionally by hydroxide ions and some multivalent ions (e.g. SO_4^{2-} in case of SiO_2 surfaces, Bai et al. 2009).

3.2 Interactions Between Colloidal Particles

The particles in colloidal suspensions interact with each other through hydrodynamic coupling and by a number of attractive and repulsive forces which are related to the particles' bulk and interfacial properties. Of particular relevance are electrodynamic interactions between the atoms and molecules of the particles (van-der-Waals interaction) and those interactions that originate from the overlap of electric double layers (double layer interaction). The DLVO theory comprises these two types of interactions for pairs of interacting particles and calculates the corresponding (excess) free energy as a function of particle distance. It neglects further types of interaction, e.g., steric or hydration interaction, the significance of which depends on the properties of particles, the solvent, and solutes. Since the range and the strength of all these interactions considerably affect the macroscopic behaviour of colloidal suspensions (e.g. stability, rheology, turbidity), it is useful to know the crucial factors and to have appropriate models for their calculation.

3.2.1 Van-der-Waals Interaction

Van-der-Waals interactions are interactions between atoms and molecules with fixed or induced polarisation, which result in a strong attractive force for short distances. Particles obey such an attractive force as well. Hamaker (1937) calculated this macroscopic interaction energy by assuming an undisturbed contribution from each elementary pair of atoms or molecules. For two spheres of diameter x_1 and x_2 he obtained:

$$V_{\mathrm{vdW}} = -\frac{A_{132}}{12}\left[\frac{x_1x_2}{h\cdot(x_1+x_2+h)} + \frac{x_1x_2}{(x_1+h)\cdot(x_2+h)} + 2\ln\left(1-\frac{x_1x_2}{(x_1+h)\cdot(x_2+h)}\right)\right], \tag{3.23}$$

where h is the minimum surface distance and A_{132} is a coefficient which depends on the materials of the particles (1 and 2) as well as on the medium 3 that separates them. Even though this approach neglects the many-body aspect of the problem, it gives a good estimate for the dependency on the geometrical parameters. In general, the van-der-Waals interaction is short-ranging and decays rapidly with surface distance. For this reason, it usually suffices to calculate the interaction for minute surface gaps ($h \ll x_i$). In that range, the interaction potential is inversely proportional to the minimum surface distance:

$$V_{\mathrm{vdW}} = -\frac{A_{132}x_1x_2}{12h(x_1+x_2)}. \tag{3.24}$$

The Hamaker-de-Boer theory[2] ignores the quantum physical aspects of the van-der-Waals interaction, which are responsible for a retardation effect, i.e. a lowering of the interaction coefficient A_{132} with increasing surface distance. An alternative, more elaborate calculation scheme was developed by Lifshitz (1956); it is based on the dielectric response of the involved substances to an electromagnetic field (e.g. Lyklema 1991, pp. 4.75; Hunter 1993, pp. 194-227). This theory takes account of the decreasing contribution of the induced dipoles to the interaction energy with growing distance between the macro-bodies (retardation effect). As a result, the Hamaker coefficient A_{132} is replaced by the Lifshitz-Hamaker function

[2] de Boer (1936) used the approach of pairwise sums for the interaction force between two flat surfaces.

$A_{132}(h)$, which can be approximated by (Bowen and Jenner 1995; Viravathana and Marr 2000):

$$A_{132}(h) = A_{132,s} + \left(A_{132,0} - A_{132,s}\right) \cdot \left(1 + \left(h/h_{cr}\right)^{3/2}\right)^{-2/3}, \tag{3.25}$$

where $A_{132,0}$ is the non-retarded Hamaker constant (vanishing gap) and $A_{132,s}$ reflects the static contribution of permanent dipoles (Viravathana and Marr 2000). These parameters can be computed from the dielectric function which is known for several materials (Bergström 1997; Nguyen 2000; cf. Appendix B.3.3). The parameter h_{cr} represents a characteristic distance of the retardation and is obtained by fitting the exact Lifshitz-Hamaker function (Bergström 1997).

Strictly speaking, Eq. (3.25) is valid for two half-spaces only. In the case of two spherical particles, there is a complex interrelation between geometry parameter and material properties. The exact analytical solution of the Lifshitz approach was presented by Langbein (1974, as cited by Thennadil and Garcia-Rubio 2001). However, the necessary computations are rather laborious and converge slowly. For that reason, it seems appropriate to approximate the exact analytical solution by combining the Lifshitz-Hamaker function of two half-spaces (Eq. (3.25)) with the Hamaker geometry function for two spherical particles (Eq. (3.23)).

Pailthorpe and Russel (1982) and Arunachalam et al. (1998) compared this approximation with the exact solution by Langbein for colloidal polystyrene spheres in an aqueous solution and for aerosol molecule clusters of tetrachloromethane (CCl_4), respectively. In both cases the approximate approach underestimates the van-der-Waals interaction energy by approx. 10 to 20 %. Pailthorpe and Russel (1982) conclude that this deviation is comparable with the uncertainty due to inaccurate dielectric values.

The static contribution (of permanent dipoles) to the van-der-Waals interaction $A_{132,s}$ is reduced when the separating medium 3 contains mobile ions. In contrast, the dispersion contribution $A_{132,d} = A_{132,0} - A_{132,s}$ remains *unaffected because the fluctuations of* electrons are much faster than the Brownian motion of ions. Mahanty and Ninham (1976, as cited in Israelachvili 1992) describe the screening for $\kappa h \gg 1$ as follows:

$$A_{132}(h) = A_{132,s} \cdot 2\kappa h e^{-2\kappa h} + A_{132,d}(h), \tag{3.26}$$

where κ is the Debye-Hückel parameter.

3.2.2 Double Layer Interaction

When two particles approach each other so that their EDLs overlap, they experience a repulsive or attractive interaction that results from the electro-static and osmotic forces between the ions and the surfaces. Consider, for instance, two particles with identical surface charge and potential. Then, the overlap of the diffuse layers means a local increase in counter-ion concentration, which produces a disjoining osmotic pressure. In general, the interaction energy is determined by the concentrations of all ionic species c_i (or mobile charges) inside the diffuse layer as well as by the corresponding potential field $\psi(\mathbf{r})$. In the overlapping region, however, the distributions of potential and ion concentration are changed compared to isolated particles, and the surface charging then locally adapts to this new environment. This so-called *charge and potential regulation* has a major influence on the magnitude of the double layer interaction.

3.2.2.1 Charge and Potential Regulation

The overlap of the diffuse layers of two neighbouring particles inevitably affects the dissociation and adsorption equilibria. Hence, surface charge σ_0 and surface potential ψ_0 as well as diffuse charge σ_d and diffuse layer potential ψ_d are locally changed in the overlapping region. The way and the extent of this charge and potential regulation is material-specific. In general, a non-linear relationship between the surface charge and the surface potential has to be established. For that reason, a lot of papers refer to the two limiting cases of a constant surface potential and a constant surface charge (Hogg et al. 1966; Chan et al. 1976; McCormack et al. 1995; Sader et al. 1995).

In order to calculate the interaction energy, it is necessary to assume a model for the charge regulation. Carnie and Chan (1993) proposed a linear approach for the absence of a Stern layer:

$$\sigma_0 = S_{\text{reg}} - K_{\text{reg}}\psi_0, \tag{3.27}$$

which is appropriate, when the linearised PBE (Eq. (3.17)) is used. The regulation parameters S_{reg} and K_{reg} allow for a broad variation of charge regulation regimes: When the regulation capacity K_{reg} vanishes, the constant charge (CC) case is obtained. In contrast, the surface potential remains constant (CP) for very large regulation capacities ($K_{\text{reg}} \gg \varepsilon_m\varepsilon_0\kappa$). Additionally, Carnie and Chan (1993) introduced a "canonical intermediate" case of charge regulation which is achieved when $K_{\text{reg}} = \varepsilon_m\varepsilon_0\kappa$.

The set of basic equations is completed by the following boundary condition:

$$\sigma_0 = \varepsilon_m\varepsilon_0 \cdot \nabla\psi \bullet \mathbf{n}. \tag{3.28}$$

This equation holds true for small permittivity of the particles ($\varepsilon_p \ll \varepsilon_m$). If this restriction is not fulfilled, the potential distribution inside the particles has to be described as well. According to Carnie and Chan (1993), the error in using Eq. (3.28) for particles with a relatively large permittivity ε_p is relevant only for constant charge regulation with small particle sizes. However, in the case of a rather rough surface, the penetration of the electric field inside the particle cannot be ignored even for small ratios $\varepsilon_p/\varepsilon_m$ (Dukhin and Lyklema 1989).

For an isolated sphere with linear charge regulation, the boundary condition may be written as:

$$S_{\mathrm{reg}} = \psi_0^{\mathrm{iso}} \varepsilon_{\mathrm{m}} \varepsilon_0 \left(\kappa + \frac{1}{a} \right) + K_{\mathrm{reg}} \cdot \psi_0^{\mathrm{iso}}. \tag{3.29}$$

This approach describes all relevant properties of the surface and the Stern layer in the four parameters S_{reg}, K_{reg}, σ_0^{iso}, and ψ_0^{iso}, and does not require an explicit description of the Stern layer. More accurate models of charge regulation, however, have to account for the mechanisms of surface charging and ion adsorption and, thus, rely on the corresponding equilibrium constants (Usui 2004; Lyklema and Duval 2005). Such models are necessary in order to explain why the overlap of two dissimilar EDLs may even result in re-charging of one of the surfaces (Chan et al. 1976; McCormack et al. 1995; Chan et al. 2006; Lyklema and Duval 2005; Usui 2008).

Consider two dissimilar oxide surfaces with PZCs of 2 and 9 in an aqueous solution of pH 9.5. That means that both surfaces are negatively charged, but one has a high charge density and the other has a low one. Excluding any specific adsorption, we find a large and a weak excess of H^+-ions in the diffuse layer of surfaces 1 and 2, respectively. Hence, the overlap is dominated by surface 1, which means that the pH in the joint diffuse layer is now lowered compared to the pH of the isolated diffuse layer 2. For a sufficiently large ΔpH, the pH falls below the PZC of surface 2, which consequently becomes positively charged. Hence, instead of surface repulsion, as expected from the isolated surfaces, the surfaces experience an attractive interaction.

3.2.2.2 Interaction Potential

The interaction energy of overlapping double layers can be calculated from the osmotic pressure and the Maxwell stress tensor (Bell et al. 1970; Hsu 1999) or via the free energy of the double layers (Chan and Mitchell 1983; Chan 1987; Carnie and Chan 1993; Carnie et al. 1994a). The free energy of an electric double layer results from the spontaneous surface charging and from the subsequent formation of the diffuse layer, i.e. from the charge separation. Since the overlap of double layers alters the diffuse layer and surface charge, the corresponding free energy changes as compared to the case of isolated particles, hence (Chan and Mitchell 1983; Carnie and Chan 1993):

$$V_{\mathrm{DL}}(h) = \int (J_1(h) - J_1(\infty))\mathrm{d}S_1 + \int (J_2(h) - J_2(\infty))\mathrm{d}S_2, \tag{3.30}$$

where $J_i(h)$ is the free energy per unit area of particle i separated from the other particle by the distance h. For the linearised PBE and the linear regulation model of Carnie (Eq. (3.27)), the interaction energy can be computed from (Carnie and Chan 1993):

$$V_{\mathrm{DL}}(h)=\frac{1}{2}S_{\mathrm{reg},1}\oint\left(\psi_{0,1}-\psi_{0,1}^{\mathrm{iso}}\right)\mathrm{d}S_1+\frac{1}{2}S_{\mathrm{reg},2}\oint\left(\psi_{0,2}-\psi_{0,2}^{\mathrm{iso}}\right)\mathrm{d}S_2 \tag{3.31}$$

for the general case of charge regulation (including CC) and from

$$V_{\mathrm{DL}}(h)=\frac{1}{2}\psi_{0,1}^{\mathrm{iso}}\oint\left(\sigma_{0,1}-\sigma_{0,1}^{\mathrm{iso}}\right)\mathrm{d}S_1+\frac{1}{2}\psi_{0,2}^{\mathrm{iso}}\oint\left(\sigma_{0,2}-\sigma_{0,2}^{\mathrm{iso}}\right)\mathrm{d}S_2 \tag{3.32}$$

for constant potential (CP) regulation. The integration is conducted over the whole surface S_i of each particle and the index "iso" refers to isolated particles.

In general, there is no simple analytical solution for the integrals in Eq. (3.30)—even for spherical particles and simple charge regulation (e.g. constant potential). That is related to the fact that there are no simple expressions for the potential distribution in the diffuse layer of a particle. Numerical schemes for deriving the interaction energy on the basis of the non-linearised Poisson-Boltzmann equation (PBE) have been discussed, e.g., by Carnie et al. (1994b) and Lyklema and Duval (2005). Most authors, however, use the Debye-Hückel approximation ($e_0\zeta/k_BT \leq 1$), which leads to the linearised PBE (Eq. (3.17)). In this form, the potential distribution $\psi(\mathbf{r})$ in the diffuse layer can be described by means of a multipole expansion (Glendinning and Russel 1983; Krozel and Saville 1992; Carnie et Chan 1993; Carnie et al. 1994a). Still, this approach requires numerical tools for calculating the interaction energy. A further approximation is possible by summation over the interaction energy of pairwise surface elements, which are considered to be parallel plates (Derjaguin approximation). Based on this, Hogg, Healy, and Fuerstenau obtained the following equation for the interaction energy between two dissimilar spheres (HHF-solution, Hogg et al. 1966)

$$V_{\mathrm{DL}}=\frac{\pi}{2}\varepsilon_{\mathrm{m}}\varepsilon_0\frac{x_1x_2}{x_1+x_2}\left[2\cdot\psi_{0,1}\psi_{0,2}\ln\left(\frac{1+\mathrm{e}^{-\kappa h}}{1-\mathrm{e}^{-\kappa h}}\right)+\left(\psi_{0,1}^2+\psi_{0,2}^2\right)\cdot\ln\left(1-\mathrm{e}^{-2\kappa h}\right)\right], \tag{3.33}$$

where $\psi_{0,1}$ and $\psi_{0,2}$ denote the surface potentials, and x_1 and x_2 denote the particle diameters. For a pair of identical particles, Eq. (3.33) reduces to:

$$V_{\mathrm{DL}}=\pi\varepsilon_{\mathrm{m}}\varepsilon_0 x\psi_0^2\ln\left[1+\exp(-\kappa h)\right]. \tag{3.34}$$

The Eq. (3.33) is based on the Gouy-Chapman model (i.e. the Stern layer is ignored) and on charge regulation with constant surface potential (CP). It applies to small surface potentials ($|\psi_{0,i}|$ < 25 mV, to thin double layers ($\kappa a \gg 1$) and to small surface distances ($h \ll x_i$). Equation (3.33) provides only rough estimates for moderate and thick double layers ($\kappa a < 5$). Yet, Sader et al. (1995) showed that by replacing the term x_1+x_2 with $x_1+x_2+2\cdot h$ (i.e. twice the particle centre distance r_{12}), the HHF-equation is applicable to arbitrary values of h.

Since the HHF-solution is based on the Debye-Hückel approximation, it can be combined with a linear charge regulation model (Eq. (3.27)). That yields the more general expression (Carnie et al. 1994a; here combined with the mentioned modification of Sader et al. 1995):

$$V_{\mathrm{DL}} = \frac{\pi}{4}\varepsilon_{\mathrm{m}}\varepsilon_0 \frac{x_1 x_2}{r_{12}} \left[4 \cdot \psi_{0,1}\psi_{0,2} \frac{artanh\left(\sqrt{\Delta_1\Delta_2}\mathrm{e}^{-\kappa h}\right)}{\sqrt{\Delta_1\Delta_2}} + \left(\frac{\psi_{0,1}^2}{\Delta_1} + \frac{\psi_{0,2}^2}{\Delta_2}\right) \cdot \ln\left(1 - \Delta_1\Delta_2\mathrm{e}^{-2\kappa h}\right) \right], \tag{3.35}$$

where:

$$\Delta_i = \frac{K_{\mathrm{reg},i} - \varepsilon_{\mathrm{m}}\varepsilon_0\kappa}{K_{\mathrm{reg},i} + \varepsilon_{\mathrm{m}}\varepsilon_0\kappa}, \; r_{12} = \frac{1}{2} \cdot (x_1 + x_2) + h. \tag{3.36}$$

Equation (3.35) reduces to the HHF-solution (Eq. (3.33), constant potential) for high regulation capacities $K_{\mathrm{reg},i} \gg \varepsilon_{\mathrm{m}}\varepsilon_0\kappa$ (i.e. $\Delta_i = 1$), while it provides the constant charge solution when the regulation capacity vanishes $K_{\mathrm{reg},i} \ll \varepsilon_{\mathrm{m}}\varepsilon_0\kappa$ (d.h. $\Delta_i = -1$). In general, both parameters, surface charge and potential, are regulated. For Carnie's "canonical intermediate" case, i.e., for $\Delta_\mathrm{i} = 0$, the interaction energy between two spherical double layers obeys:

$$V_{\mathrm{DL}} = \pi\varepsilon_{\mathrm{m}}\varepsilon_0 \frac{x_1 x_2}{r_{12}} \psi_{0,1}\psi_{0,2}\mathrm{e}^{-\kappa h}. \tag{3.37}$$

This solution agrees with the approximate solution for thick double layers ($\kappa a < 1$), when regulation is not relevant (cf. Hunter 1993, p. 418). It may, thus, provide a good approximation for regulating EDLs of arbitrary surface distance and arbitrary Debye length.

3.2.3 DLVO-Theory

The interaction between particles usually results from different forces and is quantified as total interaction energy V_t for a pair of approaching particles in dependence on the minimum surface distance h. Based on this function, one can estimate how fast colloidal particles aggregate or how the particles spatially arrange in the solvent. Assuming that the single interaction mechanisms act independently from each other, one can calculate the total interaction energy as the sum of all interaction energies. This approach was first employed by Derjaguin and Landau (1941), who considered the van-der-Waals interaction and the double layer interaction, for which they developed an approximate expression for the case of similar double layers. No other types of interaction were regarded. Based on the functional dependency $V_\mathrm{t}(h)$, they derived criteria that allow for an evaluation of the suspension stability and compared them with empirical correlations: the Schulze-Hardy rule (Hardy 1900) and Ostwald's rule (Ostwald 1935). Shortly after, Verwey and Overbeek arrived independently at similar results (Verwey and Overbeek 1948). Hence, it is customary to name the evaluation of the suspension stability via double layer and van-der-Waals interaction *DLVO theory* (e.g. Lyklema 1991, p. 1.21; Israelachvili 1992, p. 247).

Nowadays, it is rather common to extend the DLVO theory with the *Born repulsion* or *hardcore interaction*, which is a short-ranging repulsion that excludes the interpenetration of approaching surfaces (Ruckenstein and Prieve 1976; Lyklema 1991, p. 4.5). The calculation is usually based on a power-law approach for the Born repulsion of atoms and molecules as used in the Lennard-Jones (12, 6) potential (e.g. Feke et al. 1984).

The shape of the function $V_t(h)$ for the total interaction energy depends on the Hamaker function $A_{132}(h)$, the diffuse layer potentials ψ_d (approximated with the zeta-potential), and the ionic strength I (or the Debye-Hückel-Parameter κ). Ideally, one can distinguish two local minima and one local maximum (Fig. 3.13), while at vanishing surface distance ($h \ll 1$ nm), the total interaction energy tends to positive infinity due to the Born repulsion. The first or *primary minimum* originates from the van-der Waals attraction, which exceeds the double layer interaction in value for small surface distances. The primary minimum indicates that adhesion is an energetically favourable configuration for two interacting particles.

The second or *secondary minimum* at moderate surface distances is comparably feeble. It can only be observed for strong van-der-Waals interaction and small Debye lengths. It further requires a significant repulsion from the double layer interaction that becomes manifest by the local energy maximum V_{max}. This maximum is called *energy barrier* because it corresponds to an energetically unfavourable configuration and may, thus, prevent two particles from adhesion (i.e. from attaining the primary minimum). The energy barrier is effective if its value exceeds the kinetic energy of the colliding particles. In the absence of shear flow

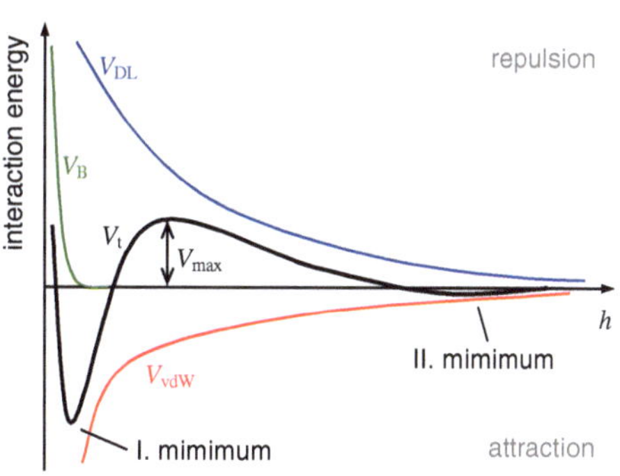

Fig. 3.13 Principal dependency of the total interaction energy V_t between two particles from the surface distance; curves for EDL interaction (V_{DL}), van-der-Waals interaction (V_{vdW}) and Born repulsion (V_B) are shown additionally; example for repelling EDLs and two local minima

and particle migration, Brownian motion prevails and the kinetic energy is a distributed quantity with an average value of 1 k_BT. In order to prevent coagulation for all particles over a sufficiently long period, i.e., to warrant suspension stability, minimum values of 10–20 k_BT are required for the energy barrier (Healy et al. 1978; Bleier and Westmoreland 1991). If the energy barrier has been overcome and the primary minimum is arrived, the particles form agglomerates or aggregates with strong interparticle bonds, which are difficult to re-disperse. In contrast, agglomerates or clusters that are related to the secondary minimum are more easy to disperse ("reversible agglomeration"), although such a suspension is also unstable.

The height of the energy barrier and the depth of the secondary minimum are affected by the electrolyte. An increase in the electrolyte concentration, for instance, compresses the diffuse layer, weakens the double layer repulsion and lowers the energy barrier. Above a critical concentration of electrolyte—the *critical coagulation concentration* (CCC)—fast coagulation is observed (Mulvaney et al. 2000; Kralchevsky et al. 2002; Pagel 2005; cf. Sect. 5.2.2). Strong repulsion, thus good suspension stability, is achieved for low electrolyte and high zeta-potential values.

3.2.4 Non-DLVO Interactions

The DLVO theory is a good starting point for calculating particle interactions and for estimating the suspension behaviour. However, it relies on an accurate characterisation of the particle and interfacial properties. Moreover, it does not explain the origin of surface charge, nor does it considers the results of particle interaction (e.g. coagulation) or aging effects, i.e. temporal changes in the suspension properties (e.g. due to dissolution or Ostwald ripening). Even more, it is based on a simplified view on the surface (e.g. with regard to roughness) and the double layer; it neglects any interdependence between van-der-Waals and double layer interaction and ignores all other mechanisms of particle interaction (e.g. steric or hydrophobic interaction). Deviations of the "real world" from DLVO assumptions or results are frequently called "non-DLVO effects", which are usually relevant for small surface distances. Of particular importance for colloidal suspensions are interaction mechanisms that are related to a second colloidal phase, which consists of much finer colloids than the particles under consideration. Such mechanisms are the steric and the depletion interaction, as well as interaction mechanisms that are somehow related to the hydration of surfaces and ions in aqueous media.

3.2.4.1 Steric Interaction

Steric interactions are interactions between two surfaces (particles) in the presence of adsorbed macromolecules (polymers) or other molecular or particular species. Such adsorbate layers prevent the direct contact of surfaces and, thus, restrain the attractive van-der-Waals forces and can, therefore, inhibit the coagulation of particles (van der Waarden 1950; Heller and Pugh 1954). Steric interactions by polymers have been systematically investigated by Donald H. Napper in a series of papers (e.g. Napper 1969; Napper and Hentschey 1971; Napper 1977).

The stabilising effect of macromolecules on colloidal suspensions has already been employed in ancient times, e.g., for stabilising carbon particles in inks. In his famous Baker lecture to the Royal Society, Faraday demonstrated the stabilisation of colloidal Au sols in the presence of electrolyte by gelatine (Faraday 1857). Some years later, Zsigmondy (1901) presented a comprehensive study on the stability of colloidal Au sols which included several natural macromolecules (i. a. gum arabic, albumin). In order to quantify their stabilising effect, Zsigmondy introduced the *Goldzahl* (gold number), which describes the critical mass (in mg) of polymer corresponding to the coagulation of the colloidal Au particles after adding 1 cm^3 of 10 wt% NaCl solution to 10 cm^3 suspension with approx. 0.0055 wt% Au.

Even though *steric interaction* refers to the interaction between two surfaces or particles, it cannot be explained without a proper understanding of the interactions of the adsorbed species with each other and with the solvent. This is particularly true for polymers which consist of large flexible molecules with numerous potential adsorption sites and various possible conformations. Their adsorption on particle surfaces depends on surface properties (charge, hydrophobicity), solvent properties (polarity, proticity), polymer properties (molecular weight, charged groups), and polymer concentration. The prerequisite for a stabilising effect of adsorbed polymers is that they completely cover the particles' surface; otherwise, bridging flocculation may occur. Moreover, the interaction between adsorbed polymer layers of two approaching particles can be purely repulsive (due to charged groups or to elastic compression) or promote weak adhesion (interpenetration of the polymer layers).

All the same, steric interaction is related to the presence of at least one additional component (adding to continuous and dispersed phase). In that regard, it depends much more on the specific material properties than on the double layer interaction. Steric interaction is relevant for the stabilisation of colloidal suspensions with high electrolyte background or with organic solvents. A good introduction is, e.g., given by Napper (1983).

3.2.4.2 Depletion Interaction

Depletion interactions are interactions between two surfaces (particles) in the presence of free, i.e., non-adsorbed, macromolecules, micelles, or very fine particles. Asakura and Oosawa (1954) first pointed out that if the distance between two surfaces h is smaller than the diameter of solute molecules $d_{M,}$ this region will contain pure solvent (*depletion zone*, cf. Fig. 3.14, left). Thus, an attractive force corresponding to the osmotic pressure of the bulk solution is acting on the two surfaces. Agglomeration caused by this effect is called *depletion flocculation*. In a second paper, the authors calculated the potential energy of this interaction for

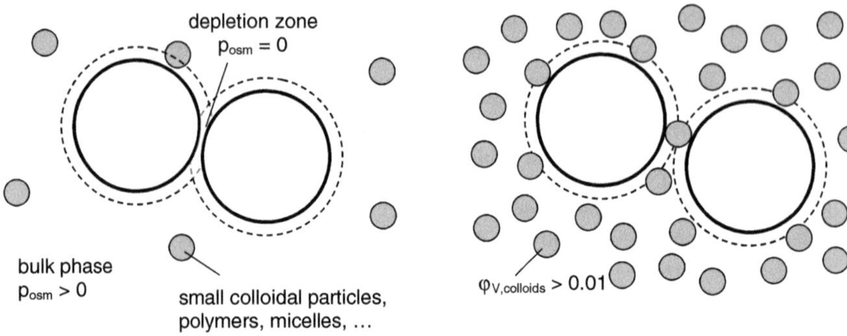

Fig. 3.14 Depletion interaction between two particles in the presence of a non-adsorbing colloidal species (e.g. nanoparticles, polymers, micelles); depletion flocculation (*left*) for low colloid concentrations and depletion stabilisation (*right*) for large ones

differently shaped macromolecules (particles) in a suspension of spheres (Asakura and Oosawa 1958).

While this attractive depletion interaction prevails for very small surface distances ($h \ll d_M$), there is a repulsive interaction when the distance between the approaching surfaces is in the order of magnitude of the macromolecules ($h \approx d_M$) and the depletion has just begun (Fig. 3.14, right). This is because the displacement of macromolecules from the gap to the bulk solvent means that they are "less evenly distributed throughout the system, which must be disfavoured in a good solvent" (Feigin and Napper 1980). The effect is of particular relevance for high polymer concentrations (≥1 vol.-%) when the volume of displaced macromolecules significantly affects the macromolecule arrangement in the bulk solvent. It is then more favourable to maintain monolayers or multiple layers of macromolecules between the two surfaces. Hence, the free macromolecules cause an oscillatory interaction, which is repulsive for each macromolecule layer (e.g. for $h \approx d_M$) but otherwise attractive (e.g. for close surface distance $h \ll d_M$). For sufficiently high polymer concentration, one, therefore, observes *depletion* or *structural stabilisation*. The balance between the repulsive and attractive effects is considerably affected by the presence of adsorbed polymers (Feigin and Napper 1980) and by the size and polydispersity of the free macromolecules (Chu et al. 1996).

3.2.4.3 Hydration Forces

Several experimental studies indicate the existence of a short-ranging repulsive interaction that cannot be explained by DLVO theory or by the presence of polymers and micelles. Since it is observed in particular for hydrophilic surfaces and polar solvents, its origin is traditionally attributed to the solvation of the surface and is called *solvation* or *hydration interaction*. However, the experimental findings seem partly contradictory and the theoretical explanations are manifold. Hydration interactions are reported for different materials, e.g. latex (e.g. Healy et al.1978), mica (Pashley and Israelachvili 1984), titania (Yotsumoto and Yoon 1993a), or silica (e.g. Grabbe and Horn 1993), but they are absent after hydrophobisation of silica (Grabbe 1993). Some authors stress the influence of ion hydration (Paunov et al. 1996), others relate the hydration interaction with gel layer formation on silica surfaces (Vigil et al. 1994; Yaminsky et al. 1998). There are several indications that silica surfaces behave differently as far as the hydration interaction is concerned: While the background electrolyte has a direct effect on this type of interaction for most surfaces (Pashley and Israelachvili 1984; Molina-Bolívar et al. 1997), there is no such clear dependency for silica surfaces (e.g. Grabbe and Horn 1993). Be that as it may, most experimental studies suggest that at high ionic strength (above 0.1 M), the hydration forces are clearly affected by the electrolyte (e.g. Yotsumoto and Yoon 1993b; Grabbe and Horn 1993) and that this influence is ion-specific (Colic et al. 1998). A huge difficulty in interpretating these studies is that the hydration forces are either directly measured, e.g., via AFM (Valle-Delgado et al. 2005), or calculated from the suspension stability (e.g. Molina-Bolivar et al. 1997; Binks and Lumdson 1999).

Hitherto, there has not been any universally accepted explanation for the hydration interaction. Frequently, it is related to the hydration of the surface, which is due to oriented adhesion of water molecules on surface groups via hydrogen bonding and may comprise several layers of water molecules (Kovalchuk et al. 2008; Zhou et al. 2009). Such hydration layers have different properties than bulk water (Basu and Sharma 1994; Belaya et al. 1986), interact repulsively with each other due to dipole interaction (Ruckenstein and Manciu 2003), require energy to be destroyed (Attard and Batchelor 1988), and have a steric effect. However, several other effects are also made responsible for short-ranging repulsion between two particles; e.g., surface gel layers that reduce the van-der-Waals attraction (Vigil et al. 1994), adsorption of hydrated cations (Yotsumoto and Yoon 1993a; Colic et al. 1998), or volume exclusion by the (hydrated) ions (Paunov et al. 1996). Boström et al. (2006) argue that the "hydration forces" can be partly attributed to non-electrostatic, i.e. van-der Waals, ion-surface interaction (cf. Sect. 3.1.6.1). That does not result in a qualitatively new type of particle-particle interaction, as, e.g., depletion interaction, but rather affects the ion distribution in the diffuse layer and, thus, the double layer interaction. This idea at least partially explains ion-specific effects and differences between SiO_2 and Al_2O_3 surfaces.

According to Grabbe and Horn (1993), the hydration interaction decays bi-exponentially with surface distance:

$$V_{\mathrm{hyd}} = \pi \cdot x \cdot \left(L_1 f_1 \cdot e^{-h/L_1} + L_2 f_2 \cdot e^{-h/L_2}\right). \tag{3.38}$$

This empiric approach was derived from force-measurements between two silica surfaces. Other authors confirmed the exponential decay but gave different values for the coefficients L_i and f_i (Yotsumoto and Horn 1993a, b; Molina-Bolívar et al. 1997).

References

General Matter

A.W. Adamson, A.P. Gast, *Physical chemistry of surfaces*, 6th edn. (Wiley, New York, 1997). ISBN 0-471-14873-3

H.-D. Dörfler, *Grenzflächen- und Kolloidchemie* (Wiley, Weinheim, 1994). ISBN 3-527-29072-9

R.J. Hunter, *Zeta potential in colloid science: principles and applications*. In series: *Colloid science*, vol. 2, 3rd edn. (Academic Press, London, 1988). ISBN 0-12-361961-0

R.J. Hunter, *Foundations in Colloid Sciences*, vol. I (Oxford University Press, Oxford, 1993). ISBN 0-19-855187-8

R.J. Hunter, *Foundations in Colloid Sciences*, vol. II (Oxford University Press, Oxford, 1995). ISBN 0-19-855392-7

J.N. Israelachvili, *Intermolecular and surface forces* (Academic Press, London, 1992). ISBN 0-12-375181-0

J. Lyklema, *Fundamentals of Interface and Colloid Science I: Fundamentals* (Academic Press, San Diego, 1991). ISBN 0-12-460525-7

J. Lyklema, *Fundamentals of Interface and Colloid Science II: Solid-Liquid Interfaces* (Academic Press, San Diego, 1995). ISBN 0-12-460524-9

J. Lyklema, *Fundamentals of Interface and Colloid Science III: Liquid-fluid Interfaces* (Academic Press, San Diego, 2000). ISBN 0-12-460523-0

Scattering

G. Bryant, H. Schotman, J.C. Thomas, The estimation of multiple light scattering in turbid colloidal suspensions using a 2-probe detector. Part. Part. Syst. Charact. **15**(4), 170–173 (1998). doi:10.1002/(SICI)1521-4117(199808)15:4<170:AID-PPSC170>3.0.CO;2-G

P. Debye, Molecular-weight determination by light scattering. J. Phys. Colloid Chem. **51**(1), 18–32 (1947). doi:10.1021/j150451a002

A. Einstein, Theorie der Opaleszenz von homogenen Flüssigkeiten und Flüssigkeitsgemischen in der Nähe des kritischen Zustandes. Ann. Phys. IV, **33**(16), 1275–1298 (1910). doi: 10.1002/andp.19103381612

M. Faraday, Experimental relations of gold (and other metals) to light. In: *Experimental Researches in Chemistry and Physics*, (Taylor and Francis, London, pp. 391–443, 1857); as cited by: J.W. Gentry, The aerosol science contributions of Michael Faraday. J. Aerosol Sci. **26** (2), 341–349 (1995)

L.L. Foldy, The Multiple Scattering of Waves. I. General theory of isotropic scattering by randomly distributed scatterers. Phys. Rev. **67**(3–4), 107–119 (1945). doi:10.1103/PhysRev.67.107

A. Ishimaru, Optical multiple scattering by particles. Part. Part. Syst. Charact. **11**(3), 183–188 (1994). doi:10.1002/ppsc.19940110303

M. Lax, Multiple scattering of waves. II. The effective field in dense systems. Phys. Rev. **85**(4), 621–629 (1952). doi:10.1103/PhysRev.85.621

L. Lorenz, in *Sur la lumière réfléchie et réfractée par une surface transparente*, ed. by H. Valentiner, Oeuvres scientifiques de L. Lorenz. (Tome Premier, Libraire Lehmann & Stage, Copenhague, 1898), pp. 403–529

L.I. Mandelstam, Über optisch homogene und trübe Medien. Ann. Phys. IV, **23**(9), 626–642 (1907). doi: 10.1002/andp.19073280904

G. Mie, Beiträge zur Optik trüber Medien, speziell kolloidaler Goldlösungen. Ann. Phys. IV, **25** (3), 377–445 (1908). doi: 10.1002/andp.19083300302

I. I. Sobel'man, On the theory of light scattering in gases. Phys. Usp. **45**(1):75–80, 2002. doi: 10.1070/PU2002v045n01ABEH001115

J.W. Strutt (Lord Rayleigh), On the light from the sky, its polarization and colour. Phil. Mag. S. 4. **41**(271), 107–120 (1871a). doi: 10.1080/14786447108640452

J.W. Strutt (Lord Rayleigh), On the light from the sky, its polarization and colour. Phil. Mag. S. 4. **41**(273), 274–279 (1871b). doi: 10.1080/14786447108640479

J.W. Strutt (Lord Rayleigh), On the scattering of light by small particles. Phil. Mag. S. 4. **41**(275), 447–454 (1871c). doi: 10.1080/14786447108640507

J.W. Strutt (Lord Rayleigh), *Die Theorie des Schalles, 2ter Bd.* (Verlag Friedrich Vieweg und Sohn, Braunschweig, 1880)

J. Tyndall, On the blue colour of the sky, the polarization of skylight, and on the polarization of light by cloudy matter generally. Phil. Mag. S. 4. **37**(250), 384–394 (1869). doi: 10.1080/14786446908640164

M. von Smoluchowski, Molekular-kinetische Theorie der Opaleszenz von Gasen im kritischen Zustande, sowie einiger verwandter Erscheinungen. Ann. Phys. IV, **25**(2), 205–226 (1908). doi: 10.1002/andp.19083300203

P.C. Waterman, R. Truell, Multiple scattering of waves. J. Math. Phys. **2**(4), 512–537 (1961). doi:10.1063/1.1703737

Diffusion, Interfacial Tension, and Bulk Properties

L. Belloni, Yes, pair correlations alone do determine sedimentation profiles of highly charged colloids. J. Chem. Phys. **123**(20), 204705 (2005). doi:10.1063/1.2121527

W.C. Hinds, *Aerosol Technology: Properties, Behaviour, and Measurement of Airborne Particles*, 2nd edn. (John Wiley & Sons, Hosoken, 1999). ISBN 0-471-19410-7

R.K. Iler, *The Chemistry of Silica* (Wiley, New York, 1979). ISBN 0-471-02404-X

M. Kreth, *Numerische Simulation von strukturellen Änderungen in Festkörpern und Nanopartikeln*. Ph.D. thesis, (Gerhard-Mercator-Universität Duisburg, 2001)

J. Lyklema, *Fundamentals of Interface and Colloid Science III: Liquid-Fluid Interfaces* (Academic Press, San Diego, 2000). ISBN 0-12-460523-0

J. Perrin, L'agitation molèculaire et le mouvement brownien. CR Hebd. Seance Acad. Sci. **146** (19), 970 (1908)

J. Perrin, Mouvement brownien et realité moléculaire. Ann. Chim. Phys. **8**(18), 1–114 (1909)

A.P. Philipse, G.H. Koenderink, Sedimentation–diffusion profiles and layered sedimentation of charged colloids at low ionic strength. Adv. Colloid Interface Sci. **100–102**, 613–639 (2003). doi:10.1016/S0001-8686(02)00078-7

S. Ripperger, J. Altmann, Crossflow microfiltration–state of the art. Sep. Purif. Technol. **26**(1), 19–31 (2002). doi:10.1016/S1383-5866(01)00113-7

Electric Double Layer and Zeta-Potential

M.D. Afonso, G. Hagmeyer, R. Gimbel, Streaming potential measurements to assess the variation of nanofiltration membranes surface charge with the concentration of salt solutions. Sep. Purif. Technol. **22–23**, 529–541 (2001). doi:10.1016/S1383-5866(00)00135-0

C. Bellmann, C. Klinger, A. Opfermann, F. Böhme, H.-J.P. Adler, Evaluation of surface modification by electrokinetic measurements. Prog. Org. Coat. **44**(2), 93–98 (2002). doi: 10.1016/S0300-9440(01)00248-X

L. Belloni, Yes, pair correlations alone do determine sedimentation profiles of highly charged colloids. J. Chem. Phys. **123**(20), 204705 (2005). doi:10.1063/1.2121527

F. Carrasco, P. Mutjé, M.A. Pelach, Control of retention in paper-making titration and zeta potential techniques by colloid. Wood Sci. Technol. **32**(2), 145–155 (1998). doi:10.1007/BF00702595

D.L. Chapman, A contribution to the theory of electrocapillarity. Philos. Mag. S. 6, **25**(148), 475–481 (1913). doi: 10.1080/14786440408634187

A.V. Delgado, F. González-Caballero, R.J. Hunter, L.K. Koopal, J. Lyklema, Measurement and interpretation of electrokinetic phenomena. J. Colloid Interface Sci. **309**(2), 194–224 (2007). doi:10.1016/j.jcis.2006.12.075

L.G. Gouy, Sur la constitution de la charge électrique à la surface d'un électrolyte. CR Hebd. Seance Acad. Sci. **149**(17), 354–357 (1909)

R. Greenwood, Review of the measurement of zeta potentials in concentrated aqueous suspensions using electroacoustics. Adv. Colloid Interface Sci. **106**(1–3), 55–81 (2003). doi:10.1016/S0001-8686(03)00105-2

R.J. Hunter, *Zeta potential in colloid science: principles and applications*. In series: *Colloid science*, vol. 2, 3rd edn. (Academic Press, London, 1988). ISBN 0-12-361961-0

R.J. Hunter, *Foundations in Colloid Sciences*, vol. I (Oxford University Press, Oxford, 1993). ISBN 0-19-855187-8

M. Kosmulski, pH-dependent surface charging and points of zero charge. J. Colloid Interface Sci. **253**(1), 77–87 (2002). doi:10.1006/jcis.2002.8490

M. Kosmulski, pH-dependent surface charging and points of zero charge–II. Update. J. Colloid Interface Sci. **275**(1), 214–224 (2004). doi:10.1016/j.jcis.2004.02.029

M. Kosmulski, pH-dependent surface charging and points of zero charge. III. Update. J. Colloid Interface Sci. **298**(2), 730–741 (2006). doi:10.1016/j.jcis.2006.01.003

M. Kosmulski, pH-dependent surface charging and points of zero charge. IV. Update and new approach. J. Colloid Interface Sci. **337**(2), 439–448 (2009). doi: 10.1016/j.jcis.2009.04.072

J. Lyklema, *Fundamentals of Interface and Colloid Science II: Solid-liquid interfaces* (Academic Press, San Diego, 1995). ISBN 0-12-460524-9

J. Lyklema, Molecular interpretation of electrokinetic potentials. Curr. Opin. Colloid Interface Sci. **15**(3), 125–130 (2010). doi:10.1016/j.cocis.2010.01.001

S. Mende, F. Stenger, W. Peukert, J. Schwedes, Production of sub-micron particles by wet comminution in stirred media mills. J. Mat. Sci. **39**(16–17), 5223–5226 (2004). doi:10.1023/B:JMSC.0000039214.12131.58

R. Nitzsche, H. Friedrich, G. Boden, W. Hermel, in *Elektrokinetic Surface Investigations—An Important Technique for Ceramic Powder Processing*. ed by R.A. Williams, N.C. De Jaeger. Advances in Measurement and Control of Colloidal Processes. (Butterworth-Heinemann, 1991), pp. 280–291

J.-S. Park, H.-J. Lee, S.-J. Choi, K.E. Geckeler, J. Cho, S.-H. Moon, Fouling mitigation of anion exchange membrane by zeta potential control. J. Colloid Interface Sci. **259**(2), 293–300 (2003). doi:10.1016/S0021-9797(02)00095-4

C. Schnitzer, S. Ripperger, Influence of surface roughness on streaming potential method. Chem. Eng. Technol. **31**(11), 1696–1700 (2008). doi:10.1002/ceat.200800180

E. Skwarek, S. Khalameida, W. Janusz, V. Sydorchuk, N. Konovalova, V. Zazhigalov, J. Skubi-szew-ska-Zięba, R. Leboda, Influence of mechanochemical activation on structure and some properties of mixed vanadium–molybdenum oxides. J. Therm. Anal. Calorim. **106**(3), 881–894 (2011). doi:10.1007/s10973-011-1744-x

J. Sonnefeld, M. Lobbus, W. Vogelsberger, Determination of electric double layer parameters for spherical silica particles under application of the triple layer model using surface charge density data and results of electrokinetic sonic amplitude measurements. Colloids Surf. A **195**(1–3), 215–225 (2001). doi:10.1016/S0927-7757(01)00845-7

G. Téllez, T. Biben, Equilibrium sedimentation profiles of charged colloidal suspensions. Eur. Phys. J. E **2**(2), 137–143 (2000). doi:10.1007/s101890050047

D.M.E. Thies-Weesie, A.P. Philipse, G. Nägele, B. Mandl, R. Klein, Nonanalytical concentration dependence of sedimentation of charged silica spheres in an organic solvent: experiments and calculations. J. Colloid Interface Sci. **176**(1), 43–54 (1995). doi:10.1006/jcis.1995.0006

B. van Lent, B. Klinksiek, Korrelation physikalischer mit anwendungstechnischen Größen bei Dispersionen. Chem. Ing. Tech. **69**(6), 793–798 (1997). doi:10.1002/cite.330690605

P. Wilhelm, D. Stephan, On-line tracking of the coating of nanoscaled silica with titania nanoparticles via zeta-potential measurements. J. Colloid Interface Sci. **293**(1), 88–92 (2006). doi:10.1016/j.jcis.2005.06.047

Electro-Viscous Effects

F. Booth, The electroviscous effect for suspensions of solid spherical particles. Proc. Roy. Soc. A **203**(1075), 533–551 (1950). doi:10.1098/rspa.1950.0155

F. Booth, Sedimentation potential and velocity of solid spherical particles. J. Chem. Phys. **22**(11), 1968 (1954). doi: 10.1063/1.1739975

H.B. Bull, Die Bedeutung der Kapillarenweite für das Strömungspotential. Kolloid Z. **60**(2), 130–132 (1932). doi:10.1007/BF01428359

R.J. Hunter, *Zeta potential in colloid science: principles and applications*. In series: *Colloid science*, vol. 2, 3rd ed. (Academic Press, London, 1988). ISBN 0-12-361961-0

H.J. Keh, J.M. Ding, Sedimentation velocity and potential in concentrated suspensions of charged spheres with arbitrary double-layer thickness. J. Colloid Interface Sci. **227**(2), 540–552 (2000). doi:10.1006/jcis.2000.6918

S. Levine, J.R. Marriott, K. Robinson, Theory of electrokinetic flow in a narrow parallel-plate channel. J. Chem. Soc. Faraday Trans. **2**(71), 1–11 (1975). doi:10.1039/F29757100001

J. Lyklema, Molecular interpretation of electrokinetic potentials. Curr. Opin. Colloid Interface Sci. **15**(3), 125–130 (2010). doi:10.1016/j.cocis.2010.01.001

J. Lyklema, J.T.G. Overbeek, On the interpretation of electrokinetic potentials. J. Colloid Sci. **16** (5), 501–512 (1961). doi:10.1016/0095-8522(61)90029-0

H. Ohshima, T.W. Healy, L.R. White, R.W. O'Brien, Sedimentation velocity and potential in a dilute suspension of charged spherical colloidal particles. J. Chem. Soc. Faraday Trans. 2 **80** (10), 1299–1317 (1984). doi:10.1039/f29848001299

E. Overbeck, C. Sinn, M. Watzlawek, Enhanced structural correlations accelerate diffusion in charge-stabilized colloidal suspensions. Phys. Rev. E **60**(2), 1936–1939 (1999). doi:10.1103/PhysRevE.60.1936

D. Quemada, C. Berli, Energy of interaction in colloids and its implications in rheological modeling. Adv. Colloid Interface Sci. **98**(1), 51–85 (2002). doi:10.1016/S0001-8686(01)00093-8

F.J. Rubio-Hernandez, F. Carrique, E. Ruiz-Reina, The primary electroviscous effect in colloidal suspensions. Adv. Colloid Interface Sci. **107**(1), 51–60 (2004). doi:10.1016/j.cis.2003.09.001

E. Ruiz-Reina, F. Carrique, F.J. Rubio-Hernández, A.I. Gómez-Merino, P. Garcia-Sánchez, Electroviscous effect of moderately concentrated colloidal suspensions. J. Phys. Chem. B **107** (35), 9528–9534 (2003). doi:10.1021/jp034795i

W.B. Russel, Rheology of suspensions of charged rigid spheres. J. Fluid Mech. **85**(2), 209–232 (1978). doi:10.1017/S0022112078000609

M. von Smoluchowski, Przyczynek do teorji endosmozy elektrycznej i kilku pokrewnych zjawisk. *Rozprawy Wydziału matematyczno-przyrodiczego Akademji Umiejętnosci w Krakowie*, T. XLIII, Serja A, 110–127, 1903; reprint in French: Contribution à la théorie de l'endosmose électrique et de quelques phénomènes corrélatifs. Bull. Int. Acad. Sci. Cracovie, Cl. Sci. Math. Nat. 8, 182–200 (1903)

M. Watzlawek, G. Nägele, Self-diffusion coefficients of charged particles: Prediction of nonlinear volume fraction dependence. Phys. Rev. E **56**(1), 1258–1261 (1997). doi:10.1103/PhysRevE.56.1258

M. Watzlawek, G. Nägele, Sedimentation of strongly and weakly charged colloidal particles: Prediction of fractional density dependence. J. Colloid Interface Sci. **214**(2), 170–179 (1999). doi: 10.1006/jcis.1999.6181

Surface Charging

M. Boström, V. Deniz, G.V. Franks, B.W. Ninham, Extended DLVO theory: Electrostatic and non-electrostatic forces in oxide suspensions. Adv. Colloid Interface Sci. **123–126**, 5–15 (2006). doi:10.1016/j.cis.2006.05.001

M. Colic, M.L. Fisher, G.V. Franks, Influence of ion size on short-range repulsive forces between silica surfaces. Langmuir **14**(21), 6107–6112 (1998). doi:10.1021/la980489y

J.A. Davis, R.O. James, J.O. Leckie, Surface ionization and complexation at the oxide/water interface: I. Computation of electrical double layer properties in simple electrolytes. J. Colloid Interface Sci. **63**(3), 480–499 (1978). doi:10.1016/S0021-9797(78)80009-5

K.J. Farley, D.A. Dzombak, F.M.M. Morel, A surface precipitation model for the sorption of cations on metal oxides. J. Colloid Interface Sci. **106**(1), 226–242 (1985). doi:10.1016/0021-9797(85)90400-X

L.E. Firment, H.E. Bergna, D.G. Swartzfager, P.E. Bierstedt, L. van Kavelaar, Silica coatings on α-alumina particles: Analysis and deposition mechanism. Surf. Interface Anal. **14**(1–2), 46–52 (1989). doi:10.1002/sia.740140111

G.V. Franks, S.B. Johnson, P.J. Scales, D.V. Boger, T.W. Healy, Ion-specific strength of attractive particle networks. Langmuir **15**(13), 4411–4420 (1999). doi:10.1021/la9815345

G.V. Franks, Zeta potentials and yield stresses of silica suspensions in concentrated monovalent electrolytes: Isoelectric point shift and additional attraction. J. Colloid Interface Sci. **249**(1), 44–51 (2002). doi:10.1006/jcis.2002.8250

R.J. Hunter, *Zeta potential in colloid science: principles and applications*. In series: *Colloid science*, vol. 2, 3rd ed. (Academic Press, London, 1988). ISBN 0-12-361961-0

J. Jabłoński, W. Janusz, M. Reszka, R. Sprycha, J. Szszypa, Mechanism of adsorption of selected monovalent and divalent inorganic ions at the alumina/electrolyte interface. Pol. J. Chem. **74**(10), 1399–1409 (2000)

R.O. James, T.W. Healy, Adsorption of hydrolyzable metal ions at the oxide–water interface. II. Charge reversal of SiO_2 and TiO_2 colloids by adsorbed Co(II), La(III), and Th(IV) as model systems. J. Colloid Interface Sci. **40**(1), 53–64 (1972). doi:10.1016/0021-9797(72)90173-7

J.-Q. Jiang, N.J.D. Graham, Pre-polymerised inorganic coagulants and phosphorus removal by coagulation-a review. Water SA **24**(3), 237–244 (1998)

S.B. Johnson, P.J. Scales, T.W. Healy, The binding of monovalent electrolyte ions on α-alumina. I. electroacoustic studies at high electrolyte concentrations. Langmuir **15**(8), 2836–2843 (1999). doi:10.1021/la980875f

M. Kosmulski, P. Eriksson, J. Gustafsson, J.B. Rosenholm, Specific adsorption of nickel and z potential of silica at various solid-to-liquid ratios. J. Colloid Interface Sci. **220**(1), 128–132 (1999a). doi:10.1006/jcis.1999.6520

M. Kosmulski, J. Gustafsson, J.B. Rosenholm, Correlation between the zeta potential and rheological properties of anatase dispersions. J. Colloid Interface Sci. **209**(1), 200–206 (1999b). doi:10.1006/jcis.1998.5884

W.H. Kuan, S.L. Lo, M.K. Wang, Modeling and electrokinetic evidences on the processes of the Al(III) sorption continuum in SiO2(s) suspension. J. Colloid Interface Sci. **272**(2), 489–497 (2004). doi:10.1016/j.jcis.2003.12.034

Y.K. Leong, Yield stress and zeta potential of nanoparticulate silica dispersions under the influence of adsorbed hydrolysis products of metal ions—Cu(II), Al(III) and Th(IV). J. Colloid Interface Sci. **292**(2), 557–566 (2005). doi:10.1016/j.jcis.2005.06.004

S. Levine, D. Calvert, G.M. Bell, Discreteness-of-charge effect in electric double layer theory. Can. J. Chem. **40**(3), 518–538 (1962). doi:10.1139/v62-080

J. Lyklema, Quest for ion–ion correlations in electric double layers and overcharging phenomena. Adv. Colloid Interface Sci. **147–148**, 205–213 (2009). doi:10.1016/j.cis.2008.12.002

H.J. Modi, D.W. Fuerstenau, Streaming potential studies on corundum in aqueous solutions of inorganic electrolytes. J. Phys. Chem. **61**(5), 640–643 (1957). doi:10.1021/j150551a029

B.W. Ninham, V. Yaminsky, Ion binding and ion specificity: the Hofmeister effect and Onsager and Lifshitz theories. Langmuir **13**(7), 2097–2108 (1997). doi:10.1021/la960974y

K. Osseo-Asare, Etching kinetics of silicon dioxide in aqueous fluoride solutions: a surface complexation model. J. Electrochem. Soc. **143**(4), 1339–1347 (1996). doi:10.1149/1.1836640

K. Osseo-Asare, Surface chemical processes in chemical mechanical polishing: Relationship between silica material removal rate and the point of zero charge of the abrasive material. J. Electrochem. Soc. **149**(12), 651–655 (2002). doi:10.1149/1.1516777

K.M. Paciejewska, *Untersuchung des Stabilitätsverhaltens von binären kolloidalen Suspensionen*. Ph.D. thesis, (Technische Universität Dresden, 2010). http://nbn-resolving.de/urn:nbn:de:bsz:14-qucosa-65050

M. Quesada-Pérez, E. González-Tovar, A. Martín-Molina, M. Lozada-Cassou, R. Hidalgo-Álvarez, Overcharging in colloids: Beyond the Poisson-Boltzmann approach. Chem. Phys. Chem. **4**(3), 234–248 (2003). doi:10.1002/cphc.200390040

M. Quesada-Pérez, E. González-Tovar, A. Martín-Molina, M. Lozada-Cassou, R. Hidalgo-Álvarez, Ion size correlations and charge reversal in real colloids. Colloids Surf. A **267**(1–3), 24–30 (2005). doi:10.1016/j.colsurfa.2005.06.034

W. Szczepaniak, H. Kościelna, Specific adsorption of halogen anions on hydrous gamma-Al2O3. Anal. Chim. Acta **470**(2), 263–276 (2002). doi:10.1016/S0003-2670(02)00661-X

G.R. Wiese, R.O. James, T.W. Healy, Discreteness of charge and solvation effects in cation adsorption at the oxide/water interface. Discuss. Faraday Soc. **52**, 302–311 (1971). doi:10.1039/DF9715200302

Surfactants

R. Atkin, V.S.J. Craig, E.J. Wanless, S. Biggs, Mechanism of cationic surfactant adsorption at the solid–aqueous interface. Adv. Colloid Interface Sci. **103**(3), 219–304 (2003). doi:10.1016/S0001-8686(03)00002-2

N.K. Dimov, V.L. Kolev, P.A. Kralchevsky, L.G. Lyutov, G. Broze, A. Mehreteab, Adsorption of ionic surfactants on solid particles determined by zeta-potential measurements: Competitive binding of counterions. J. Colloid Interface Sci. **256**(1), 23–32 (2002). doi:10.1006/jcis.2001.7821

A. Fan, P. Somasundaran, N.J. Turro, Adsorption of alkyltrimethylammonium bromides on negatively charged alumina. Langmuir **13**(3), 506–510 (1997). doi:10.1021/la9607215

K. Holmberg, B. Jönsson, B. Kronberg, B. Lindman, *Surfactants and polymers in aqueous solution*, 2nd edn. (Wiley, Chichester, 2002). ISBN 0-471-49883-1

D. Myers, *Surfaces, interfaces, and colloids: Principles and applications*, 2nd edn. (Wiley, New-York, 1999). ISBN 0-471-33060-4

S. Paria, K.C. Khilar, A review on experimental studies of surfactant adsorption at the hydrophilic solid–water interface. Adv. Colloid Interface Sci. **110**(3), 75–95 (2004). doi:10.1016/j.cis.2004.03.001

Surface Dissolution

Z. Adamczyk, B. Jachimska, M. Kolasińska, Structure of colloid silica determined by viscosity measurements. J. Colloid Interface Sci. **273**(2), 668–674 (2004). doi:10.1016/j.jcis.2004.01.008

C.F. Baes, R.E. Mesmer, *The Hydrolysis of Cations*, (Wiley, New York, 1976), p. 58. ISBN 0-471-03985-3

S. Bai, S. Urabe, Y. Okaue, T. Yokoyama, Acceleration effect of sulfate ion on the dissolution of amorphous silica. J. Colloid Interface Sci. **331**(2), 551–554 (2009). doi:10.1016/j.jcis.2008.11.076

B. Bonelli, P. Palmero, F. Lomello, M. Armandi, M. Lombardi, Study of the effect of prolonged magnetic stirring on the physico-chemical surface properties of nanometric transition alumina. J. Mater. Sci. **45**(22), 6115–6125 (2010). doi:10.1007/s10853-010-4698-7

X. Carrier, E. Marceau, J.F. Lambert, M. Che, Transformations of γ-alumina in aqueous suspensions. 1. Alumina chemical weath-ering studied as a function of pH. J. Colloid Interface Sci. **308**(2), 429–437 (2007). doi:10.1016/j.jcis.2006.12.074

I. Gunnarson, S. Arnórsson, Amorphous silica solubility and the thermodynamic properties of H4SiO4 in the range of 0° to 350°C at Psat. Geochim. Cosmochim. Acta **64**(13), 2295–2307 (2000). doi:10.1016/S0016-7037(99)00426-3

R.K. Iler, *The Chemistry of Silica* (Wiley, New York, 1979). ISBN 0-471-02404-X

J.-Q. Jiang, N.J.D. Graham, Pre-polymerised inorganic coagulants and phosphorus removal by coagulation-a review. Water SA **24**(3), 237–244 (1998)

W.H. Kuan, S.L. Lo, M.K. Wang, Modeling and electrokinetic evidences on the processes of the Al(III) sorption continuum in SiO_2(s) suspension. J. Colloid Interface Sci. **272**(2), 489–497 (2004). doi:10.1016/j.jcis.2003.12.034

G. Lefèvre, M. Duc, P. Lepeut, R. Caplain, M. Fédoroff, Hydration of γ-alumina in water and its effects on surface reactivity. Langmuir **18**(20), 7530–7537 (2002). doi:10.1021/la025651i

W.L. Lindsay, *Chemical Equilibria in Soils* (Wiley, New York, 1979), pp. 35–49. ISBN 0-471-02704-9

K.M. Paciejewska, *Untersuchung des Stabilitätsverhaltens von binären kolloidalen Suspensionen*. Ph.D. thesis, (Technische Universität Dresden, 2010). http://nbn-resolving.de/urn:nbn:de:bsz:14-qucosa-65050

D. Rickert, *Dissolution kinetics of biogenic silica in marine environments*. Ph.D. thesis, (Geomar Forschungszentrum für marine Geowissenschaften Kiel, 2000)

F. Roelofs, W. Vogelsberger, Dissolution kinetics of nanodispersed γ-alumina in aqueous solution at different pH: Unusual kinetic size effect and formation of a new phase. J. Colloid Interface Sci. **303**(2), 450–459 (2006). doi:10.1016/j.jcis.2006.08.016

W. Stumm, R. Wollast, Coordination chemistry of weathering: Kinetics of the surface-controlled dissolution of oxide minerals. Rev. Geophys. **28**(1), 53–69 (1990). doi:10.1029/RG028i001p00053

G. Vigil, Z. Xu, S. Steinberg, J. Israelachvili, Interactions of silica surfaces. J. Colloid Interface Sci. **165**(2), 367–385 (1994). doi:10.1006/jcis.1994.1242

W. Vogelsberger, M. Löbbus, J. Sonnefeld, A. Seidel, The influence of ionic strength on the dissolution process of silica. Colloids. Surf. A **159**(2–3), 311–319 1999. doi: 10.1016/S0927-7757(99)00268-X

V.V. Yaminsky, B.W. Ninham, R.M. Pashley, Interaction between surfaces of fused silica in water. Evidence of cold fusion and effects of cold plasma treatment. Langmuir **14**(12), 3223–3235 (1998). doi:10.1021/la9713762

Van-der-Waals Interaction

V. Arunachalam, W.H. Marlow, J.X. Lu, Development of a picture of the van der Waals interaction energy between clusters of nanometre-range particles. Phys. Rev. E **58**(3), 3451–3457 (1998). doi:10.1103/PhysRevE.58.3451

L. Bergström, Hamaker constants of inorganic materials. Adv. Colloid Interface Sci. **70**, 125–169 (1997). doi:10.1016/S0001-8686(97)00003-1

W.R. Bowen, F. Jenner, The calculation of dispersion forces for engineering applications. Adv. Colloid Interface Sci. **56**, 20–243 (1995). doi:10.1016/0001-8686(94)00233-3

J.H. de Boer, The influence of van der Waals' forces and primary bonds on binding energy, strength and orientation, with special reference to some artificial resins. Trans. Faraday Soc. **32** (1), 10–37 (1936). doi:10.1039/TF9363200010

H.C. Hamaker, The London—van der Waals attraction between spherical particles. Physica **4**(10), 1058–1072 (1937). doi:10.1016/S0031-8914(37)80203-7

R.J. Hunter, *Foundations in Colloid Sciences*, vol. I (Oxford University Press, Oxford, 1993). ISBN 0-19-855187-8

J.N. Israelachvili, *Intermolecular and Surface Forces* (Academic Press, London, 1992). ISBN 0-12-375181-0

D. Langbein, Non-retarded dispersion energy between macroscopic spheres. J. Phys. Chem. Solids **32**(7), 1657–1667 (1971). doi:10.1016/S0022-3697(71)80059-8

D. Langbein, *Theory of Van der Waals Attraction*. In series: *Springer Tracts in Modern Physics*, vol. 72. (Springer, Berlin, 1974). doi: 10.1007/BFb0042407

E.M. Lifshitz, The theory of molecular attractive forces between solids. Sov. Phys. JETP **2**(1), 73–83 (1956)

J. Lyklema, *Fundamentals of Interface and Colloid Science I: Fundamentals* (Academic Press, San Diego, 1991). ISBN 0-12-460525-7

J. Mahanty, B.W. Ninham, *Dispersion Forces* (Academic Press, London, 1976). ISBN 0-124-65050-3

A.V. Nguyen, Improved approximation of water dielectric permittivity for calculation of Hamaker constants. J. Colloid Interface Sci. **229**, 648–651 (2000). doi:10.1006/jcis.2000.7010

B.A. Pailthorpe, W.B. Russel, The retarded van der Waals interaction between spheres. J. Colloid Interface Sci. **89**(2), 563–566 (1982). doi:10.1016/0021-9797(82)90208-9

S.N. Thennadil, L.H. Garcia-Rubio, Approximations for calculating van der Waals interaction energy between spherical particles—a comparison. J. Colloid Interface Sci. **243**(1), 136–142 (2001). doi:10.1006/jcis.7851

P. Viravathana, D.W.M. Marr, Optical trapping of titania/silica core-shell colloidal particles. J. Colloid Interface Sci. **221**, 301–307 (2000). doi:10.1006/jcis.1999.6603

Double Layer Interaction

G.M. Bell, S. Levine, L.N. McCartney, Approximate methods of determining the double-layer free energy of interaction between two charged colloidal spheres. J. Colloid Interface Sci. **33**(3), 335–359 (1970). doi:10.1016/0021-9797(70)90228-6

S.L. Carnie, D.Y.C. Chan, Interaction free energy between plates with charge regulation: a linearized model. J. Colloid Interface Sci. **161**(1), 260–264 (1993). doi:10.1006/jcis.1993.1464

S.L. Carnie, D.Y.C. Chan, J.S. Gunning, Electrical double-layer interaction between dissimilar spherical colloidal particles and between a sphere and a plate: The linearized Poisson-Boltzmann theory. Langmuir **10**(9), 2993–3009 (1994a). doi:10.1021/la00021a024

S.L. Carnie, D.Y.C. Chan, J. Stankovich, Computation of forces between spherical colloidal particles: Nonlinear Poisson-Boltzmann theory. J. Colloid Interface Sci. **165**(1), 116–128 (1994b). doi:10.1006/jcis.1994.1212

D. Chan, T.W. Healy, L.R. White, Electrical double layer interactions under regulation by surface ionization equilibria–dissimilar amphoteric surfaces. J. Chem. Soc. Faraday Trans. **1**(72), 2844–2865 (1976). doi:10.1039/f19767202844

D.Y.C. Chan, D.J. Mitchell, The free energy of an electrical double layer. J. Colloid Interface Sci. **95**(1), 193–197 (1983). doi:10.1016/0021-9797(83)90087-5

D.Y.C. Chan, in *Free energies of electrical double layers at the oxide-solution interface*. ed. by J.A. Davis, K.F. Hayes Geochemical Processes at Mineral Surfaces, (ACS Symposium Series 323, *American* Chemical Society, 1987), Chapter 6, pp 99–112. doi: 10.1021/bk-1987-0323.ch006

D.Y.C. Chan, T.W. Healy, T. Supasiti, S. Usui, Electrical double layer interactions between dissimilar oxide surfaces with charge regulation and Stern-Grahame layers. J. Colloid Interface Sci. **296**(1), 150–158 (2006). doi:10.1016/j.jcis.2005.09.003

S.S. Dukhin, J. Lyklema, Electrostatic interaction of colloidal particles and deviations of their double-layers from electrical neutrality (in Russian: Elektrostatićeskoe Vsaïmodejstwie kolloidnich ćastic i otklonenie ich dwoinich sloew ot elektronejtralnosti). Colloid J. USSR **51** (2), 212–221 (1989)

A.B. Glendinning, W.B. Russel, The electrostatic repulsion between charged spheres from exact solutions to the linearized Poisson-Boltzmann equation. J. Colloid Interface Sci. **93**(1), 95–104 (1983). doi:10.1016/0021-9797(83)90388-0

R. Hogg, T.W. Healy, D.W. Fuerstenau, Mutual coagulation of colloidal dispersions. Trans. Faraday Soc. **62**, 1638–1651 (1966). doi:10.1039/TF9666201638

J.-P. Hsu, *Interfacial forces and fields: theory and applications*. In series: *Surfactant science*, vol. 85. (Dekker, New York, 1999). ISBN 0-8247-1964-6

R.J. Hunter, *Foundations in Colloid Sciences*, vol. I (Oxford University Press, Oxford, 1993). ISBN 0-19-855187-8

J.W. Krozel, D.A. Saville, Electrostatic interactions between two spheres: solutions of the Debye-Hückel equation with a charge regulation boundary condition. J. Colloid Interface Sci. **150**(2), 365–373 (1992). doi:10.1016/0021-9797(92)90206-2

J. Lyklema, J.F.L. Duval, Hetero-interaction between Gouy–Stern double layers: Charge and potential regulation. Adv. Colloid Interface Sci. **114–115**, 27–45 (2005). doi:10.1016/j.cis.2004.05.002

D. McCormack, S.L. Carnie, D.Y.C. Chan, Calculations of electric double-layer force and interaction free energy between dissimilar surfaces. J. Colloid Interface Sci. **169**(1), 177–196 (1995). doi:10.1006/jcis.1995.1019

J.E. Sader, S.L. Carnie, D.Y.C. Chan, Accurate analytic formulas for the double-layer interaction between spheres. J. Colloid Interface Sci. **171**(1), 46–54 (1995). doi:10.1006/jcis.1995.1149

S. Usui, Interaction between dissimilar double layers with like signs under charge regulation on the basis of the Gouy–Chapman–Stern–Grahame model. J. Colloid Interface Sci. **280**(1), 113–119 (2004). doi:10.1016/j.jcis.2004.07.014

S. Usui, Electrical double-layer interaction between oppositely charged dissimilar oxide surfaces with charge regulation and Stern-Grahame layers. J. Colloid Interface Sci. **320**(1), 353–359 (2008). doi:10.1016/j.jcis.2007.12.016

DLVO Theory (Including Born Repulsion)

A. Bleier, C.G. Westmoreland, Effects of pH and particle size on the processing of and the development of microstructure in alumina-zirconia composites. J. Am. Ceram. Soc. **74**(12), 3100–3111 (1991). doi:10.1111/j.1151-2916.1991.tb04307.x

B. Derjaguin, L. Landau, Theory of the stability of strongly charged lyophobic sols and of the adhesion of strongly charged-particles in solutions of electrolytes. Prog. Surf. Sci. **43**(1–4), 30–59 (1993); reprinted from: Acta phys.-chim. **14**(6), 633–662 (1941). doi: 10.1016/0079-6816(93)90013-L

D.L. Feke, N.D. Prabhu, J.A. Mann Jr, J.A. Mann III, A formulation of the short-range repulsion between spherical colloidal particles. J. Phys. Chem. **88**(23), 5735–5739 (1984). doi:10.1021/j150667a055

W.B. Hardy, Preliminary investigation of the conditions which determine the stability of irreversible hydrosols. J. Phys. Chem. **4**(4), 235–253 (1900). doi:10.1021/j150022a001

T.W. Healy, A. Homola, R.O. James, R.J. Hunter, Coagulation of amphoteric latex colloids: reversibility and specific ion effects. Faraday Discuss. Chem. Soc. **65**, 156–163 (1978). doi:10.1039/dc9786500156

J.N. Israelachvili, *Intermolecular and Surface Forces* (Academic Press, London, 1992). ISBN 0-12-375181-0

P.A. Kralchevsky, K.D. Danov, N.D. Denkov, in *Chemical Physics of Colloid Systems and Interfaces*. ed. by K.S. Birdi, *Handbook of Surface and Colloid Chemistry*, 2nd edn, Chapter 5, pp. 137–344. (CRC Press, New York, 2002). ISBN 0-8493-1079-2

J. Lyklema, *Fundamentals of Interface and Colloid Science I: Fundamentals* (Academic Press, San Diego, 1991). ISBN 0-12-460525-7

P. Mulvaney, L.M. Liz-Marzan, M. Giersig, T. Ung, Silica encapsulation of quantum dots and metal clusters. J. Mater. Chem. **10**(6), 1259–1270 (2000). doi:10.1039/b000136h

W. Ostwald, Elektrolytkoagulation schwach solvatisierter Sole und Elektrolytaktivität. Kolloid Z. **73**(3), 301–323 (1935). doi:10.1007/BF01428785

R. Pagel, *Laserpulsinduzierte Deaggregation von TiO2-Nanopartikeln in wässriger Suspension*. Ph.D. thesis, (Freie Universität Berlin, 2005)

E. Ruckenstein, D.C. Prieve, Adsorption and desorption of particles and their chromatographic separation. AIChE J. **22**(2), 276–283 (1976). doi:10.1002/aic.690220209

E.J.W. Verwey, J.T.G. Overbeek, *Theory of stability of lyophobic colloids* (Elsevier, New York, 1948)

Steric and Depletion Interaction

S. Asakura, F. Oosawa, On interaction between two bodies immersed in a solution of macromolecules. J. Chem. Phys. **22**(7), 1255–1256 (1954). doi:10.1063/1.1740347

S. Asakura, F. Oosawa, Interaction between particles suspended in solutions of macromolecules. J. Polym. Sci. **33**(126), 183–192 (1958). doi:10.1002/pol.1958.1203312618

X.L. Chu, A.D. Nikolov, D.T. Wasan, Effects of particle size and polydispersity on the depletion and structural forces in colloidal dispersions. Langmuir **12**(21), 5004–5010 (1996). doi:10.1021/la960359u

M. Faraday, Experimental relations of gold (and other metals) to light. In: *Experimental Researches in Chemistry and Physics*, pp. 391–443. (Taylor and Francis, London, 1857); as cited by: J.W. Gentry, The aerosol science contributions of Michael Faraday. J. Aerosol Sci. **26** (2), 341–349 (1995)

R.I. Feigin, D.H. Napper, Depletion stabilization and depletion flocculation. J. Colloid Interface Sci. **75**(2), 525–541 (1980). doi:10.1016/0021-9797(80)90475-0

W. Heller, T.L. Pugh, "Steric protection" of hydrophobic colloidal particles by adsorption of flexible macromolecules. J. Chem. Phys. **22**(10), 1778 (1954). doi:10.1063/1.1739900

D.H. Napper, The steric stabilization of hydrosols by nonionic macromolecules. J. Colloid Interface Sci. **29**(1), 168–170 (1969). doi:10.1016/0021-9797(69)90363-4

D.H. Napper, A. Netschey, Studies of the steric stabilization of colloidal particles. J. Colloid Interface Sci. **37**(3), 528–535 (1971). doi:10.1016/0021-9797(71)90330-4

D.H. Napper, Steric stabilization. J. Colloid Interface Sci. **58**(2), 390–407 (1977). doi:10.1016/0021-9797(77)90150-3

D.H. Napper, *Polymeric stabilization of colloidal dispersions*. In series: *Colloid science*, vol. 3. (Academic Press, London, 1983). ISBN 0-12-513980-2

M. van der Waarden, Stabilization of carbon-black dispersions in hydrocarbons. J. Colloid Sci. **5** (4), 325 (1950). doi: 10.1016/0095-8522(50)90056-0

R. Zsigmondy, Die hochrothe Goldlösung als Reagens auf Colloide. Z. Anal. Chem. **40**(11), 697–719 (1901). doi:10.1007/BF01334022

Hydration Interaction

P. Attard, M.T. Batchelor, A mechanism for the hydration force demonstrated in a model system. Chem. Phys. Lett. **149**(2), 206–211 (1988). doi:10.1016/0009-2614(88)87223-3

S. Basu, M.M. Sharma, Effect of dielectric saturation on disjoining pressure in thin films of aqueous electrolytes. J. Colloid Interface Sci. **165**(2), 355–366 (1994). doi:10.1006/jcis.1994.1241

M.L. Belaya, M.V. Feigel'man, V.G. Levadny, Hydration forces as a result of non-local water polarizability. Chem. Phys. Lett. **126**(3–4), 361–364 (1986). doi:10.1016/S0009-2614(86)80099-9

B.P. Binks, S.O. Lumsdon, Stability of oil-in-water emulsions stabilised by silica particles. Phys. Chem. Chem. Phys. **1**(12), 3007–3016 (1999). doi:10.1039/a902209k

M. Boström, V. Deniz, G.V. Franks, B.W. Ninham, Extended DLVO theory: electrostatic and non-electrostatic forces in oxide suspensions. Adv. Colloid Interface Sci. **123–126**, 5–15 (2006). doi:10.1016/j.cis.2006.05.001

M. Colic, M.L. Fisher, G.V. Franks, Influence of ion size on short-range repulsive forces between silica surfaces. Langmuir **14**(21), 6107–6112 (1998). doi:10.1021/la980489y

A. Grabbe, Double layer interactions between silylated silica surfaces. Langmuir **9**(3), 797–801 (1993). doi:10.1021/la00027a032

A. Grabbe, R.G. Horn, Double-Layer and hydration forces measured between silica sheets subjected to various surface treatments. J. Colloid Interface Sci. **157**(2), 375–383 (1993). doi:10.1006/jcis.1993.1199

T.W. Healy, A. Homola, R.O. James, R.J. Hunter, Coagulation of amphoteric latex colloids: reversibility and specific ion effects. Faraday Discuss. Chem. Soc. **65**, 156–163 (1978). doi:10.1039/dc9786500156

N. Kovalchuk, V. Starov, P. Langston, N. Hilal, V. Zhdanov, Colloidal dynamics: Influence of diffusion, inertia and colloidal forces on cluster formation. J. Colloid Interface Sci. **325**(2), 377–385 (2008). doi:10.1016/j.jcis.2008.06.017

J.A. Molina-Bolívar, F. Galisteo-González, R. Hidalgo-Álvarez, Colloidal stability of protein-polymer systems: a possible explanation by hydration forces. Phys. Rev. E **55**(4), 4522–4530 (1997). doi:10.1103/PhysRevE.55.4522

R.M. Pashley, J.N. Israelachvili, Molecular layering of water in thin films between mica surfaces and its relation to hydration forces. J. Colloid Interface Sci. **101**(2), 511–523 (1984). doi:10.1016/0021-9797(84)90063-8

V.N. Paunov, R.I. Dimova, P.A. Kralchevsky, G. Broze, A. Mehreteab, The hydration repulsion between charged surfaces as an interplay of volume exclusion and dielectric saturation effects. J. Colloid Interface Sci. **182**(1), 239–248 (1996). doi:10.1006/jcis.1996.0456

E. Ruckenstein, M. Manciu, Specific ion effects via ion hydration: II. Double layer interaction. Adv. Colloid Interface Sci. **105**(1–3), 177–200 (2003). doi:10.1016/S0001-8686(03)00068-X

J.J. Valle-Delgado, J.A. Molina-Bolívar, F. Galisteo-González, M.J. Gálvez-Ruiz, A. Feiler, M.W. Rutland, Hydration forces between silica surfaces: Experimental data and predictions from different theories. J. Chem. Phys. **123**(3), 034708 (2005). doi:10.1063/1.1954747

G. Vigil, Z. Xu, S. Steinberg, J. Israelachvili, Interactions of silica surfaces. J. Colloid Interface Sci. **165**(2), 367–385 (1994). doi:10.1006/jcis.1994.1242

V.V. Yaminsky, B.W. Ninham, R.M. Pashley, Interaction between surfaces of fused silica in water. Evidence of cold fusion and effects of cold plasma treatment. Langmuir **14**(12), 3223–3235 (1998). doi:10.1021/la9713762

H. Yotsumoto, R.-H. Yoon, Application of extended DLVO-theory: 1. Stability of rutile suspensions. J. Colloid Interface Sci. **157**(2), 426–433 (1993a). doi:10.1006/jcis.1993.1205

H. Yotsumoto, R.-H. Yoon, Application of extended DLVO-theory: 2. Stability of silica suspensions. J. Colloid Interface Sci. **157**(2), 434–441 (1993b). doi:10.1006/jcis.1993.1206

J. Zhou, J. Ralston, R. Sedev, D.A. Beattie, Functionalized gold nanoparticles: Synthesis, structure and colloid stability. J. Colloid Interface Sci. **331**(2), 251–262 (2009). doi:10.1016/j.jcis.2008.12.002

Chapter 4
Suspensions of Colloidal Aggregates

Aggregation is a generic term describing any process that leads to the formation of contiguous clusters of colloidal particles called aggregates. Aggregation is an intrinsic step in precipitation and is also observed in gas phase particle synthesis. In suspensions, aggregation can be induced by adding electrolytes or polymers (coagulation and flocculation, respectively) and is affected by external fields (e.g. gravity) and the flow regime. Indeed, aggregation is an inherent behaviour of any particle system, though the extent to which it prevails may largely differ. Aggregation is of particular relevance for colloidal particles, because their adhesion forces frequently exceed the dispersion forces driving two particles apart.

Due to the multitude of aggregation processes and aggregating particles, there is a large variety in the morphology and physical properties of aggregates. However, aggregates possess some general properties that are remarkable from an engineer's point of view:

(i) Aggregates have a larger surface area than compact particles of the same geometric size. That means a higher adsorption capacity, better catalyst performance, and faster dissolution.
(ii) Aggregates are porous, i.e., relatively little solid material is required to occupy a certain space which is employed for thickening agents. Besides this, the pores may be used for the storage of other substances.
(iii) Aggregates are larger than the primary particles they are made of. This enhances solid-fluid-separation and powder handling. It affects optical properties and concerns toxicological aspects.
(iv) Aggregates have certain mechanical properties, which determine the aggregate breakage during dispersion or the erosion of primary particles under mechanical stress. The mechanical properties are also relevant for the elasticity of particle gels.

F. Babick, *Suspensions of Colloidal Particles and Aggregates*,
Particle Technology Series 20, DOI 10.1007/978-3-319-30663-6_4

Each of these properties may be beneficial or unwanted. The large size of aggregates, when compared to the primary particles, improves powder handling, but may deteriorate the optical properties (colour/opacity) of the final product or may require intensive dispersion when, e.g., used in polishing slurries. Sometimes just the presence or absence of aggregates is of interest, e.g., when evaluating the stability of colloidal suspensions.

All of the above mentioned technologically relevant properties are related to the aggregate morphology, which usually impedes a simple characterisation because aggregates are generally neither spherical, homogeneous, nor isotropic. Moreover, in the course of aggregation, the porosity of the evolving aggregate typically increases. That means that the mass of the aggregates do not increase with the third power of their size, but follow some non-integer power law (with exponents between 1 and 3). Such aggregates are called *fractal aggregates*.

The characterisation of such aggregates requires a theoretical framework that goes beyond the traditional description of particle systems. Its main tool is the *fractal dimension* introduced by Bênoit Mandelbrot in the 1970s. Some of the characterisation techniques presented in Chap. 2 can be employed to quantify the fractal dimension, i.e., aggregate structure, and most of them can be used for aggregate sizing. However, meaningful measurements are only possible when appropriate structure-property-relationships are established. More generally, the knowledge of these relationships is needed for designing and controlling aggregation processes.

This chapter briefly reviews the modelling of aggregation processes and the quantification of aggregate morphology. It further addresses the scattering properties and the hydrodynamic behaviour of aggregates and shows how this knowledge facilitates the characterisation of suspensions of pyrogenic powders. Finally, the interaction forces between aggregates are considered.

4.1 Aggregation Mechanisms and Kinetics

The morphology (i.e. size and structure) of aggregates is closely related to the microprocesses that affect the aggregation process (e.g. particle diffusion and migration, particle adhesion, and aggregate breakage or consolidation). Thus, it depends on a lot of microprocess and material parameters. However, for a principal look at the physical behaviour of aggregates, this manifold reality can be reduced to a few idealised cases. The two aggregation concepts of the *diffusion-limited cluster aggregation* (DLCA) and of the *reaction-limited cluster aggregation* (RLCA) have been proven to be good approximations for the aggregation of colloidal particles. This section introduces both concepts, while the corresponding aggregates are used for exemplary discussion in the subsequent sections.

4.1.1 Origin of Aggregation

Aggregation is the result of collision and adhesion. Collision means a relative motion between the particles. A major cause for the collision of colloidal particles is their Brownian motion. Additionally, the relative motion can be caused by laminar shear flow, turbulence, or particle migration (e.g. settling). Hence, the collision is primarily governed by the flow regime and by external fields. Additionally, collision is affected by long-ranging interparticle forces, i.e. by double layer interaction for colloidal suspensions. In the case of very strong repulsive forces, the collision rate can be reduced by several orders of magnitude (cf. Sect. 5.2.3). A third factor is the particle concentration: The higher the concentration, the higher the collision frequency.

When looking on the aggregation of small colloids in a viscous medium, the particle collision is mostly due to Brownian motion. This concerns, e.g., the aggregation of primary particles during the precipitation of SiO_2 (Schlomach and Kind 2004) or the formation of fumed oxides in gas phase reactions (Schaefer and Hurd 1990). If, in such a case, there are no or just weak repulsive particle interactions, then colliding particles and/or aggregates adhere at first contact. This kind of aggregation is called *diffusion limited cluster aggregation.*[1] On the other hand, when there is significant repulsive interaction, the collision of two approaching particles is very unlikely. Therefore, it takes more trials and more time before collision and subsequent adhesion is successfully “finished”. This type of aggregation is named *reaction limited cluster aggregation.*

Adhesion requires attractive interparticle forces between two particles in contact. Usually, such an attraction inevitably exists as a result of the London-van-der-Waals interaction (cf. Sect. 3.2.1). However, short-range repulsive forces, e.g. originating from adsorbed molecule layers (cf. Sect. 3.2.4), may outweigh the van-der-Waals forces and impede the adhesion even in the case of small surface distances.

In general, aggregation coincides with further processes that affect growth and morphology of aggregates: the breakage of particle bonds, the re-structuring within the aggregate (consolidation), the alteration of contact zones due to sintering (Al’Zaitone et al. 2009), or the (slow) formation of covalent bonds between particle surfaces in contact (e.g. siloxan-bonds for SiO_2, Iler 1979, pp. 223–224). However, these processes do not possess the same universal character as the sheer aggregation of diffusing, interacting particles, because they depend to a large extent on process parameters (e.g. temperature, velocity gradient) and material properties (e.g. solubility).

[1]The term *cluster* states that aggregation occurs between individual clusters incl. single particles, whereas in a *particle aggregation* mechanism only single particles adhere at a central aggregate (Schaefer and Hurd 1990).

4.1.2 Aggregation Kinetics

For the sake of simplicity and clarity, the kinetics of aggregation processes are usually examined for monodisperse particles in the absence of nucleation, dissolution, and aggregate breakage. In that case, clusters of dimensionless mass m (i.e. the number of particles in the aggregate) result from the aggregation of two sub-clusters with masses i and $j = m - i$. At the same time, clusters of m join with other clusters. Hence, the mass balance of aggregates m takes the form:

$$\frac{\mathrm{d}c_{N,m}}{\mathrm{d}t} = \frac{1}{2}\sum_{i,j=m-i}^{m-1} K_{ij}c_{N,i}c_{N,j}E_{ij} - c_{N,m}\sum_{n=1}^{\infty} K_{mn}c_{N,n}E_{mn}. \tag{4.1}$$

This is the population balance equation (PBE). Its principal form was first given by Smoluchowski (1916). In Eq. (4.1), c_N denotes the aggregate number concentration, while K_{ij} is a kinetic constant related with the collision between clusters of mass i and j, and E_{ij} is the sticking probability of the sub-clusters. Commonly, it is assumed that any successful collision (i.e. surface contact) leads to a stable particle bond and that E_{ij} can be, thus, set to 1.

The collision kernels K_{ij} depend on the particle trajectories and the particle interaction. For purely diffusing particles and clusters without any particle interaction (i.e. diffusion-limited cluster aggregation—DLCA), Smoluchowski (1916) derived:

$$K_{ij}^{\mathrm{diff}} = 4\pi \cdot r_{ij,\min} \cdot \left(D_i + D_j\right) = 4\pi \cdot \left(R_{\mathrm{C},i} + R_{\mathrm{C},j}\right) \cdot \left(D_i + D_j\right), \tag{4.2}$$

where $D_{i/j}$ are the diffusion coefficients and $R_{\mathrm{C},i/j}$ are the geometric (collision) radii.

In the presence of long-ranging repulsive interactions, Brownian collision is considerably hampered and the kernel K_{ij} is reduced by a factor W_{ij}, which can be calculated from the energy potential of the acting conservative forces (Fuchs 1934) and the hydrodynamic hindrance functions $B(r)$, which accounts for viscous lubrication (Honig et al. 1971; cf. Sect. 5.2.3):

$$W_{ij} = \int_{r_{ij,\min}}^{\infty} B(r)\exp\left(-\frac{V_{\mathrm{tot}}}{\mathrm{k_B}T}\right)\frac{\mathrm{d}r}{r^2}. \tag{4.3}$$

Repulsive interactions cannot be ignored for electrically charged aerosols or for suspended colloids with high surface charges. For colloidal suspensions, it is frequently assumed that the double layer is relatively thin ($\kappa a > 1$). Thus, the interaction between two sub-clusters is determined by the closest pairs of primary particles and the kernel K_{ij} is approximated by:

$$K_{ij} = K_{ij}^{\mathrm{diff}} P_{ij} \Big/ W_{11}. \tag{4.4}$$

The parameter P_{ij} is regarded as the number of potential particle contacts between two clusters of mass i and j. For DLCA, the parameter P_{ij} is set to 1, while for RLCA it is frequently approximated by a power law: $P_{ij} \approx (i{\cdot}j)^{\lambda}$ (e.g. $\lambda = 1$—Asnaghi et al. 1992, $\lambda = 0.5$—Lattuada et al. 2004, $\lambda \approx 0.25$–0.5—Sandkühler et al. 2005a). However, since a reactivity P_{ij}/W_{11} above 1 has no physical relevance for RLCA, it is recommended to employ $P_{ij} \approx (i{\cdot}j)^{\lambda}/(1 + ((i{\cdot}j)^{\lambda} - 1)/W_{11})$ with $\lambda \approx 0.45$ (Sandkühler et al. 2005b).

A consequence of Eq. (4.4) is that the impact of repulsive particle interactions is vanishing as the aggregates grow. An RLCA regime should, thus, steadily transmute into a DLCA regime. Indeed, such a transition could be experimentally confirmed (e.g. Asnaghi et al. 1992; di Biasio et al. 1994; Sandkühler et al. 2005b). Even more, ideal RLCA obviously exists only during the initial stages of aggregation.

The solutions of the PBE for the DLCA and RLCA regimes have been extensively investigated by several authors (Klein et al. 1990; Lin et al. 1990; Asnaghi et al. 1992; di Biasio et al. 1994). Recent research in this field has been conducted by the Morbidelli group (e.g. Lattuada et al. 2003; Sandkühler et al. 2005b; Gauer et al. 2009). For that reason, there is a fairly consistent view on the differences between these two fundamental aggregation mechanisms. Important features are summarised in Table 4.1.

Apart from Brownian motion, fluid flow or particle migration can be responsible for particle collision. For an interception-like aggregation process of particles moving on two different streamlines in a laminar shear flow, Smoluchowski (1916) calculated the following kernel:

$$K_{ij}^{\mathrm{sh}} = \frac{4\dot{\gamma}}{3} R_{ij}^{3} = \frac{4\dot{\gamma}}{3} \left(R_i + R_j\right)^3. \tag{4.5}$$

For particle collision due to turbulent eddy motion, Saffman and Turner (1956) obtained:

Table 4.1 Comparison between reaction-limited and diffusion-limited cluster aggregation (RLCA and DLCA)

Parameter	RLCA	DLCA
Energy barrier	$V_{\max} >\approx k_B T$	$V_{\max} \ll k_B T$
Reactivity	$P_{ij}/W_{11} \ll 1$	$P_{ij}/W_{11} \approx 1$
Kinetic	Slow, $\overline{N}(t) \sim \exp(t)$	Fast, $\overline{N}(t) \sim t^z,\ z \approx 1$
Aggregate morphology	Relatively dense	Very porous
Fractal dimension	$d_f \approx 2.1$	$d_f \approx 1.8$
Kernel: $K(ai, aj) = a^{\lambda} K(i, j)$	$\lambda > 1$, (i.e. exponent. growth)	$\lambda = 0$, (i.e. linear growth)
Cluster mass distribution	Aggregated particle mass decays with cluster size i	Most particle mass in cluster with $i = \overline{N}$

$$K_{ij}^{\mathrm{tu}} = \sqrt{\frac{8\pi}{15}}(r_i + r_j)^3 \left(\frac{P_V}{\eta}\right)^{1/2}, \tag{4.6}$$

where P_V denotes the volume-specific dissipation rate and η the dynamic viscosity.

Modifications of the kernels have been proposed for special types of particle interaction, e.g. for bridging flocculation (Smellie and La Mer 1958; Lu and Pelton 2001; Ahmad et al. 2008) or for short range structural forces (Runkana et al. 2005). The PBE itself has to be changed when fragmentation is included (Hogg 1992; Tolpekin et al. 2004), when heteroaggregation is examined (Lin et al. 2006), when a continuous expression is employed (Ahmad et al. 2008), or when the consolidation of aggregates is considered (Gruy 2011).

4.1.3 Conceptual Aggregation Mechanisms

The population balance equation is an important tool for understanding the kinetic behaviour of aggregation and for predicting the aggregate (or cluster) mass distribution. However, it does not reveal details on the morphology of the freshly formed aggregates. This requires simulation tools that adequately consider the two essential factors of aggregation: the particle trajectories (e.g. Brownian) and the particle interaction (e.g. strongly repulsive).

The development of such algorithms started in the late 1970s and yielded several papers on theoretical aggregation formalisms in two, three, and even higher dimensions as well as on the corresponding aggregate properties. These algorithms can be distinguished with regard to the particle trajectories (Brownian, ballistic, none), the aggregating objects (particles, monodisperse or polydisperse clusters), the sticking probability (reaction limited, diffusion limited) and the employment of a lattice.

A diffusion-limited aggregation algorithm for two-dimensional particles was first presented by Witten and Sander (1981) who simulated Brownian motion of a single particle in the vicinity of a fixed particle cluster. This algorithm is comparatively fast, but leads to relatively dense aggregates because the diffusing particles can easily enter the interior of the growing aggregate. Shortly afterwards, the simulation of a DLCA process was introduced independently by both Meakin (1983) and Kolb et al. (1983). The algorithm starts with a certain number of particles randomly distributed in a periodic box, and subsequently allows for a random walk of all particles and clusters. A new cluster is formed as soon as two objects collide. These two papers as well as subsequent studies explored the impact of the algorithm parameters. It could be shown that the initial concentration should be sufficiently low (Kolb et al. 1983; Kusaka et al. 2011) and that the adjusted mobility (either independent from aggregate mass or not) has no effect on the aggregate structure (Meakin 1983; Jullien 1987), yet influences the resulting mass distribution of the clusters (Meakin 1984b). Moreover, the rotational diffusion appeared to have a vanishing impact on the resulting aggregate morphology (Meakin 1984a). In conclusion, the numeric DLCA algorithms adequately reproduce real DLCA processes.

First algorithms for RLCA processes were published by Jullien and Kolb (1984), Brown and Ball (1985), and Meakin and Familiy (1987). The basic idea behind the algorithms is that in RLCA the collision trajectory has no influence on the aggregation. Any possible configuration of two sub-clusters in contact is equally likely. One way to realise this idea is to choose a pair of primary particles belonging to different clusters, to join the two clusters at this pair in random orientation, and to accept this aggregation if there is no overlap between particles. In comparison to the DLCA algorithm, RLCA simulations are relatively fast and produce more compact aggregates (Fig. 4.1).

Both types of algorithms, DLCA and RLCA, are intended to imitate real aggregation processes. In particular, they conserve the stochastic nature of aggregate formation, even though the aggregation conditions are controlled. As a result of this, the generated aggregates are not uniform with regard to configuration and morphological parameters.

A different approach is the concept of deterministic aggregates with a tuneable morphology (mostly fractal dimension). Such aggregation algorithms warrant certain morphology parameters without referring to any real aggregation mechanism. Deterministic aggregates can be generated by particle-cluster aggregation (PCA, e.g. Mackowski 1995; Kätzel et al. 2008a) or by cluster-cluster aggregation (CCA, e.g. Thouy and Jullien 1994; Filippov et al. 2000). Tuneable PCA leads to isometric aggregates with a dense central part, whereas tuneable CCA yields asymmetric morphologies similar to those obtained by DLCA or RLCA and should be preferred for even more reasons (Filippov et al. 2000). Tuneable aggregates allow for studying structure-property-relations for arbitrary values of fractal dimensions;

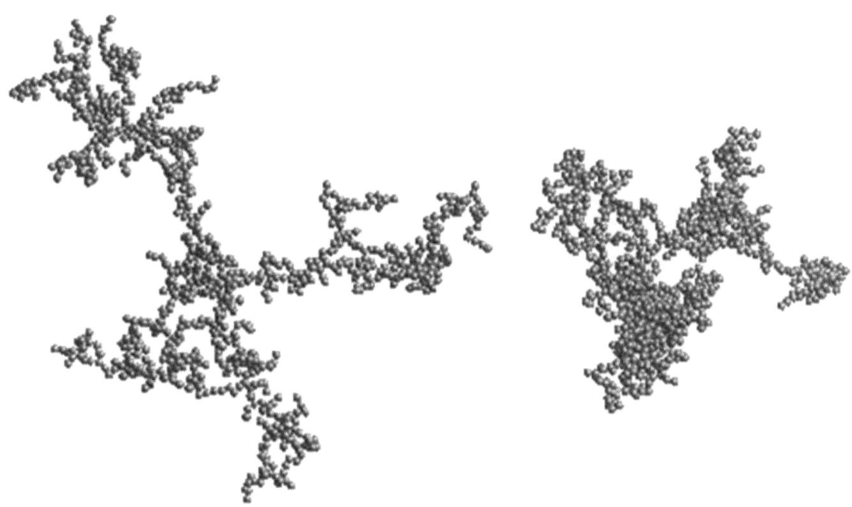

Fig. 4.1 Numerically designed DLCA (*left*) and RLCA (*right*) aggregates, $N = 1000$

uncertainty due to the variation in structure is eliminated and the algorithms are relatively fast. Their employment is meaningful when the structure (and polydispersity) of the relevant aggregates are already known.

4.2 Structure of Colloidal Aggregates

4.2.1 Concept of Fractality

4.2.1.1 General Remarks

Fractals are geometric (or other kind of mathematical) objects that are composed of small, at least approximate copies of the whole object. That means that these copies can be further split into smaller copies of themselves; a property which is called self-similarity.

A famous example of fractals is the Koch curve (von Koch 1904), which is continuous everywhere, and not differentiable anywhere (Fig. 4.2). The curve has a rough appearance which is typical to fractal bodies.[2] In contrast to non-fractal objects this roughness is infinitely fine, which leads to qualitatively new properties.

In the case of the Koch curve, we can state that the curve length increases with each step of iteration by a factor of 4/3. The final curve is, thus, of infinite length. Hence, length is not a meaningful parameter to describe this curve. A more appropriate description of the curve is the power law relationship between the number N of basic elements used in curve design and their size L.

$$N \propto 1/L^{d_f}. \tag{4.7}$$

For instance, the first iteration step of the Koch curve employs $N = 4$ elements of length $L = 1/3$, while the second one uses $N = 16$ elementary lines of length $L = 1/9$, etc. The exponent d_f in Eq. (4.7) takes the value 1.26; it is called Hausdorff-Besicovitch dimension or just *fractal dimension*.

The principal concept of fractality and its relevance for real phenomena were elaborated by Benoît Mandelbrot. He impressively showed that fractality does not only apply to some strange “pathological” curves in mathematical analysis, but is a general tool to describe irregularities in shapes and processes or to handle any fragmented phenomenon. In his seminal book “Fractals, form, chance and dimension” from 1977, he exemplarily studied fractality in turbulent flow, as well as showed the relevance of fractal structures for the spatial distribution of holes in Swiss cheese and even for phenomena in economics and linguistics.

A very intriguing example for the “reality” of fractal phenomena in nature is the coastline problem, which was first studied by Lewis Fry Richardson (1961). It says

[2]The term *fractal* is derived from the latin *fractus*, i.e. broken.

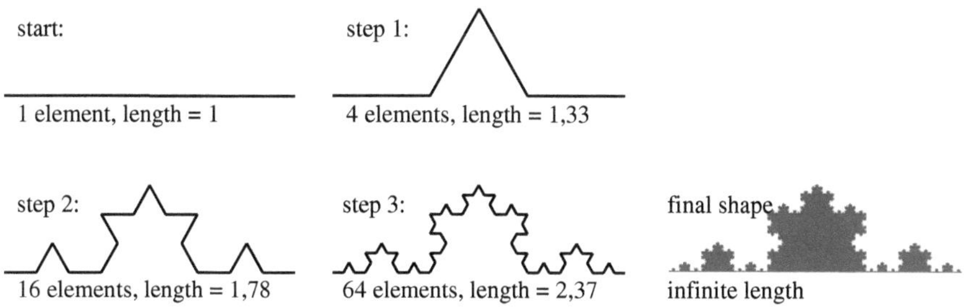

Fig. 4.2 Koch curve, construction, and final shape

that the length of a coastline when determined from a map depends on the map's magnification and obeys a power law $L \propto l_{\mathrm{scale}}^{1-d_\mathrm{f}}$, where l_{scale} is the elementary scale length and d_f is the fractal dimension of the coastline (Mandelbrot 1967). The power-law is a representation of the statistic self-similarity of natural coastlines.

> …when a bay or peninsula first noticed on a map scaled to 1/100.000 is re-examined on a map at 1/10.000, innumerable subbays and subpeninsulas become visible. On a 1/1.000 scale map, sub-sub-bays and sub-sub-peninsulas appear…
>
> (Mandelbrot 1977).

4.2.1.2 Fractal Aggregates

Soon after the publication of Mandelbrot's textbook, Forrest and Witten (1979) showed that density correlations in TEM images of smoke aggregates obey a power law:

$$g \sim r^{d_\mathrm{f} - d_\mathrm{eucl}}, \tag{4.8}$$

where d_eucl is the dimension of Euclidean space and d_f can be identified as the fractal dimension. This means that particle aggregates are self-similar in a statistical sense and that they can be characterised by means of fractality.

In 1981, Witten and Sander used the fractal dimension to describe their numerically designed aggregates of a diffusion-limited particle-cluster aggregation scheme. They derived the fractal dimension on to ways: (i) via the density-correlation function (Eq. (4.8)) and (ii) via the correlation between the aggregate radii of gyration R_g and the dimensionless mass, i.e. the number of primary particles N:

$$N \sim R_\mathrm{g}^{d_\mathrm{f}}. \tag{4.9}$$

The two values of d_f are similar but not identical. A thorough study should, therefore, distinguish between d_α—related to the density correlation (Eq. (4.8))—and between d_g—related to the fractal growth (Eq. (4.9)). For instance,

3-dimensional numeric DLCA aggregates showed a correlation dimension $d_\alpha = 1.8 \pm 0.1$ and a gyration dimension $d_g = 1.75 \pm 0.05$ (Meakin 1984c).

Since the publication of theses papers, numerous studies have confirmed the fractal nature of real and numeric aggregates of colloidal particles. Fractal dimension was verified as a powerful concept for quantifying the structure of large and medium sized aggregates. This is primarily due to its simplicity. Moreover, the power-law dependency of the density correlation $g(r)$, i.e., the statistic self-similarity of aggregates, results in corresponding power-laws for hydrodynamic and scattering properties. However, real aggregates are of finite size and the important fractal property of self-similarity is strictly found only at length scales above the primary particle size and below the aggregate size ($a \ll r \ll R_{agg}$). It is, therefore, rather questionable to apply the concept of fractality to small aggregates. Consequently, some authors treat very small aggregates (doublets, triplets, quartets) separately when investigating aggregation processes and aggregate properties (e.g. Lattuada et al. 2003; Gruy and Cugniet 2004).

4.2.2 *Morphological Parameters of Fractal Aggregates*

4.2.2.1 Aggregate Mass

The mass of fractal aggregates is defined as the total mass of all primary particles[3] that they are composed of. It is a crucial parameter for the physical aggregate properties (e.g. settling or scattering). For scale-independent modelling and simulation, it is quite common to use the normalised aggregate mass m_{agg}/m_p, which is identical with the number N of primary particles in the aggregate (i.e. the *aggregation number*).

In general, the aggregation number N is a distributed quantity. The corresponding number weighted distributions are called *cluster mass distribution* (CMD). They can be calculated for ideal aggregation processes by solving the PBE (Eq. (4.1)). For DLCA processes, the absolute frequency $n(N)$ of aggregates with mass N can be estimated as (Lin et al. 1989, 1990a):

$$n^{\mathrm{DLCA}}(N) = \frac{n_\mathrm{p}}{\overline{N}^2}\left(1 - \frac{1}{\overline{N}}\right)^{N-1}, \tag{4.10}$$

where n_p is the initial number of primary particles and $\overline{N}$ is the average aggregation number of the aggregate population. For large $\overline{N}$, the power function on the r.h.s. of

[3]This definition deviates from the one some authors use for compact aggregates, where the fluid captured in the voids are considered as part of the aggregate. However, in the case of fractal aggregates it is difficult to define a meaningful aggregate surface, which is necessary for calculating a total volume or mass.

Eq. (4.10) can be replaced by an exponential function and the relative frequency $p(N)$ is given by:

$$\lim_{\overline{N}\to\infty} p^{\mathrm{DLCA}}(N) = \frac{1}{\overline{N}(t)} \cdot \mathrm{e}^{-N/\overline{N}}. \tag{4.11}$$

Hence, the fundamental shape of the CMD does not change with proceeding aggregation; the CMD during DLCA is, therefore, a self-preserving distribution. The only parameter is the average aggregate mass $\overline{N}$, which was shown to increases linearly with time $\overline{N} \propto t/t_0$ (Lin et al. 1989, $t_0 = 3\eta/(8k_B T c_{N,\mathrm{p}})$).

Aggregation within the RLCA regime results in a power-law-shaped CMD with exponential cut-off (van Dongen and Ernst 1985):

$$p^{\mathrm{RLCA}}(N) \propto N^{-\tau} \cdot \mathrm{e}^{-N/N_c}, \tag{4.12}$$

where the parameter τ is approximately 1.5 (Lin et al. 1989; Sandkühler et al. 2005b). The cut-off mass Nc is the scaling parameter of this approach; it decays exponentially with time ($\overline{N} \propto N_c \propto \exp^{-t/t_M}$; Ball et al. 1987; Lin et al. 1989). Normalisation transforms Eq. (4.12) into (cf. Lin et al. 1990b):

$$p^{\mathrm{RLCA}}(N) = \frac{\overline{N}/N_c}{N_c \Gamma(2-\tau)} \cdot (N/N_c)^{-\tau} \cdot \mathrm{e}^{-N/N_c} \propto \frac{1}{N_c(t)} \cdot \mathrm{funct.}(N/N_c), \tag{4.13}$$

which implies that the CMD of an RLCA processes is—to first approximation—a self-preserving distribution as well (cf. Lattuada 2003, pp. 11–17).

4.2.2.2 Primary Particle Size

The elementary length scale of particle aggregates is the diameter of the primary particles[4] x_p (or equivalently their radius a). It is inversely proportional to the specific surface area of the aggregate and, thus, affects adsorption capacity, dissolution kinetics, or even solubility. Moreover, the size of the primary particles determines their scattering strength, thus on the scattering regime. It is the minimum interparticle distance and, therefore, a scale for the range of particle interaction (e.g. double layer thickness).

Primary particles are often regarded as monodisperse, when simulating aggregation processes. This assumption is based on the frequent finding that within the aggregates of precipitation or combustion products there is only little variation of particle size (Schlomach and Kind 2004; Boldridge 2010). However, when considered for the total aggregate population and for general aggregation processes,

[4]Note that the primary particles can change their shape and size after aggregation, e.g. due to sintering. In this case, the term *constituent particles* would be more appropriate.

polydispersity of primary particles cannot be neglected. Distinction has to be made between the polydispersity within the individual aggregates, among all aggregates, and of all primary particles, which is expressed in different size averages:

- $x_{\mathrm{p,av}}^{(k)}$ = the average primary particle size x_{p} of the kth aggregate,
- $x_{\mathrm{p,av}}$ = the average of all $x_{\mathrm{p,av}}^{(k)}$, $x_{\mathrm{p,av}} = \frac{1}{n_{\mathrm{Cluster}}}\sum x_{\mathrm{p,av}}^{(k)}$
- $x_{\mathrm{p,mean}}$ = the arithmetic mean size of all primary particles, $x_{\mathrm{p,mean}} = \frac{1}{n_0}\sum x_{\mathrm{p,av}}^{(k)} \cdot N(k)$

Differences between $x_{\mathrm{p,av}}$ and $x_{\mathrm{p,mean}}$ indicate some kind of correlation between the average particle size $x_{\mathrm{p,av}}^{(k)}$ and the aggregation number $N(k)$.

4.2.2.3 Aggregate Size

The term *aggregate size* refers to a number of different geometric and physical length scales that grow with the aggregate mass and that are rough estimates of the aggregates' outer dimensions. The most important size parameter is probably the radius of gyration R_{g}, which is defined by the local density distribution around the aggregate centre ($\mathbf{r}_{\mathrm{c}}$):

$$R_{\mathrm{g}}^2 = \frac{1}{m_{\mathrm{tot}}}\int_V \rho_m(\mathbf{r})|\mathbf{r}-\mathbf{r}_{\mathrm{c}}|^2 \mathrm{d}V. \tag{4.14}$$

For clusters of infinitely small particles ($a \ll r_k$, r_{ij}), Eq. (4.14) transmutes into two equivalent definitions:

$$R_{\mathrm{g,c}}^2 = \frac{1}{N}\sum_k |\mathbf{r}_k - \mathbf{r}_{\mathrm{c}}|^2 = \left\langle r_k^2 \right\rangle \text{ and} \tag{4.15}$$

$$R_{\mathrm{g,c}}^2 = \frac{1}{2N^2}\sum_{i,j} \left|\mathbf{r}_i - \mathbf{r}_j\right|^2 = \frac{1}{2}\left\langle r_{ij}^2 \right\rangle. \tag{4.16}$$

This particle-centre-based radius of gyration $R_{\mathrm{g,c}}$ is the rms of the particle distance from the aggregate centre and, at the same time, it is $\sqrt{1/2}$ times the rms of the interparticle distance.

According to Eqs. (4.15) and (4.16), the centre-based radius of gyration $R_{\mathrm{g,c}}$ would be zero for $N = 1$. This is because $R_{\mathrm{g,c}}$ ignores the spatial extension of the primary particles. The total radius of gyration is:

$$R_{\mathrm{g}}^2 = R_{\mathrm{g,p}}^2 + R_{\mathrm{g,c}}^2,\ R_{\mathrm{g,p}}^2 = \frac{3}{5}a^2. \tag{4.17}$$

This total radius of gyration is commonly used to evaluate the "size" of aggregates and can be experimentally determined. The centre-based radius of gyration $R_{g,c}$ only has some significance for the characterisation of numeric aggregates. However, the difference between R_g and $R_{g,c}$ rapidly vanishes with increasing aggregate size.

The radius of gyration can be measured by scattering experiments. Alternatively, it may be derived from 2D images of the particle aggregate. However, in the case of image analysis the aggregate size is preferentially associated with the diameter of the circumscribing sphere x_C (Hess et al. 1986; Gmachowski 1998; Tang et al. 2002; Lee and Kramer 2004), which is the same as or slightly larger than the maximum Feret diameter $x_{Feret,max}$. This geometric aggregate size is approximately proportional to the radius or diameter of gyration (R_g or x_g, respectively). For aggregates of spherical symmetry having a radial density distribution $\sim r^{d_f-3}$, the proportionality factor can be easily calculated (Hess et al. 1986):

$$\frac{x_C}{x_g} = \sqrt{\frac{d_f+2}{d_f}}. \tag{4.18}$$

This solution is consistent with the result for compact spheres in Euclidean space ($x_C/x_g = \sqrt{1+2/d}$), but it fails for the anisometric clusters obtained by DLCA and RLCA processes, as can be seen in Fig. 4.3, which depicts the size ratio $x_{Feret,max}/x_g$ as a function of the aggregation number N for DLCA, RLCA, and hexagonal close-packed (hcp) aggregates.

The data points in Fig. 4.3 are obtained from numeric aggregate populations with 100 aggregates per each mass fraction, which leads to some statistic scattering in the curves. Regarding this uncertainty, the ratio x_C to x_g can be considered virtually universal for DLCA aggregates ($x_C/x_g \approx 1.66$). That means that the geometric size obeys the same fractal law as the radius of gyration: $N \propto R_C^{d_f}$ (cf. Eq. (4.9)).

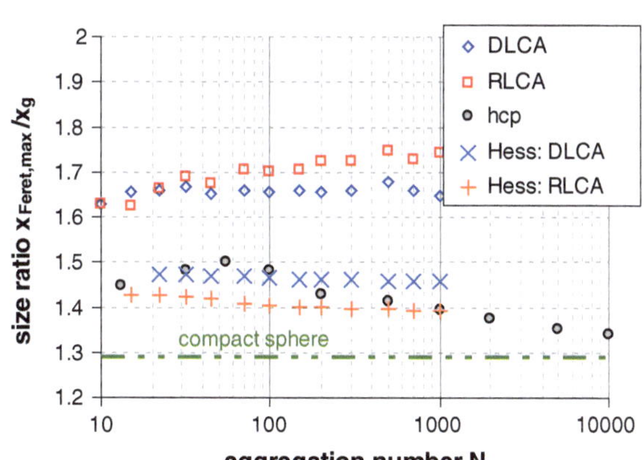

Fig. 4.3 Ratio of the geometric aggregate size $x_{Feret,max}$ to the diameter of gyration x_g for DLCA, RLCA, and hexagonal close-packed (hcp) aggregates; determined at numeric aggregate populations and calculated acc. to Hess et al. (Eq. (4.18)); for comparison, the compact sphere value is indicated

In contrast, the ratio x_C/x_g slightly increases with the aggregation number N for RLCA aggregates. This is related to the densification of growing RLCA aggregates (cf. Sect. 4.2.3.3).

4.2.2.4 Aggregate Structure

Although *mass* and *size* are fundamental parameters in particle technology, the major difference between particle aggregates and compact particles is their internal structure. *Structure* refers to the spatial distribution of primary particles. This distribution can be described by the radial density function:

$$\rho(r) = \frac{1}{4\pi r^2} \cdot \left\langle \sum_k \delta(r - |\mathbf{r}_k - \mathbf{r}_c|) \right\rangle, \quad \int \rho \mathrm{d}V = N. \tag{4.19}$$

The radial distribution function can be used to calculate the frequency of interparticle distances and related physical phenomena (Hess et al. 1986; Naumann and Bunz 1993; Lattuada et al. 2006). However, it implicitly assumes spherical symmetry and, thus, ignores the anisometry and the density variations inside real aggregates.

A better description of the aggregate structure is achieved by the pair-correlation function (also density auto correlation, particle-particle-correlation or two-point-correlation):

$$g = \frac{1}{4\pi |\mathbf{r}|^2 \bar{\rho} N} \left\langle \sum_{k \neq l} \delta(\mathbf{r} - \mathbf{r}_{kl}) \right\rangle, \quad \int g \mathrm{d}V = N - 1. \tag{4.20}$$

Note that some authors use a different normalisation, due to which the volume integral in (4.20) is independent of the aggregation number N (e.g. Jullien 1986; Sorensen 2001; Khlebtsov 1996; Gruy 2009). In this case, however, the values of g—e.g. for the first coordination shells—depend on N, and, thus, defy an immediate comparison. This is avoided with the convention used here (cf. Lattuada et al. 2004; Sandkühler et al. 2005b), which has the additional appeal of being consistent with the radial distribution function $g(r)$ of colloidal suspensions.

The pair correlation function of fractal aggregates reflects their self-similarity for medium interparticle distances ($a \ll r \ll R_g$) by a power-law relationship (Eq. (4.8)). At large arguments ($r > R_g$), the pair correlation function decays steeply, which can be modelled by a cut-off function. For medium and large arguments, one obtains:

$$g(r) = A \cdot r^{d_f - 3} \cdot h\left(\frac{r}{\xi}\right) (r \gg a), \tag{4.21}$$

with $h(x) = \exp(-x^{\varphi})$. For $\varphi = 1$, the function $h(x)$ describes an exponential cut-off, whereas $\varphi = 2$ leads to a Gaussian cut-off function, which seems to be more realistic (Sorensen 2001).

At small interparticle distances r, the pair correlation function $g(r)$ describes the structured vicinity (coordination shells) of the individual particles. Figure 4.4 shows the pair correlation functions of DLCA and RLCA aggregates (N = 10–1000) for both distance regions. They are depicted as scaled functions $g \times R_g^3/(N-1)$ versus r/R_g on a log-log plot for medium and large interparticle distances, while unscaled functions $g(r)$ are plotted versus r/a on linear axes for small arguments. The log-log-presentation confirms the existences of a power-law region at an intermediate distance ($a \ll r \ll R_g$) that allows the determination of the fractal dimension (cf. Eq. (4.8)). For large interparticle distances ($r > R_g$), the correlation function decreases rapidly but without considerable difference between the two aggregate types, DLCA and RLCA. This decrease can be approximated well by a Gaussian cut-off function, whereas an exponential cut-off overestimates the local density in the outer aggregate regions.

The right diagram of Fig. 4.4 is zoomed into the first three coordination shells around the reference particle. In that region, pair-correlation functions do not steadily decrease with r, but show a clear discontinuous behaviour as can be expected for aggregates of monodisperse particles (Dimon et al. 1986). The function starts at $r = 2a$ with a delta function that corresponds to the contacting neighbours and cannot be shown in the graph. The volume integral of this delta function equals the average coordination number N_C, which is approximately 2 for DLCA and RLCA aggregates when $N \geq 20$ (cf. Lattuada 2003, p. 97). A second discontinuity is observed at $r = 4a$; it reflects the transition from the slightly ordered, second coordination shell to the random, self-similar structure on medium interparticle distances. The figure shows that RLCA aggregates are more densely packed than DLCA aggregates in the second coordination shell ($2a < r \leq 4a$). A quantitative analysis reveals that the average number of neighbouring particles within $r \leq 4a$ converges to approx. 6.3 and 8.3 for infinitely large DLCA and

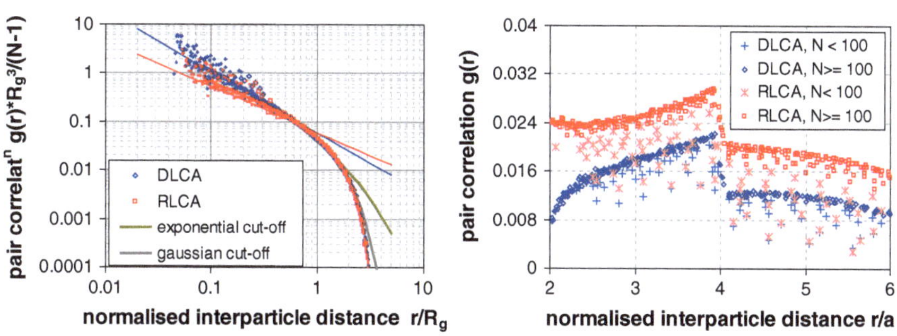

Fig. 4.4 Pair-correlation functions of DLCA and RLCA aggregates of varying aggregation numbers N (10–1000); averaged from 100 aggregates per aggregation number N; *left* scaled correlation function $g \times R_g^3/(N-1)$ versus r/R_g, *right* correlation function g versus r/a

RLCA aggregates, respectively. Note that these values lie considerably below the corresponding one for hexagonal close-packing (=54).

The pair correlation function can be used to calculate various morphological or physical properties of fractal aggregates. Of particular importance is their scattering behaviour because it can be employed to measure the fractal dimension. For aggregates of small colloidal particles, the scattering of waves can be traced back to an interference problem (cf. Sect. 4.3.1). That means that the scattering response of the aggregate is proportional to the Fourier-transform of the pair correlation function. In the scattering domain, the aggregate structure is described by the *static structure factor* (or *static structure function*):

$$S_{\mathrm{agg}}(\mathbf{q}) = \frac{1}{N}\left(1 + \int \mathrm{e}^{\mathrm{i}\mathbf{qr}} g(\mathbf{r})\mathrm{d}V\right), \tag{4.22}$$

After orientation averaging, one gets:

$$S_{\mathrm{agg}}(q) = \frac{1}{N}\left(1 + \int \frac{\sin qr}{qr} g(r)\mathrm{d}V\right). \tag{4.23}$$

The relevance of the function $S_{\mathrm{agg}}(q)$ for the structural properties of the aggregate depends on the argument q: At small arguments q ($q < 1/R_{\mathrm{g}}$), the aggregate structure is not resolved, but the aggregate size can be derived. In this so-called Guinier regime, the structure factor is approximated by $S_{\mathrm{agg}}(q) \approx \exp\left(\frac{1}{3}q^2R_{\mathrm{g,c}}^2\right)$, where the particle-centre-based radius of gyration $R_{\mathrm{g,c}}$ has to be used (Guinier 1938). With increasing argument q, more details of the structure are resolved. For intermediate arguments ($1/R_{\mathrm{g}} \ll q \ll 1/a$), the self-similarity of fractal aggregates is reflected by a power-law dependency $S_{\mathrm{agg}}(q) \propto (qR_{\mathrm{g}})^{-d_{\mathrm{f}}}$, where the total radius of gyration R_{g} is employed for normalisation. At large arguments ($q \gg 1/a$), only the individual primary particles contribute to $S_{\mathrm{agg}}(q)$, which then yields $S_{\mathrm{agg}}(q) = 1/N$.

Expressions for $S_{\mathrm{agg}}(q)$ can be derived from analytical integration of the pair-correlation function $g(r)$ (Eq. (4.23)). They are rather sophisticated and are usually replaced by simple approximations like the Fisher-Burford equation (cf. Sorensen 2001):

$$S_{\mathrm{agg}} = \left(1 + \frac{2}{3d_{\mathrm{f}}} q^2 R_{\mathrm{g}}^2\right)^{-d_{\mathrm{f}}/2}, \tag{4.24}$$

which corresponds to a pair correlation function with exponential decay at a fractal dimension $d_{\mathrm{f}} = 2$. An alternative is the polynomial regression function of Lin et al. (1989):

$$S_{\mathrm{agg}} = \left(1 + \sum_{s=1}^{4} c_s (qR_{\mathrm{g}})^{2s}\right)^{-d_{\mathrm{f}}/8}, \tag{4.25}$$

where the first coefficient is $c_1 = 8/(3\ d_f)$, and the others depend on the aggregate type (DLCA with $d_f = 1.8$: $c_{2/3/4} = 2.50/-1.52/1.02$, RLCA with $d_f = 2.10$: $c_{2/3/4} = 3.13/-2.58/0.95$).

In Fig. 4.5 (left), the two analytical formulations are compared with the averaged structure factor of DLCA aggregates consisting of 1000 monodisperse particles. Additionally, Guinier's approximation for small arguments and the solution of Eq. (4.23) for a fractal pair correlation function with Gaussian cut-off are shown. Apparently, the Gaussian solution and Lin's regression fit very well with the real data, whereas the Fisher-Burford Eq. (4.24) clearly overestimates $S_{agg}(q)$ in the fractal regime. It should be mentioned that a similar picture is obtained for RLCA aggregates.

All of the analytical solutions in Fig. 4.5 (left) fail to describe the transition from the fractal regime to large arguments (i.e. when $S_{agg}(q) = 1/N$). For this purpose, slightly more elaborate solutions have to be employed. A possible approximation for DLCA and RLCA aggregates was proposed by Babick et al. (2012a):

$$S_{agg}(q) = S_{G+f}(q) \cdot \left(1 - w_p(q)\right) + S_p(q) \cdot w_p \tag{4.26}$$

with

$$S_{G+f}(q) = \exp\left(-\tfrac{1}{3}\left(qR_{g,c}\right)^2\right) + \frac{1}{1000 \cdot \left(qR_{g,c}\right)^{-5} + \left(qR_{g,c}\right)^{d_f}}, \tag{4.27}$$

$$S_p(q) = \frac{1}{N}\left(1 + 1.5 \cdot \frac{\sin 2qa}{2qa}\right) \tag{4.28}$$

$$\text{and}\quad w_p = 1 - \exp\left(-\left(\tfrac{3}{5}qa\right)^4\right). \tag{4.29}$$

For $N \geq 10$, the parameter $R_{g,c}$ can be replaced by R_g without significant effect. The corresponding curves are compared with the averaged structure factor of DLCA

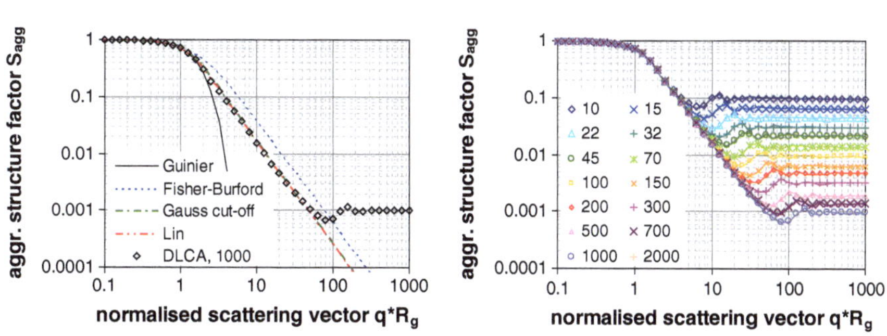

Fig. 4.5 Static structure factor $S_{agg}(q)$ of DLCA aggregates; *left* model predictions and averaged for $N = 1000$; *right* $S_{agg}(q)$ for $N = 10$–1000 (symbols) and data fit (solid lines) with Eq. (4.26)

aggregates (d_f = 1.80) in Fig. 4.5(right). These equations even provide a good approximation for the static structure factor of RLCA aggregates when the fractal dimension in Eq. (4.27) is set to 2.1.

4.2.2.5 Aggregate Shape

The two examples of numeric DLCA and RLCA aggregates in Fig. 4.1 illustrate that the particles are not homogenously distributed in space and that there is no spherical symmetry. Even more, a systematic deviation from spherical (macro) shape can generally be observed for both aggregate types. This asphericity was first examined by Thouy and Jullien (1997), who introduced a gyration tensor $R_{ij}^2 = \langle (x_k^{(i)} - x_c^{(i)}) \cdot (x_k^{(j)} - x_c^{(j)}) \rangle$, where $x_k^{(i)}$ is a Cartesian coordinate of the kth particle. The sum of the eigenvalues R_m^2 (m = 1–3) of this tensor is equal to the squared centre-based radius of gyration $R_{g,c}$, while the ratio of these eigenvalues describe the aggregate's anisometry. Gyration tensors were calculated for deterministic aggregates of tuneable fractal dimension in the range from 1.0 to 2.5. The ratio between the largest and smallest eigenvalue were quantified to $\langle R_1^2 \rangle / \langle R_3^2 \rangle \approx$ 12 for d_f = 1.8 and to 6 for d_f = 2.1. These values are in (approximate) accordance with the average ratios that can be found for numeric DLCA and RLCA aggregates (12.0 and 8.2, respectively). The corresponding ratios for the first two eigenvalues are 4.5 (DLCA) and 3.6 (RLCA). Hence, aggregates evolving from a cluster-cluster aggregation do not expand uniformly in space, but rather grow in two dimensions.

Although the gyration tensor offers a way to quantify the anisotropy of aggregates, its experimental determination is bound to the analysis of 2D images and then relies on the resolution of all primary particles. A more practical way to describe the macroshape by means of image analysis is the comparison of characteristic lengths, e.g. the minimum and maximum Feret diameters. A completely different approach of evaluating the aggregate anisotropy is the multipole expansion of the modified static structure factor $S^*(q)$ (Lindsay et al. 1989):

$$S^*(q) = N^2 \cdot S_{agg}(q) = \left| \sum e^{i\mathbf{q}\mathbf{r}_j} \right|^2 = 4\pi \sum_l S_l(q) \tag{4.30}$$

with

$$S_l = \sum_m \left| \sum_i j_l(qb_i) Y_l^m(\Omega_i) \right|^2, \tag{4.31}$$

where j_l are the spherical Bessel functions of first kind and Y_l^m are the spherical harmonics (cf. Appendix C.1). The functions $S_l(q)$ determine the impact of the aggregate rotation on its scattering behaviour and can, thus, be indirectly measured with dynamic light scattering. However, there is no real application of these functions to shape characterisation. They are solely used when predicting the impact of rotational diffusion on DLS signals for a given particle configuration (cf. Sect. 4.5.2.1).

Summarising, it can be stated that aggregates obtained from a cluster-cluster aggregation mechanism cannot be considered spherically symmetric. For this reason, any calculation of aggregate properties that is based on spherical shape and isotropic density distribution can only provide a rough estimation. It is not, however, easy to determine the macroshape of aggregates experimentally. Therefore, its impact on aggregate properties has only been implicitly studied; that is, by assuming a certain type of aggregate.

4.2.3 Fractal Dimension of Aggregates

The previous section showed that the various morphological aggregate parameters are specifically related to the fractal dimension, which itself is nothing else than a morphological parameter. Yet, there is no simple and unambiguous definition of the fractal dimension of aggregates. This is due to the fact that real aggregates are finite in size and show stochastic variations in their properties. In these respects, they differ from mathematical fractals (cf. Sect. 4.2.1). The two scaling laws that are characteristic for fractals—mass scaling $N \propto R_{\mathrm{agg}}^{d_\mathrm{f}}$ and self-similarity $g \propto r^{d_\mathrm{f}-3}$—do not strictly apply to single aggregates, but rather quantify empirical correlations of population averaged properties. Moreover, for finite-sized aggregates, these scaling laws are related to different length scales (mass scaling to the aggregate size, e.g. R_g or R_C, self-similarity to length scales in the range $a \ll r \ll R_\mathrm{g}$).

Another reason for the difficulties in defining fractal dimension is that there is no experimental technique that is able to reveal the exact configuration of colloidal aggregates. Instead, the structural properties have to be derived from images (e.g. obtained with electron microscopes) or from physical properties that reflect the aggregate structure to a certain degree. Indeed, there are different experimental methods which are employed for the determination of the fractal dimension. Hence, they should always be indicated when stating values of d_f.

4.2.3.1 Experimental Determination of the Fractal Dimension

The most descriptive way to quantify the structural properties of aggregates is probably the evaluation of aggregate images, from which the projected coordinates of the primary particles and/or size-related properties of the aggregate can be extracted. There are various ways to derive the fractal dimension from such parameters. For very porous aggregates, when the positions of all primary particles are extractable from the image, it is possible to calculate the pair-correlation function of the projected coordinates, which shows a power-law behaviour acc. to Eq. (4.8). However, this procedure does not work for large and compact aggregates when the single particles cannot be resolved due to multiple overlap. In that case, geometric properties, e.g. the diameter of the enveloping sphere, the Feret diameter, the perimeter, or the projected area, are quantified and related to the aggregation number N or to each other (Weitz et al. 1985; Park et al. 2004; Shin et al. 2009). Nevertheless, fractal dimensions that are obtained by image analysis do not reflect the 3-dimensional structure and, thus, have to be converted when used for further analysis (Lee and Kramer 2004). A completely different approach for calculating the fractal dimension of aggregates from their images is the box-counting method, which was introduced by Mandelbrot (1977): When the projected area of the aggregate is covered by non-overlapping equal boxes (basic elements) of defined size L, we find a power law relationship between the number of boxes for complete coverage and the box size (e.g. Shin et al. 2009; Ibaseta and Biscans 2010). The box-counting dimension is very close to the definition given in Sect. 4.2.1 for ideal fractal objects (Eq. (4.7)).

For coarse, micrometre-sized aggregates, images can be obtained with optical detection systems which even facilitate online measurements (Tang et al. 2002). However, colloidal aggregates can be resolved only with electron microscopy, which requires sophisticated sample preparation procedures and appropriate image analysing tools. In general, image analysis requires a fairly high number of examined aggregates for ensuring representativity. This statistical aspect can be ignored in the case of static scattering techniques. Papers that uses static light scattering (SLS), small angle X-ray scattering (SAXS), or even small angle neutron scattering (SANS) for studying the aggregation of colloidal particles or the structure of flame-made aggregates are vast (e.g. Schaefer et al. 1984; Weitz et al. 1985; Wiltzius 1987; Lin et al. 1989; Schaefer and Hurd 1990; Burns et al. 1997; Bugnicourt et al. 2007). All these studies rely on the fact that scattering on a colloidal aggregate can be approximately related to the pairwise interference of scattering fields from the single particles and, thus, reflect the pair-correlation function $g(r)$ and the static structure factor $S_{\text{agg}}(q)$. The fractal dimension is then the exponent of power-law relation between the scattering intensity and the inverse scattering vector q (see above, p. 134).

Similarly, the interaction of hydrodynamic fields within an aggregate can be approximated by a pairwise interaction of hydrodynamic disturbances at the primary particles. As a result, the hydrodynamic drag on an aggregate depends on its pair-correlation function. The drag force on fractal aggregates, thus their hydrodynamic diameter, eventually obeys a fractal scaling law that is similar to the one for the geometric aggregate size ($N \sim F_{\text{D}}^{d_{\text{f}}}$). Hence, the combined measurement of a dynamic property (e.g. diffusion coefficient) and a mass-related aggregate property (e.g. volume, settling velocity) offers a further way to quantify the fractal dimension. Such a quantification of the fractal dimension is rather common in aerosol science, where narrow fractions of defined mobility can be generated and analysed with regard to the mass (e.g. Weber et al. 1996; Park et al. 2004; Lall et al. 2006; Olfert 2007; Shin et al. 2010) or to the aerodynamic diameter, which is proportional to the Stokes diameter (Schleicher et al. 1995; Virtanen et al. 2010).

The scattering and hydrodynamic behaviour of aggregates are considered in detail in Sects. 4.3 and 4.4, respectively.

4.2.3.2 Fractal Dimension of Numeric Aggregates

Aggregates that have been generated by a numeric aggregation algorithm bear the advantage that all positions of the primary particles are known; thus, that the pair-correlation function and the static structure factor can be computed and used for a profound analysis of the morphological and physical aggregate properties. They provide, therefore, an important tool to reveal systematic differences between the fractal dimensions from the numerous experimental methods. Even more, they allow a comparison between any global, i.e. population-averaged, fractal dimension and the distribution of the individual fractal dimension.

Two methods that can be employed for the determination of individual fractal dimensions were already introduced in the context of image analysis (see above): box-counting and pair-correlation. Both methods can be applied to the 3-dimensional structure (e.g. box-counting with cubes) or to the 2-dimensional projection of the numeric aggregates (Thouy and Jullien 1997). Regardless, deriving the fractal dimension from the pair-correlation functions is meaningful only for large aggregates ($R_g/a \gg 1$, Witten and Sander 1981). A further method for computing the fractal dimension of a numeric aggregate is based on the monitoring of size (e.g. R_g) and aggregation number N during aggregation. The two parameters correlate in form of a power law acc. to Eq. (4.9) (Bedrich 2006). Last but not least, any numeric aggregate can be successively decomposed into medium-sized and small sub-aggregates—independently from the aggregation mechanism. Again, size and aggregation number of the sub-aggregates form a power-law that contains the fractal dimension. This fragmentation method can be considered as a combination of box-counting (with unequal irregular boxes) and aggregation growth monitoring (thereby implying a cluster-cluster aggregation).

Even though the individual determination of the fractal dimension gives insight into the variation of structural properties within a population of aggregates, it is only of minor relevance for the physical properties of aggregate systems. Those typically reflect the population-averaged aggregate structure. Global values for the fractal dimension of an ensemble of numeric aggregates can be derived from the averaged pair-correlation functions (Jullien 1986) or from the relationship between the aggregation number N and the averaged aggregate size.

4.2.3.3 Densification of Aggregates

One of the most popular features of fractal aggregates is the independence of the aggregate structure on size. However, Gmachowski (2002) pointed out that this does not hold true for small fractal aggregates. In particular, he examined the correlation between the aggregation number N and the geometric aggregate radius

R_C, which should appear as a straight line in a log-log plot. Yet, the slope of this curve is not constant and indicates a steady densification during the aggregate growth. The effect is especially pronounced for particle-cluster aggregation mechanisms, where free individual particles have the chance to enter the inner part of an existing aggregate even at later stages of aggregation. In contrast, the penetration of one aggregate into the inner region of a second aggregate is less likely due to steric reasons. Therefore, densification is only of minor relevance for ideal cluster-cluster aggregation processes.

Regardless of this difference, the densification of aggregates is real and cannot be completely ignored. In fact, it contradicts the concept of fractality, which assumes self-similarity, i.e. a scale-invariant morphology. Moreover, it raises the question of whether and how the fractal dimension should be defined in the presence of densification. The answer is very pragmatic: For individual methods, like box-counting and fragmentation, the history of an aggregate has no effect for the quantification of the fractal dimension. It is important only for the individual or global aggregation growth monitoring, i.e. the analysis of the correlation between aggregation number N and aggregate size (e.g. R_g); the average slope of this curve in a log-log plot is then defined as the fractal dimension d_f for aggregates of mass N. The alternative determination of d_f via the pair correlation function is meaningful only for large aggregates ($R_g/a \gg 1$) and does not help to quantify densification.

The effect of densification is illustrated in Fig. 4.6, where the fractal dimension of numeric DLCA and RLCA aggregates are plotted versus the aggregation number N. The values of the fractal dimension were derived with different methods which do no yield identical results. The data reveal a significant densification for $N < 1000$—despite the data scatter, which is due to the limited size of the aggregate population (100 aggregates for each N). Additionally, densification appears more pronounced for RLCA than for DLCA. This is related to the fact that the cluster mass distribution of an RLCA process is dominated by single particles and small aggregates, whereas during a DLCA process single particles and small aggregates disappear first (cf. Sect. 4.2.2.1). Consequently, in its initial stages, RLCA is mainly

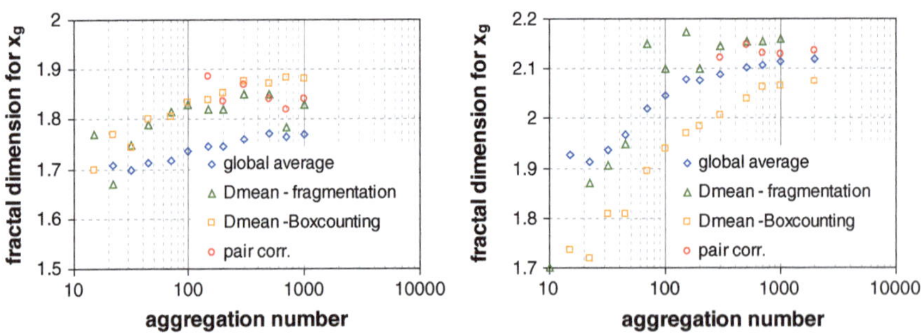

Fig. 4.6 Fractal dimension from growth of $x_g = 2 \times R_g$, from fragmentation, from box-counting, and from pair-correlation versus aggregation number, each time averaged from 100 aggregates; *left* DLCA aggregates, *right* RLCA aggregates

a particle-cluster aggregation process, and, thus, prone to considerable densification, whereas DLCA is fairly similar to an ideal cluster-cluster aggregation process.

Nevertheless, the implications of the densification have to be considered for both types of aggregates because the assumption of a size-independent fractal dimension can considerably mislead the computation of physical aggregate properties or the interpretation of experimental data.

4.2.3.4 Distribution of Structural Parameters

Densification is only one aspect in which fractal aggregates deviate from fractal mathematical objects. Another is the variation of fractal properties within a population of aggregates. This is not really surprising since the statistical meaning of their fractality has already been emphasised. The variation of structure can be examined on numeric aggregates via the individual fractal dimensions as determined with the box-counting method and the fragmentation method. For DLCA and RLCA aggregates, the cumulative frequencies of the individual fractal dimensions are shown in Fig. 4.7. The curves have been determined for aggregate populations with uniform mass (i.e. N = const.) and reveal a rather broad variation of d_f for $N < 100$ regardless of the aggregation mechanism. The correlation between the aggregation number and the spread of the structural properties is quantitatively shown in the right graph of Fig. 4.7. This graph further states an effect of the analysing method on the distribution width; it does not, however, change the nature of correlation.

In summary, the statistic variation of the fractal dimension within a population of aggregates is a function of the aggregation number: While for large N a uniform structure can be assumed, the variation of the fractal dimension is not negligible for small aggregation numbers ($N < 100$) and has to be appropriately considered when studying the physical properties of aggregates.

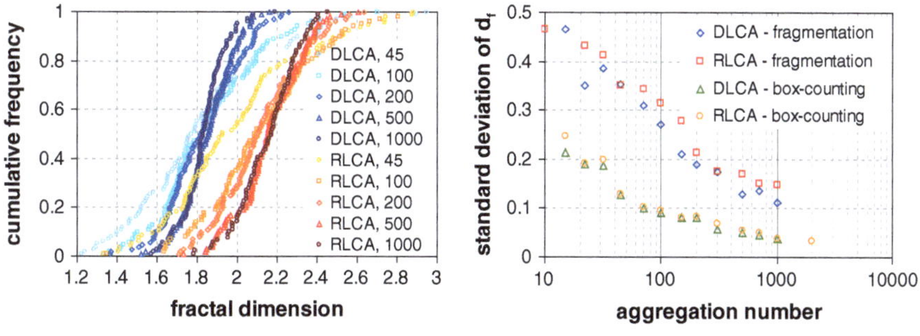

Fig. 4.7 Distribution of the fractal dimension within populations of numeric DLCA and RLCA aggregates; *left* cumulative frequencies for d_f for monodisperse aggregates determined by the fragmentation method; *right* standard deviation of the d_f-distribution determined by fragmentation and box-counting

4.2.3.5 Fractal Prefactor

Frequently, the structure of aggregates is solely described by their fractal dimension. However, this parameter does not provide a complete description of the aggregate structure.

Consider an aggregation process where particles arrange in regular lattices. These crystal-like aggregates obviously have a fractal dimension of 3 because the particles are homogenously distributed in space, i.e. the pair correlation is independent from the inter-particle distance. Nevertheless, we can imagine different kinds of lattice structures, e.g. a simple cubic (sc) system with 48 % porosity or a hexagonal close-packed (hcp) structure with 26 % porosity. That means that aggregates of the same fractal dimension and same geometric size may contain a different mass, i.e. they consist of different numbers of primary particles. For instance, hcp-aggregates contain approx. 42 % more particles than equally sized sc-aggregates.

For this reason, it is necessary to explicitly quantify the proportionality between aggregation number and scaled aggregate size in Eq. (4.9):

$$N = k_Z \cdot (R_Z/a)^{d_Z} = k_Z \cdot \left(x_Z/x_p\right)^{d_Z}. \tag{4.32}$$

The parameter k_Z is called fractal prefactor. In analogy to the fractal dimension d_Z, it is always related to a morphological property Z (e.g. "g" for gyration) that has been used to evaluate the aggregate size.

The fractal prefactor allows the quantification of the size or mass of fractal aggregates with a known fractal dimension. Moreover, it is required when numeric aggregates with a specified fractal structure are to be created in order to imitate real aggregates without the problem of structure variation (cf. Sect. 4.1.3). For these purposes, the fractal prefactors and their interrelation with aggregate size and fractal dimension have to be known.

It can be shown that the fractal prefactor correlates with the fractal dimension in a manner that is specific for the aggregation mechanism. Figure 4.8 shows the

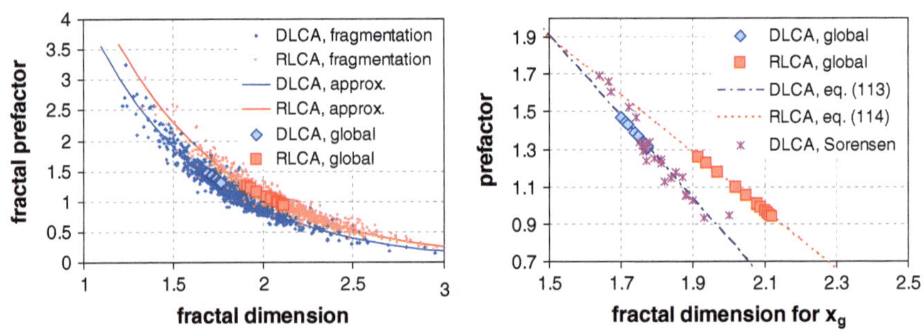

Fig. 4.8 Correlation between fractal dimension and fractal prefactor; *left* global values (from fractal growth, Eq. (4.32)) and individual prefactors (from method of fragmentation); *right* comparison with Sorensen and Roberts (1997) and linear fit

individual fractal prefactors k_g of DLCA and RLCA aggregates (N = 10–1000) obtained with the method of fragmentation. Additionally, the global prefactors are depicted, which do not differ significantly from the individual ones. The apparent anticorrelation between k_g and d_g results from geometric effects and has been discussed in detail by Sorensen and Roberts (1997). These authors could further show that, even for arbitrary aggregation processes, the prefactors anticorrelate with d_g and lie within a limited range of real numbers.

While the correlation between k_g and d_g is clearly non-linear for the individual prefactors, it can be considered linear for the narrow ranges of the global fractal properties (Fig. 4.8):

$$k_g^{DLCA} = 5.1900 - 2.1874 \cdot d_g, \tag{4.33}$$

$$k_g^{RLCA} = 4.2149 - 1.5433 \cdot d_g. \tag{4.34}$$

Note that the results of Sorensen and Roberts (1997) for different DLCA populations fit very well to the values obtained in this work (Fig. 4.8).

4.2.3.6 Final Remarks on Fractality

The fractal dimension is a powerful tool to describe and classify particle aggregates because the relevant morphological quantities are condensed in just one parameter. The subsequent sections will show that the fractal dimension is crucial for various physical properties of aggregates. That means that it affects measurable properties of aggregate suspensions and is, thus, accessible by experimental determination.

However, there is no unique definition of the fractal dimension. It can be related to the pair-correlation function or to the dependency of aggregate mass on aggregate size (the latter being an ambiguous term as well). Moreover, the fractal dimension can be derived for single aggregates or determined at an aggregate population. Finally, real aggregates are fractal in a statistical sense. They resemble mathematical fractal objects only for large aggregation numbers ($N > 1000$), while the concept of fractality may even fail for small aggregation numbers (in particular for $N < 10$).

4.3 Scattering on Aggregates

The scattering of visible light, X-rays, or neutron beams at particle aggregates is generally affected by mass and structure of the aggregates as well as by the size of the primary particles. Measuring the intensity of scattering signals as a function of scattering angle or beam energy (static scattering techniques), therefore, offers a way for aggregate characterisation. This requires a proper understanding as well as quantitative models on the phenomenon of scattering. Additionally, such knowledge is necessary if scattering signals are used to quantify the motion or concentration of particle aggregates (e.g. with dynamic light scattering).

This section gives a brief introduction into the scattering theory for aggregates with the focus being on the colloidal size range. Only diluted aggregate systems are considered. The general idea is that the scattering signal from one aggregate results from the interference of the scattered waves generated at all primary particles. These waves interact with each other depending on the interparticle distance and the scattering properties of the particles (*intracluster multiple scattering*, Sorensen 2001). For weak scatterers, this effect can be neglected (Rayleigh-Debye-Gans scattering), and the scattering at simple particle geometries and at particle aggregates can be expressed by analytical equations. If such a simplification is not possible, the scattering at aggregates has to be calculated by computationally expensive approaches (e.g. finite element method or discrete dipole approximation, cf. Jacquier and Gruy 2010). For aggregates of spherical particles, the scattering can be rigorously computed by means of the generalised multi-particle Mie theory (GMM), which decomposes the electromagnetic fields by a multipole expansion (Mackowski 1991; Xu 1995). The section closes with a look on the scattering behaviour of polydisperse aggregate populations.

4.3.1 Scattering on Aggregates in the Rayleigh-Debye-Gans Limit

4.3.1.1 Rayleigh-Debye-Gans (RDG) Scattering of Single Aggregates

The term *Rayleigh-Debye-Gans scattering* (RDG scattering) applies to a situation where each volume element experiences the same incident field, i.e. within the whole scattering volume there is no difference in amplitude and phase to the original field. Hence, the scattered fields from all volume elements are identical and independent from each other. However, when the scattered fields are received by a detector in the far field, they have an individual phase shift $e^{i\mathbf{qr}}$ because the total path length of the incident and the scattered wave depends on the location $\mathbf{r}$ of the volume element in space. That means for vanishing scattering vectors $\mathbf{q}$ (cf. Eq. (2.17)), that the amplitudes of the scattered fields simply add up (constructive interference) and the total scattering volume acts as a Rayleigh scatterer. In any other case, there is a partially destructive interference and the intensity of the detected scattering signal is obtained by:

$$I_{\text{sca}}(\mathbf{q}) = I_{\text{sca,Rayleigh}}(\mathbf{q}) \cdot P(\mathbf{q}), \tag{4.35}$$

where

$$\sqrt{P(\mathbf{q})} = \frac{1}{V} \cdot \oint e^{i\mathbf{qr}}\, d\mathbf{V} \leq 1. \tag{4.36}$$

The factor $P(\mathbf{q})$ or its orientation average $P(q)$ is called *form factor*.

The assumption of independent scattering within the scattering volume is fulfilled if the refractive indices of particles and continuum are nearly identical ($|m_r - 1| \ll 1$) and if the phase shift in the particles with respect to the incident wave is negligibly small ($ka \cdot |m_r - 1| \ll 1$). Both conditions hold true for SAXS measurements of colloidal particles because the refractive indices of any material are extremely close to 1 in the X-ray domain of electromagnetic radiation (cf. Sect. 2.3.4; Henke et al. 1993). For light scattering, the validity of the RDG approximation strongly depends on the material and particle size because the optical contrast is usually quite significant ($|m_r - 1| = 0.01$–3).

For instance, the optical contrast for silica particles suspended in water is $|m_r - 1| \approx 0.09$; the analogue value for titania particles is approx. 1. Hence, RDG scattering cannot, in principle, be applied to titania suspension. It is, however, possible to assume Rayleigh scattering when the titania particles are very fine ($ka \ll 1$, $x \leq 50$ nm). In contrast, optical scattering in aqueous silica suspensions can be approximated by RDG scattering for particle sizes up to 200 nm.

In the context of aggregates, the term RDG scattering refers to weakly scattering primary particles; thus, that the incident field on a primary particle is not affected by the scattered field of neighbouring particles (otherwise there is intracluster multiple scattering). Hence, all particles in the aggregate are exerted to the same field and scatter independently from each other as if they were isolated. The total scattering field of the aggregate $E_{sca,agg}$ results from the interference of the particle scattering fields $E_{sca,k}$:

$$E_{sca,agg} = \sum E_{sca,k} = \sum \frac{E_0}{|\Delta \mathbf{r}_k|} e^{i(\mathbf{k}_e \mathbf{r}_D - \omega t)} \cdot a_k(\theta) \cdot e^{i\mathbf{q}\mathbf{r}_k}, \tag{4.37}$$

where a_k is the field amplitude for particle k. It is proportional to the polarisability and, hence, to the particle's volume (cf. Appendix B.2). The intensity of this field is defined as the product $E_{sca,agg}E^*_{sca,agg}$; for identical primary particles, it can be expressed as:

$$\begin{aligned} I_{sca,agg} &= K \cdot a(\theta)^2 \cdot \sum \sum e^{i\mathbf{q}(\mathbf{r}_k - \mathbf{r}_l)} = I_{sca,p} \cdot \left(N + \sum_{k \neq l} e^{i\mathbf{q}\mathbf{r}_{k,l}} \right) \\ &= I_{sca,p} \cdot N^2 \cdot S_{agg}(\mathbf{q}). \end{aligned} \tag{4.38}$$

In this equation, $I_{sca,p}$ is the scattering intensity of the primary particles, which is proportional to the form factor $P(\mathbf{q})$ when RDG scattering applies. The factor $S_{agg}(\mathbf{q})$ reflects the aggregate structure and depends on the aggregate orientation.

Commonly, the aggregates are not uniformly aligned. Moreover, their orientation is steadily changing due to migration as well as shear and thermal motion (Brownian rotation). For these reasons, usually the orientation average of Eq. (4.38) is considered, which can be derived by means of the Debye equation (Debye 1915):

$$\left\langle e^{i\mathbf{qr}} \right\rangle_{\Omega} = \frac{\sin qr}{qr} \tag{4.39}$$

leading to

$$I_{\text{sca,agg}}(\theta) = I_{\text{sca,p}}(\theta) \cdot \sum_{k,l} \frac{\sin(qr_{kl})}{qr_{kl}} = I_{\text{sca,p}}(\theta) \cdot N^2 \cdot S_{\text{agg}}(q). \tag{4.40}$$

In this equation, the parameter $S_{\text{agg}}(q)$ is the *static structure factor* which was introduced in Sect. 4.2.2.4. By assuming that the particle scattering obeys the RDG approximation, the scattering intensity can be written as:

$$I_{\text{sca,agg}}(q) = K \cdot N^2 a^6 \cdot S_{\text{agg}}(q) \cdot P(q). \tag{4.41}$$

This expression reveals the main features of aggregate scattering: For vanishing q and/or vanishing aggregate size ($qx_g \ll 1$), the aggregate behaves as a Rayleigh scatterer with the scattering intensity proportional to the squared aggregate volume. However, since the factor K includes material properties and measurement parameters, the aggregate volume cannot usually be measured absolutely. What is more interesting is that the asymptotic behaviour of the scattering curve at small arguments ($qx_g < 1$) can be used to determine the aggregate's radius of gyration:

$$\lim_{q \to 0} I_{\text{sca,agg}}(q) \propto \exp\left(\frac{1}{3} q^2 R_{g,c}^2\right) \cdot \exp\left(\frac{1}{3} q^2 R_{g,p}^2\right) = \exp\left(\frac{1}{3} q^2 R_g^2\right). \tag{4.42}$$

Furthermore, at moderate q-values ($x_g^{-1} < q < x_p^{-1}$), the aggregate scattering is additionally affected by the interparticle distances r_{kl}. In the case of fractal aggregates, this leads to a power law dependency of type $I_{\text{sca,agg}}(q) \propto q^{-d_f}$. In this structure-dominated scattering regime, the primary particles have virtually no influence on the shape of the scattering curve ($P(q) \to 1$). This is different for large arguments ($qx_p \geq 1$) where the shape of the scattering function $I_{\text{sca}}(q)$ is solely affected by the size and shape of primary particles.

4.3.1.2 Light Scattering from Dilute Aggregate Suspensions in the RDG Limit

The equations above refer to the scattering of single aggregates, yet in practice one usually measures the scattering with a large number of aggregates. For dilute suspensions, the total scattering intensity I_{sca} is just the sum of all individual aggregate signals $I_{sca,agg}$ (Eq. (4.40)):

$$I_{sca} = c_{N,agg} \cdot V_{meas} \cdot I_{sca,agg} = c_{N,agg} \cdot V_{meas} \cdot N^2 I_{sca,p} S_{agg} \propto c_{N,agg} \cdot N^2 V_p^2 S_{agg} P, \tag{4.43}$$

where V_{meas} denotes the illuminated suspension volume, and $c_{N,agg}$ the number concentration of the aggregates. The latter parameter is usually unknown and changes during aggregation or dispersion procedures. It is related to the number concentration of all primary particles $c_{N,p}$ via the aggregation number N ($c_{N,agg} = N \times c_{N,p}$), which in turn is related to the aggregate size ($N \propto \left(x_{agg}/x_p\right)^{d_f}$). Note that the form factor P can be set to 1 if $q < x_p^{-1}$.

This equation yields rather intriguing results for the two limiting cases of aggregate scattering. If the aggregates are very small (Rayleigh scattering, $qx_{agg} \ll 1$), and the static structure factor S_{agg} approaches 1, then the total scattering from the suspensions is:

$$I_{sca} \propto c_{V,p} \cdot N \cdot V_p \propto c_{V,p} \cdot \left(x_{agg}/x_p\right)^{d_f} \tag{4.44}$$

with $c_{V,p} = V_p \times c_{N,p}$. That means that the relationship between scattering intensity I_{sca} and aggregate size x_{agg} allows a direct determination of the aggregates' fractal dimension d_f, provided that there is neither growth nor dissolution of the primary particles.

In the fractal regime of aggregate scattering (i.e. for $x_{agg}^{-1} < q < x_p^{-1}$), the static structure factor declines non-linearly with aggregate size ($S_{agg} \propto \left(qx_{agg}\right)^{-d_f} \propto N^{-1}$), and Eq. (4.43) takes:

$$I_{sca} \propto c_{N,p} \cdot I_{sca,p} \propto c_{V,p} \cdot V_p \cdot \left(qx_p\right)^{-d_f}, \tag{4.45}$$

which is constant as long as the primary particles are not changed (e.g. during dispersion).

These two distinct scattering regions can be observed, e.g., during the aggregation of very fine colloidal particles. Figure 4.9 plots such experimental data for a suspension of colloidal silica (approx. 10 nm, 0.1 wt%, pH = 9) which was destabilised by the addition of NaCl (9 wt%); hydrodynamic aggregate size and suspension scattering intensity were determined by means of dynamic light scattering. There is a clear power-law relationship between particle size and scattering intensity for small aggregates below 100 nm, which yields a fractal dimension of 1.85. The close similarity of this value to that of ideal DLCA aggregates (cf. Sect. 4.1.2) is not surprising, since the high salt concentration effectively reduces the electrostatic repulsion between the charged particle surfaces. Moreover, after having reached aggregate sizes above 200 nm, there is hardly any increase in the suspension scattering intensity, which confirms the prediction above (Eq. (4.45)).

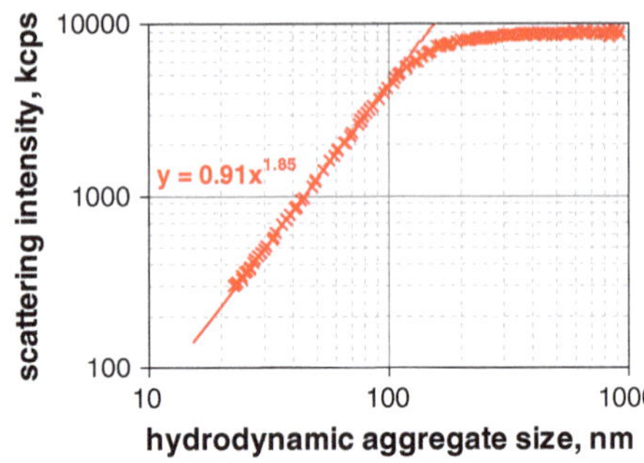

Fig. 4.9 Increase of the total scattering intensity during the aggregation of colloidal silica (approx. 10 nm, 0.1 wt% in 9 wt% NaCl, pH = 9); size and intensity were measured with dynamic light scattering (Babick et al. 2013)

4.3.2 *Light Scattering on Aggregates with Intracluster Multiple Scattering*

Although universally valid for SAXS and SANS measurements of colloidal aggregates, the RDG approximation does not necessarily apply to their light scattering, which may be significantly affected by intracluster multiple scattering. Moreover, the light scattering on the primary particles may considerably deviate from Rayleigh scattering (see above). An accurate computation of the light scattering on aggregates needs to solve the Maxwell equations for the electromagnetic fields for the specific aggregate morphology. Several numerical methods have been proposed for that problem, e.g. the finite element method or the volume integral equation method (cf. Jacquier and Gruy 2010). This section will discuss the generalised multi-particle Mie theory (GMM)—a method based on a separation of variables, similar to that of the Mie's solution for light scattering on single spheres (Mie 1908). Since the computational costs for exact solutions of light scattering on aggregates are usually high, the necessity of using numerical methods is tested for fractal aggregates in the colloidal size range.

4.3.2.1 Generalised Multi-particle Mie Theory (GMM)

The *generalised multi-particle Mie theory* (GMM) is based on a multipole expansion of the scattering fields at each primary particle (Mackowski 1991; Xu 1995). Each multipole (monopole, dipole, quadrupole, etc.) is a solution of Maxwell's equations, which describe the propagation of electromagnetic waves. The "weights" of these poles are adjusted in such a way that the boundary conditions (e.g. continuity of the tangential component of the electric field on the particle surface) hold true. The mathematical framework of the GMM relies on the simple formulation of the boundary conditions for spherical shape. This theory is, therefore, restricted to clusters and aggregates of non-sintered spheres. There is, however, no limitation with regard to particle size, aggregate structure, and material.

The concept of GMM and the details of computation were presented by Yu-Lin Xu in a series of papers (e.g. Xu 1995, 1997, 1998a). The main idea is to express the scattering field of the particle j as an infinite series of spherical vector wave functions (SVWF) $\mathbf{M}_{mn}^{(3)}$ and $\mathbf{N}_{mn}^{(3)}$ (i.e. spherical harmonics for vector fields) that are defined for spherical coordinates centred in particle j. In terms of the electric field, it reads:

$$\mathbf{E}_{\text{sca}}(j) = \sum_{n=1}^{\infty} \sum_{m=-n}^{n} \mathrm{i} \cdot E_{mn} \left[a_{mn}^{j} \mathbf{N}_{mn}^{(3)} + b_{mn}^{j} \mathbf{M}_{mn}^{(3)} \right]. \tag{4.46}$$

The factor E_{mn} is proportional to the amplitude of the incident field E_0. The parameters a_{mn}^j and b_{mn}^j are the coefficients of the multipole expansion which are to be attained. In a similar way, one can derive an expansion for the incident field on the primary particle j:

$$\mathbf{E}_{\text{inc}}(j) = -\sum_{n=1}^{\infty} \sum_{m=-n}^{n} \mathrm{i} \cdot E_{mn} \left[p_{mn}^{j} \mathbf{N}_{mn}^{(1)} + q_{mn}^{j} \mathbf{M}_{mn}^{(1)} \right]. \tag{4.47}$$

Note that the SVWFs of first kind ($\mathbf{M}_{mn}^{(1)}$ and $\mathbf{N}_{mn}^{(1)}$) have to be used for the incident field because, unlike the scattered field, it does not vanish at infinity. The parameters p_{mn}^j and q_{mn}^j are the corresponding expansion coefficients.

The relationship between the incident and scattered field for a single sphere is known as Mie solution (Mie 1908), which is based on a multipole expansion with the corresponding scattering coefficients a_n^j and b_n^j. For aggregate, this leads to:

$$a_{mn}^{j} = a_n^j \cdot p_{mn}^j \quad \text{and} \quad b_{mn}^j = b_n^j \cdot q_{mn}^j. \tag{4.48}$$

Hence, a solution for the aggregate scattering is achieved once the incident field on the primary particle j is known. This incident field, however, results from the original field E_0 and the scattered fields $E_{\text{sca}}(l)$ from all other particles $l \neq j$:

$$\mathbf{E}_{\text{inc}}(j) = \mathbf{E}_0(j) + \sum_{l \neq j} \mathbf{E}_{sca}(l,j). \tag{4.49}$$

The contribution of E_0 to $E_{\text{inc}}(j)$ can be described by an expansion of type (4.47), but p_{mn}^j and q_{mn}^j have to be replaced by p_{mn}^{jj} and q_{mn}^{jj}. The scattered field from particle l, $E_{\text{sca}}(l,j)$, when seen from particle j (i.e. spherical coordinates in j), is also expanded like an incident field of type (4.47)—quantified by the coefficients p_{mn}^{lj} and q_{mn}^{lj}.

On the other hand, $E_{\text{sca}}(l)$ can be expressed as a scattered field seen from particle l, i.e. $E_{\text{sca}}(l)$ can be described within the coordinates of l by Eq. (4.46) when a_{mn}^j and b_{mn}^j are replaced by a_{mn}^l and b_{mn}^l. That means the incident field on particle j can be expressed in terms of the scattering coefficients a_{mn}^l and b_{mn}^l of all particles $l \neq j$ when the VSWFs (the multipoles) on l can be expressed in spherical coordinates j. This is achieved by employing the vector translation theorems developed by Stein (1961) and Cruzan (1962). This transformation of coordinates is a linear operation which relates the m, n-poles of particle j with the μ, ν-poles of particle l by the coefficients $A^{lj}_{mn\mu\nu}$ and $B^{lj}_{mn\mu\nu}$ that are solely defined by the aggregate structure.

Eventually, a system of linear equations is obtained that allows for a solution of the unknown expansion coefficients a^j_{mn} and b^j_{mn}:

$$a^j_{mn} = a^j_n \left[p^{jj}_{mn} - \sum_{l \neq j}^{N} \sum_{\nu=1}^{N^l_W} \sum_{\mu=-\nu}^{\nu} \left(a^l_{\mu\nu} A^{lj}_{mn\mu\nu} + b^l_{\mu\nu} B^{lj}_{mn\mu\nu} \right) \right], \tag{4.50}$$

$$b^j_{mn} = b^j_n \left[q^{jj}_{mn} - \sum_{l \neq j}^{N} \sum_{\nu=1}^{N^l_W} \sum_{\mu=-\nu}^{\nu} \left(a^l_{\mu\nu} B^{lj}_{mn\mu\nu} + b^l_{\mu\nu} A^{lj}_{mn\mu\nu} \right) \right]. \tag{4.51}$$

In these equations, the multipole expansion is terminated at N^j_W, which is Wiscombe's termination criterion of the Mie solution for the primary particle j ($N_W = ka + 4 \cdot \sqrt[3]{ka} + 2$, Wiscombe 1980). Hence, the number of unknowns, $2 \times N \times N_W(ka) \times (N_W(ka) + 2)$, depends on the size and number of the primary particles. The size of the coefficient matrix scales with with N^2 and N^4_W.

Once the coefficients a^j_{mn} and b^j_{mn} are known, it is possible to compute the amplitudes of the scattered wave and the relevant cross sections. For instance, the total extinction cross section is calculated from:

$$C_{ext} = \sum_{l=1}^{N} C^l_{ext} = \frac{4\pi}{k^2} \sum_{l=1}^{N} \sum_{n=1}^{N^l_W} \sum_{m=-n}^{n} \mathrm{Re}\left(\left(p^l_{mn}\right)^* a^l_{mn} + \left(q^l_{mn}\right)^* b^l_{mn} \right). \tag{4.52}$$

More details are found in Appendix B.2.5.

Already in his first paper on GMM, Xu (1995) compared his theoretical calculations by experimental data from microwave scattering (ka = 3.1, 4.4) on doublets and triplets (Wang and Gustafson 1984). He found excellent agreement with regard to the scattering profile. Later publications of Xu verified the GMM with microwave scattering on compact clusters ($N \leq 27$; $ka = 5, 7$) and linear chains of spheres ($N \leq 5$; $ka \leq 4.4$) with regard to scattering profile and the orientation-averaged asymmetry parameter (Xu and Gustafson 1997; Xu 1998b). Kätzel et al. (2005) described light extinction experiments on colloidal suspensions of stable doublets and triplets (and primary particles) and found agreement with GMM calculation as well.

4.3.2.2 Relevance of Intracluster Multiple Scattering

Although numeric methods, such as GMM, facilitate the exact computation of aggregate scattering, they are not suited for routine calculations because of their high computational costs. If the intracluster multiple scattering is not huge, approximate methods like the *coupled-dipole* approach (Purcell and Pennypacker 1973; Singham and Bohren 1993) or the *mean-field-approximation* (Berry and Percival 1986; Khlebtsov 1996) can be used to compute light scattering on aggregates. From a practical point of view, however, one would prefer that the intracluster multiple scattering can be ignored, thus when the RDG approximation applies. For aggregates of fine colloidal particles, in particular for nanoparticles ($x < 100$ nm), this is not unlikely to occur because particles are smaller than or in the range of the wavelength and their scattering is weak.

The decisive parameter for any scattering phenomenon is ka, i.e. the ratio between the particle perimeter and the wavelength in the continuous phase. For light scattering ($\lambda \approx 0.5$ µm) and particles of 100 nm diameter, this parameter is $ka \approx 0.6$.

This means that the multiple scattering inside the aggregates of such particles is weak as well. For $ka < 1$, the effect of multiple scattering to the aggregate scattering can be estimated with the mean-field approach of Berry and Percival (1986). This theory assumes a uniform field inside the aggregate, which differs from the originally exciting field due to phase shift and absorption in the particles. In the notation of Eq. (4.41), the aggregate scattering in the presence of intracluster multiple scattering can be written (Khlebtsov 1996):

$$I_{\text{sca,agg}}(q) = K \cdot |d|^2 \cdot N^2 a^6 \cdot S_{\text{agg}}(q) \cdot P(q), \tag{4.53}$$

where $|d|^2$ is a factor associated with the intracluster multiple scattering. It is calculated from:

$$d^{-1} = \left(N - \frac{3}{2} \sin\delta \exp(\mathrm{i}\delta) \sum_{i \neq j} P_1(kr_{ij}) \right) \Big/ N \tag{4.54}$$

with

$$\delta = \frac{2}{3} \mathrm{i} S_1(0) = \frac{2}{3} (ka)^3 \frac{m_\mathrm{r}^2 - 1}{m_\mathrm{r}^2 + 2} \tag{4.55}$$

and

$$P_1(z) = \frac{\exp(\mathrm{i}z)}{z} \left(j_0(z) - \frac{1}{z} \cdot j_1(z) + \frac{1 - \mathrm{i}z}{z^2} \cdot j_2(z) \right), \tag{4.56}$$

where j_k are the spherical Bessel functions of first kind.

Plots of the factor $|d|^2$ are shown in Fig. 4.10 as a function of ka for different types of aggregates and for different types materials (non-absorbing: titania,

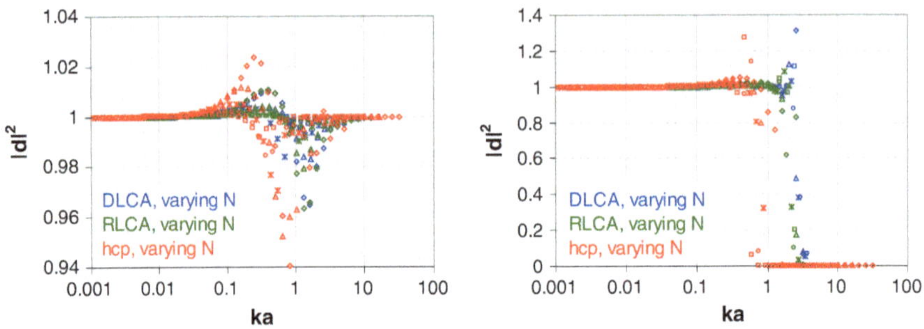

Fig. 4.10 Influence of intracluster multiple scattering on scattering intensity according to the mean field theory, correction factor $|d|^2$ from Eq. (4.54) versus scaled primary particle size *ka* for different aggregate types and varying aggregation number (N = 10, 32, 100, 300 and 1000); *left* TiO_2 in water; *right* carbon black in air

absorbing: carbon black). According to these graphs there is virtually no impact of intracluster multiple scattering for $ka \leq 0.1$—independent of the type of aggregate. Above this value there is little effect for non-absorbing materials, which is most pronounced for hcp aggregates. In contrast, there is an enormous effect for absorbing material, which becomes relevant for $ka \geq 0.4$ and $ka \geq 1$ for hcp and for fractals aggregates, respectively. There is a small, but not decisive impact of the aggregation number. Finally, it should be noted that the mean field approximation should not be applied to $ka > 1$.

Similar results were obtained by Berry and Percival (1986) and Singham and Bohren (1993). The former concluded from an analytical analysis of the mean-field theory that multiple scattering can be neglected for fractal aggregates with $d_f < 2$. The latter found that multiple scattering is virtually insignificant for carbon black aggregates with $d_f < 2.5$ ($ka = 0.3$, $N \leq 1000$).

The main conclusion from these results is that the light scattering on fractal aggregates (DLCA, RLCA type) obeys the RDG approximation when the primary particles are sufficiently small ($ka \leq 1$). However, this has to be verified with exact computation. For that purpose, the GMM algorithm was used to calculate the light scattering properties of numeric populations of fractal aggregates, which differed in size and number of primary particles (x_p and N, respectively). The calculation yields different aggregate parameters, like the partial and total scattering cross sections ($C_{psca}(\theta, \phi)$, C_{sca}, Sect. B.2.5) or the extinction cross section C_{ext} (Eq. (4.52)). In Fig. 4.11, the latter is used as reference for the corresponding values from the RDG approach. The results shown there refer to DLCA and RLCA aggregates with primary particles from 10 nm ($N \leq 500$) up to 100 nm ($N \leq 100$). Two material combinations with low and high optical contrast (silica and carbon black, respectively, dispersed in water; wavelength set to 632 nm) are considered separately.

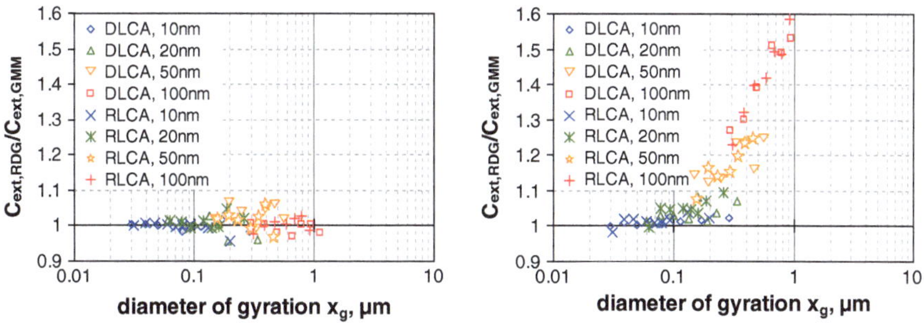

Fig. 4.11 Comparison of RDG scattering with GMM via ratio of the corresponding extinction cross sections C_{ext} for DLCA and RLCA aggregates; variation of the primary particle size and of the aggregation number N; silica (*left*) and carbon black (*right*) in water

The GMM confirms the previous statement that the intracluster multiple scattering (thus the deviation from RDG) increases with primary particle size x_p and with the optical contrast $|m_r - 1|$. Additionally, one observes that the effect of the intracluster multiple scattering steadily rises[5] with the aggregation number N, while a significant difference between the two aggregate types is not found. The most impressive result is the virtual agreement between RDG and GMM for the weakly scattering material (silica in water), in which case the deviations between both approaches is less than 10 % for aggregates up to $x_g = 1$ μm. In contrast, the differences rise to 60 % in that size range for carbon black aggregates with 100 nm primary particles. Note that similar results are obtained when the total scattering cross section is considered and that, in contrast to Fig. 4.10, the deviation between GMM and RDG for non-absorbing materials with high optical contrast (e.g. titania in water) is similar to that of carbon black Babick et al. (2012a).

In summary, the relevance of intracluster multiple scattering is mainly determined by the size of the primary particles, their optical behaviour, and the aggregate packing density. For fractal aggregates (e.g. DLCA and RLCA) of nanosized primary particles ($x_p \leq 100$ nm), the light scattering can be reliably calculated by the RDG approach if the optical contrast of the material is low (e.g. silica in water). For strongly scattering materials, the impact of intracluster multiple scattering grows with the aggregation number and, thus, limits the validity of RDG to aggregate sizes well below 1 μm (dependent on x_p). In spite of this, the RDG approach does not completely fail in predicting the light scattering behaviour of fractal aggregates with $x_p \leq 100$ nm. It thus provides a good estimate for the functional dependency of scattering on morphology parameters of such aggregates over the whole submicrometre range.

[5]The slight data scatter results from the low number of orientations (15) used for averaging in GMM calculations.

4.3.3 Scattering on Polydisperse Aggregate Populations

The previous sections considered the scattering on aggregates with defined aggregation number N and primary particle size (radius a or diameter x_p). In practice, however, one encounters aggregate systems with varying aggregation number and polydisperse primary particles; even within an aggregate the primary particles may differ in size. This results from the stochastic nature of real aggregation processes, which was already discussed in Sect. 4.2.2.1. Similarly, the dispersion of aggregates, even if we assume uniformity in the initial mass and size, yields a distribution of the aggregate parameters. Moreover, the polydispersity of the primary particles is strongly related to the context of the aggregation process. For instance, when coagulation is induced in a colloidal suspension, the polydispersity of the primary particles is determined by the employed particle system. When aggregation is part of a crystallisation process, it is affected by reaction kinetics, material properties, and process parameters.

It is obvious that a meaningful analysis of scattering experiments requires a profound understanding of the effect of aggregate polydispersity. For that purpose, this section assumes an aggregate population in which the primary particles within a given aggregate are of virtually uniform size—diameter $x_{p,av}$ or corresponding radius a_{av}. Additionally, a correlation is assumed between this average particle size and the aggregation number N. Hence, the variation of the aggregation number N can be expressed by a conditional probability function $p(N|a)$, with $\int p(N|a)\,da = 1$. Within the notation of Eq. (4.41), the scattering intensity of such an aggregate population can be calculated with:

$$I_{sca}(q) = c_{N,agg} V_{meas} \cdot K \iint N^2 a^6 \cdot S_{agg}(q) \cdot P(q) p(N|a) p(a) \mathrm{d}N \mathrm{d}a, \tag{4.57}$$

where $p(a)$ is the probability function, i.e. the number weighted distribution function q_0 of the average particle radius a. A possible correlation between a_{av} and the aggregation number N can be used to replace $p(N|a)$ with $p(N|\overline{N}(a_{av}))$, e.g. given in Eq. (4.16) for DLCA aggregates.

From Eq. (4.57), several characteristic features of the scattering curve $I_{sca}(q)$ can be derived. For instance, the intercept with the ordinate is:

$$I_{sca}(0) = c_{N,agg} V_{meas} \cdot K\overline{N^2 a_{av}^6}, \tag{4.58}$$

while the effective radius of gyration $R_{g,eff}$, which is determined from a Guinier-plot analysis of $I_{sca}(q)$, is obtained by:

$$R_{g,eff}^2 = \iint R_g^2 N^2 a^6 p(N|a) p(a) \mathrm{d}N \mathrm{d}a \Big/ \iint N^2 a^6 p(N|a) p(a) \mathrm{d}N \mathrm{d}a. \tag{4.59}$$

Equations (4.57) and (4.59) are of particular benefit for fractal aggregates for which simple expressions can be used to describe the structure factor $S_{\mathrm{agg}}(q)$ or the radius of gyration (cf. Sect. 4.2.2). In that case, the effective radius is defined as:

$$R_{\mathrm{g,eff}}^2 = k_{\mathrm{g}}^{-2/d_{\mathrm{f}}}\overline{N^{2+2/d_{\mathrm{f}}}a_{\mathrm{av}}^8}\Big/\overline{N^2a_{\mathrm{av}}^6},\ x_{\mathrm{g,eff}}^2 = k_{\mathrm{g}}^{-2/d_{\mathrm{f}}}\cdot\overline{N^{2+2/d_{\mathrm{f}}}\cdot x_{\mathrm{p,av}}^8}\Big/\overline{N^2}\cdot\overline{x_{\mathrm{p,av}}^6}. \tag{4.60}$$

Furthermore, it has already been shown that the shape of the static structure factor $S_{\mathrm{agg}}(q)$ depends on the fractal dimension d_{f} at moderate q-values ($x_{\mathrm{g}}^{-1} \ll q \ll x_{\mathrm{p,av}}^{-1}$). This also holds true for polydisperse systems:

$$I_{\mathrm{sca,fractal}}(q) = c_{N,\mathrm{agg}}V_{\mathrm{meas}}\cdot K\cdot\frac{C\cdot k_{\mathrm{g}}}{2^{d_{\mathrm{f}}}}\cdot\frac{\overline{N}\cdot\overline{a_{\mathrm{av}}^{6-d_{\mathrm{f}}}}}{q^{d_{\mathrm{f}}}} = c_{N,\mathrm{agg}}V_{\mathrm{meas}}\cdot K\cdot C\cdot k_{\mathrm{g}}\cdot\frac{\overline{N}\cdot\overline{x_{\mathrm{p,av}}^{6-d_{\mathrm{f}}}}}{q^{d_{\mathrm{f}}}}. \tag{4.61}$$

Eventually, the surface scattering regime ($q \gg x_{\mathrm{p,av}}^{-1}$) solely depends on the number and size of the primary particles. The scattering intensity decays with the 4th power of q and the effective decay coefficient is proportional to the average surface of the aggregate S_{agg}:

$$B_{\mathrm{eff}} = \lim_{q\to\infty} I_{\mathrm{sca}}(q)\cdot q^4 = c_{N,\mathrm{agg}}V_{\mathrm{meas}}\cdot\frac{9}{2}K\cdot\overline{N}\cdot\overline{a_{\mathrm{av}}^2} = c_{N,\mathrm{agg}}V_{\mathrm{meas}}\cdot 2\pi K'\cdot\overline{N}\cdot\overline{S_{\mathrm{agg}}}. \tag{4.62}$$

Hence, the transition from the fractal regime to surface scattering scales with:

$$q_{\mathrm{fs}}^{4-d_{\mathrm{f}}} = \frac{72}{Ck_{\mathrm{g}}}\cdot\frac{\overline{x_{\mathrm{p,av}}^2}}{\overline{x_{\mathrm{p,av}}^{6-d_{\mathrm{f}}}}}. \tag{4.63}$$

The transition is, thus, unaffected by the aggregation number N and its polydispersity; the crucial parameter is the size distribution of the primary particles.

A characteristic parameter for the surface scattering regime is the Porod invariant Q, which should be derived only from that part of the scattering that is due to the primary particles (cf. Sect. 2.3.4.4; Eq. (2.25)). This parameter is proportional to the average aggregate volume:

$$\begin{aligned} Q_{\mathrm{eff}} &= \int I_{\mathrm{sca}}(q)\cdot q^2\mathrm{d}q \geq c_{N,\mathrm{agg}}V_{\mathrm{meas}}\cdot\frac{3}{2}\pi K\cdot\overline{N}\cdot\overline{a_{\mathrm{av}}^3} \\ &= c_{N,\mathrm{agg}}V_{\mathrm{meas}}\cdot 2\pi^2K'\cdot\overline{N}\cdot\overline{V_{\mathrm{agg}}}. \end{aligned} \tag{4.64}$$

The ratio of B_{eff} and Q_{eff} is proportional to the volume specific surface S_V, which is the sixfold inverse of the Sauter diameter x_{ST}.

It is noteworthy that polydispersity specifically affects the different parameters of the scattering curve $I_{\mathrm{sca,eff}}(q)$. For example, $R_{\mathrm{g,eff}}$ scales with the rms of the intensity weighted distribution of the primary particles, whereas the inverse of the volume specific surface or x_{ST} scales with the harmonic mean of the volume weighted distribution. Hence, coarse primary particles have a much stronger impact on the effective radius of gyration $R_{g,eff}$ than on the Sauter diameter x_{ST}. Also, while the latter is independent on the aggregation number, the former grows approximately with $\sqrt{\overline{N^3}/\overline{N^2}}$.

Nevertheless, the most pronounced effect of polydispersity is the smoothing of the scattering curve. This is illustrated in Fig. 4.12, which depicts normalised scattering curves of DLCA aggregates. The graphs in the left diagram are obtained for aggregates consisting of $N = 200$ uniformly sized spheres, the size of which varies among the aggregates according to a log-normal distribution (average primary particle size in the aggregate $x_{p,av} = 20$ nm). Already, a low polydispersity ($\sigma_{ln} = 0.1$) suffices to smooth out the resonance peaks of the spherical surface at large q. Increasing the width of the distribution additionally leads to a shift of the curve to smaller q-values, indicating that the effective aggregate and sphere sizes are growing. In the right chart of Fig. 4.12, polydispersity in N was additionally assumed to a fixed polydispersity in $x_{p,av}$ ($\sigma_{ln} = 0.3$). A log-normal distribution (varying $\sigma_{ln}(N)$) and a self-preserving distribution as resulting from an ideal DLCA process (Eq. (4.10)) were considered. Furthermore, a negative correlation between the average primary particle size and the average aggregation number was assumed $\overline{N} \propto \bar{x}_{p,av}^{-1}$. Obviously, there is only minor variation among the "log-normal curves", but a clear distinction between these curves and that of the self-preserving distribution function.

The normalised scattering curves in Fig. 4.13 are plotted over a broad range of q-values, which cannot be covered by just one instrument (typical values for SLS are 0.002 to 0.03 nm^{-1} and for SAXS: 0.05 to 2 nm^{-1}, cf. Sect. 2.3.4). That causes tremendous difficulties in deriving reliable parameters on size and structure of particle aggregates. For the purpose of illustration, Fig. 4.13 shows the normalised

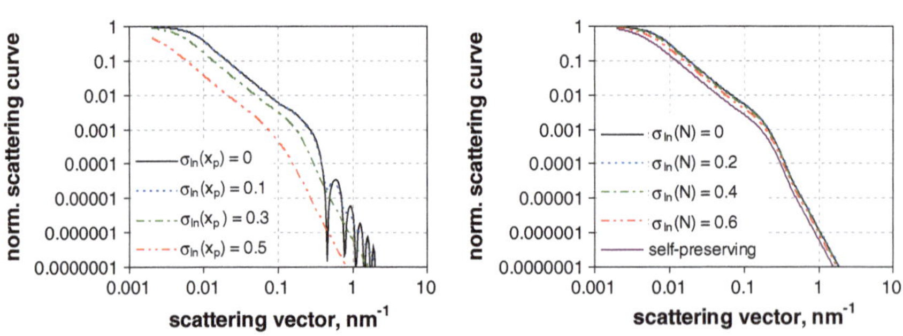

Fig. 4.12 Effect of polydispersity on the static structure factor; *left* constant aggregation number N with log-normal distribution of primary particle size, *right* variation of N with $p(N|x_{p,av})$ as log-normal distribution and as self-preserving distribution as obtained with DLCA processes (Eq. (4.10))

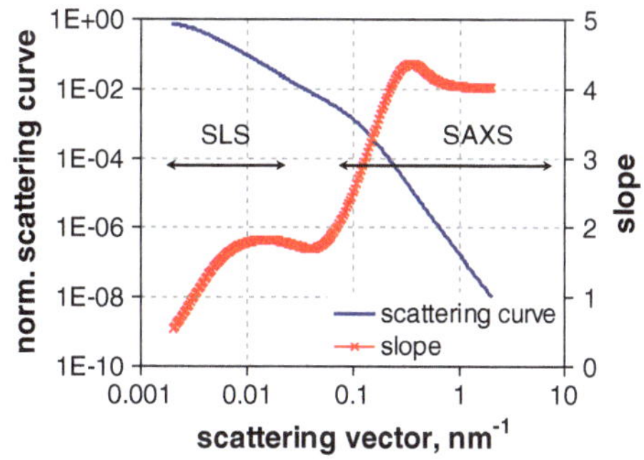

Fig. 4.13 Normalised scattering curve and its negative slope in a log-log plot for polydisperse DLCA aggregates ($x_{p,av}$ = 20 nm, σ_{ln} = 0.3, self-preserving distribution of N with $\overline{N}$ = 200)

scattering curve and its negative slope in the log-log plot for a polydisperse aggregate population (log-normal distribution of primary particles around $x_{p,av}$ = 20 nm, self-preserving distribution of N with $\overline{N}$ = 200). Additionally, the typical working ranges of SLS and SAXS are indicated. The figure clearly shows that SAXS is primarily useful when describing the primary particles, and SLS is mainly useful for the determination of the aggregate size (x_g). Structural parameters, e.g., the fractal dimension, may be obtained by either of the two methods, but are accessible only on a small q-range.

Due to such technical limitations as well as polydispersity, it is often rather difficult to directly extract single morphological parameters from static scattering experiments (e.g. d_f from the slope of the scattering curve). Alternatively, the measured scattering curves can be fitted with parametric model equations (Kammler et al. 2004; Kätzel et al. 2008b).

4.4 Hydrodynamic Properties of Aggregates

Several macroscopic properties of colloidal suspensions—like rheology, sedimentation, or stability—are affected by the diffusion of the individual particles and their migration in external fields. Measurement techniques like dynamic light scattering, field-flow-fractionation, or centrifugation analysis probe particle systems with regard to these hydrodynamic properties (cf. Chap. 2), which depend on size, shape and—in the case of aggregates—internal structure. There are several approaches for calculating the hydrodynamic behaviour of aggregates: approximate ones, which provide some rough idea on the structure-property relationship, and more rigorous ones, which commonly require high computation costs. This section reviews popular models and discusses their results for fractal aggregates.

4.4.1 Diffusion and Settling of Aggregates

Any motion of colloidal particles is driven by volume forces (e.g. gravity), stochastic Brownian forces and hydrodynamic forces. The latter act on the particles' surface and can be calculated via the hydrodynamic stress tensor $\mathbf{\Pi}$:

$$F_{\mathrm{D}} = \int_S \mathbf{\Pi} \cdot \mathbf{n} \mathrm{d}S = \int_S \mathbf{\Pi}_{\mathbf{n}} \mathrm{d}S, \quad \text{and} \quad \mathbf{M} = \int_S \mathbf{r} \times \mathbf{\Pi}_{\mathbf{n}} \mathrm{d}S \tag{4.65}$$

$$\text{with } \mathbf{\Pi} = -p\mathbf{I} + 2\eta\mathbf{D}, \tag{4.66}$$

where $\mathbf{D}$ is the rate of strain tensor ($D_{ij} = ½·(∂_j u_i + ∂_i u_j)$). The calculation of hydrodynamic forces, therefore, requires the knowledge of the flow field at the particle's surface, which in general obeys the non-linear Navier–Stokes equations. However, the next sections solely deal with the motion of colloidal aggregates at very small Reynolds numbers (i.e. for creeping flow) in an incompressible fluid. Hence, the non-linear terms of the Navier–Stokes equations can be neglected and the hydrodynamic forces depend linearly on the velocity field. In this case, the drag force on a single sphere moving with velocity $\mathbf{v}_{\mathrm{p}}$ in a quiescent fluid ($\mathbf{u}^{\infty} = 0$) can be calculated with Stokes' law:

$$\mathbf{F}_{\mathrm{D}} = -6\pi\eta a \cdot \mathbf{v}_{\mathrm{p}}. \tag{4.67}$$

This relationship is fundamental for the calculation of hydrodynamic properties of colloidal objects (e.g. the terminal settling velocity of small spheres). The factor $6\pi\eta a$ is called friction coefficient; its reciprocal is the hydrodynamic mobility μ_{hd}.

When the frame of reference is shifted to the particle centre, Eq. (4.67) describes the hydrodynamic force acting on a resting particle in a uniform flow field with $\mathbf{u}^{\infty} = -\mathbf{v}_{\mathrm{p}}$. Besides this, the resting particle causes a perturbation of the velocity field $\mathbf{v}(\mathbf{r}) = \mathbf{u}(\mathbf{r}) - \mathbf{u}^{\infty}$, which decays with $1/r$ from the particle surface. That is, the disturbance of the flow field caused by the presence of one particle is long-ranging. Consequently, the flow around two sufficiently close particles is governed by the mutual influence of the perturbation fields. To first approximation, this interdependency can be described as if the perturbation field of one particle exerts a drag force on the other particle's centre, which adds to that of the undisturbed field $\mathbf{u}^{\infty}$. Happel and Brenner (1983, p. 235) derived analytical expressions for this kind of interaction. Their results suggest that the total drag on two particles is reduced when they approach each other (depending on their alignment in the velocity field).

This may be understood by regarding the total flow field. In the region between the two particles, the external flow field is screened considerably (i.e. the velocities are smaller than in the far field) and the local shear rates are lowered accordingly (cf. Fig. 4.14). As a result, the momentum flux from the fluid phase to the particles is not as high as for two completely separated particles.

The analytical approach of Happel and Brenner provides first insight into the physics of the hydrodynamic interaction between particles. However, it fails in

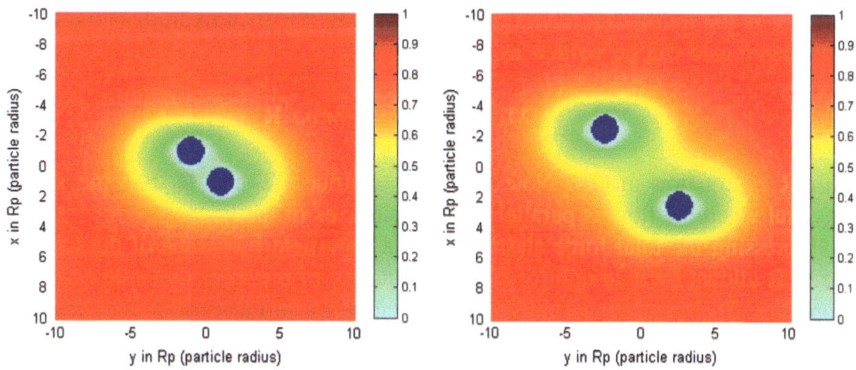

Fig. 4.14 Magnitude of the velocity field around two spherical particles; creeping flow along the horizontal axis; *left* centre distance = 2.8 × *a*, *right* centre distance = 7.1 × *a*

describing the hydrodynamics of closely spaced particles where the net effect of perturbation fields can no longer be approximated by central forces but leads to a distinct force distribution on the (other) particle surface. Moreover, if two freely moving particles approach each other, one cannot ignore the unsteady nature of the flow field, which results in strong retarding interactions (*lubrication*).

For clusters or aggregates of particles, the total drag in a uniform velocity field results from the drag each individual particle experiences. Similar to the 2-sphere problem, this drag can be split into contributions from the far field $\mathbf{u}^{\infty}$ and the perturbation fields of all primary particles. By doing so, one benefits from the fact that the creeping flow is governed by linear equations (Stokes equations) and, thus, all perturbation superpose linearly. Nevertheless, since the perturbation fields decay rather slowly the hydrodynamic interaction within the aggregate are principally many-body interactions.

Once the total velocity field inside an aggregate is known, the drag on each particle can be computed with Eq. (4.65) and the total drag acting on the aggregate is obtained by summation:

$$F_{\mathrm{D}} = \sum F_{\mathrm{D},i}. \tag{4.68}$$

The total force is not necessarily parallel to the direction of particle velocity (exciting flow field). Indeed, the direction of the total hydrodynamic force on an aggregate depends on its configuration and orientation. Force components parallel and perpendicular to the direction of flow (drag and drift, respectively) can be distinguished. For an aggregate of monodisperse particles, the total drag may be expressed as:

$$\mathbf{F}_{\mathrm{D}} = -6\pi\eta a \mathbf{T} \cdot \mathbf{v}_{\mathrm{p}}, \tag{4.69}$$

where **T** is the translation matrix. In a similar way, the total hydrodynamic torque on an aggregate can be written as

$$\mathbf{M} = \sum \mathbf{M}_i + \mathbf{r}_i \times \mathbf{F}_i = -8\pi\eta a^3 \mathbf{R} \cdot \boldsymbol{\omega}, \tag{4.70}$$

with **R** as rotation matrix. Note that both matrices, **T** and **R**, become strictly diagonal when the flow is parallel to the main axes of inertia.

Expressing the total force and torque in a flow field by means of the matrices **T** and **R** allows for a straightforward computation of the hydrodynamic aggregate properties. The translational diffusion coefficient, for instance, is derived from the Langevin equation, which reads in x-direction (without volume forces):

$$m\frac{\mathrm{d}^2x}{\mathrm{d}t^2} = -6\pi\eta a T_{xx} \cdot \frac{\mathrm{d}x}{\mathrm{d}t} + F_{\mathrm{stoch}} \tag{4.71}$$

and has the result $\langle x^2 \rangle = \mathrm{k_B}T/(3\pi\eta a T_{xx}) \cdot t$. The total mean square displacement $\langle r^2 \rangle$ is simply the sum of $\langle x^2 \rangle$, $\langle y^2 \rangle$ and $\langle z^2 \rangle$. By comparison with the Stokes-Einstein Eq. (2.4) and after orientation averaging, one obtains for the translational diffusion coefficient:

$$D_{\mathrm{t}} = \frac{\mathrm{k_B}T}{6\pi\eta a} \cdot \left\langle \frac{1}{T_{ii}} \right\rangle. \tag{4.72}$$

A similar analysis of Brownian rotation yields for the rotational diffusion coefficient:

$$D_{\mathrm{r}} = \frac{\mathrm{k_B}T}{8\pi\eta a^3} \cdot \left\langle \frac{1}{R_{ii}} \right\rangle. \tag{4.73}$$

For fractal aggregates, the diagonal elements of the matrices **T** and **R** are considerably larger than the off-diagonal ones ($T_{i \neq j} \ll T_{ii}$) and the orientation averages can be replaced by harmonic sums of the matrices' eigenvalues. These eigenvalues reflect the hydrodynamic anisotropy (in R_{ii} more dramatically than in T_{ii}).

The hydrodynamic anisotropy is described by the ratio of the maximum and minimum eigenvalues of **T** and **R**. For hcp aggregates, these ratios are both 1.0, while for DLCA aggregates they take the values of 1.1 and 2.2, respectively. The respective ratios for RLCA aggregates are 1.1 and 1.9 ($N \geq 10$). Obviously, the drag is not influenced much by the aggregates anisotropy and, thus, also not by their orientation. The aspherical macroshape chiefly affects the torque. Yet, when converted to hydrodynamic sizes, the effect of orientation is still less than 30 %. That may be surprising, since the ratio of the maximum and minimum eigenvalues of the gyration tensor are significantly larger (12 for DLCA and 8 for RLCA, cf. p. 136). However, the hydrodynamic behaviour of the aggregate is strongly related to the screening of the velocity field in the inner core of the aggregates, whereas the geometric anisotropy (e.g. quantified with the gyration tensor) is related to their outer regions.

The diffusion coefficients are used to define the hydrodynamic equivalent diameters for translation and rotation:

$$x_{\mathrm{h,t}} = x_{\mathrm{p}} \Big/ \left\langle \frac{1}{T_{ii}} \right\rangle \quad \text{and} \quad x_{\mathrm{h,r}} = x_{\mathrm{p}} \Big/ \left\langle \frac{1}{R_{ii}} \right\rangle^{1/3}. \tag{4.74}$$

These two equivalent diameters only differ a little and in such a way that frequently only one hydrodynamic diameter is considered. Nevertheless, they are decisive for the diffusive behaviour of aggregates and affect any motion in viscous fluids. Of particular relevance for the processing of colloidal suspensions and for their stability is the settling of aggregates (i.e. their migration in a gravitational or centrifugal field). This motion can be considered as a quasi-steady flow. Moreover, the creeping flow condition holds for colloidal particles even at high centrifugal acceleration. The sedimentation of spherical particles is always in the direction of the exerting field. However, in the general case of non-spherical particles and aggregates, a horizontal drift may additionally be observed. A force balance based on gravity and drag reveals the relationship between the drift velocity v_x and the settling velocity v_z ($T_{xz}v_z + T_{xx}v_x = 0$, when z is parallel to the field and x shows in drift direction) as well as their dependence on gravitation:

$$v_z = \frac{2g\Delta\rho}{9\eta} \cdot \frac{R_V^3}{a \cdot \tilde{T}_{zz}}, \quad \text{with } \tilde{T}_{zz} = T_{zz} - \frac{T_{xz}^2}{T_{xx}}. \tag{4.75}$$

As already discussed above, the off-diagonal elements of the translation matrix **T** are considerably smaller than those on the diagonal (i.e. $\tilde{T}_{zz} \approx T_{zz}$). Moreover, for a huge variety of fractal aggregates—including DLCA and RLCA—approximate hydrodynamic isotropy with regard to translation is observed ($T_{xx} \approx T_{yy} \approx T_{zz}$, see above). Even more, Brownian rotation leads to an orientation averaging during the sedimentation process. The relevance of this effect can be estimated by comparing the characteristic time scales of diffusion (τ_{diff}) and settling (τ_{sett}). Their ratio is a Péclet number, which is computed from $Pe = v_{\mathrm{sett}}/D_r R_{\mathrm{h}}$ and depends on the $(1 + d_{\mathrm{f}})$ th power of the hydrodynamic diameter. For gravitational settling of fractal aggregates of nanosized particles, the Pe-number is usually <1 over the complete colloidal size range (≤1 μm). Hence, it is permissible to approximate the settling velocity of such aggregates with its orientation average (i.e. $T_{zz}^{-1} = \langle T_{ii}^{-1} \rangle$):

$$v_{\mathrm{sett}} = \frac{2g\Delta\rho}{9\eta} \cdot \frac{R_V^3}{R_{\mathrm{h,t}}} = \frac{g\Delta\rho}{18\eta} \cdot \frac{x_V^3}{x_{\mathrm{h,t}}}. \tag{4.76}$$

Hence, the sedimentation equivalent diameter, the Stokes diameter x_{Stokes}, is given by:

$$x_{\mathrm{Stokes}}^2 = v_{\mathrm{sett}} \cdot \frac{18\eta}{g\Delta\rho} = \frac{N \cdot x_{\mathrm{p}}^3}{x_{\mathrm{h,t}}}, \tag{4.77}$$

where N is the number and x_{p} the diameter of the primary particles.

This equation defines the Stokes diameter x_{Stokes} on the basis of the density contrast $\Delta\rho = \rho_s - \rho_m$ between solid and fluid phase. It should not be confused with the density contrast $\Delta\rho$ between the porous "particle phase" (with $\rho_{p,eff}$) and the fluid. Such a definition might be meaningful for dense agglomerates with defined porosity, which can be measured by nitrogen adsorption (Staiger et al. 2002), or may be deduced from the powder bulk density (Barthel et al. 1998) or from sediment properties (Kätzel 2007, pp. 183–187). However, it should be avoided for fractal aggregates, where the concept of uniform porosity fails for several reasons (no defined surface, size-dependent and locally varying porosity).

Finally, it should be mentioned that gas-borne aggregates require different or modified models for their dynamic properties than those suspended in liquids, to which the concepts of continuum mechanics apply. The drag on colloidal aerosol particles is not just a result from viscous friction, but is considerably affected by impacts of the gas molecules on the particle surface. This is because gas molecules interact via elastic collisions with mean free paths much larger than molecule size (approx. 66 nm at atmospheric pressure), whereas the interaction of liquid molecules occurs over distances in the Ångström range. This difference can be regarded by appropriate correction functions (e.g. Cunningham 1910) or may require completely different solutions (Epstein 1924). Since aerosol aggregates are beyond the scope of this book, the readers are referred to the literature (Dahneke 1982; Wang and Sorensen 1999; Filippov 2000; Lall and Friedlander 2006).

4.4.2 Calculation of the Hydrodynamic Drag on Aggregates

The necessity to understand the relationship between the structural properties of aggregates and their dynamic behaviour, e.g., for accomplishing dynamic simulations or for analysing experimental data, has led to various approaches for calculating the hydrodynamic drag on aggregates. Besides purely numerical methods (Coelho et al. 1997; Binder et al. 2006), several analytical calculation schemes on different levels of approximation are employed. One way to simplify the physical problem is to consider aggregates as porous particles (e.g Veerapaneni and Wiesner 1996; Vanni 2000), which are well-studied with regard to their hydrodynamic behaviour. While this approach seems appropriate for compact aggregates ($d_f > 2$), loose aggregates ($d_f < 2$) may be better approximated by straight chains of spheres (Dahneke 1982; Lall and Friedlander 2006).

More rigorous analytical approaches are based on the individual perturbation fields of the primary particles within the aggregate and the corresponding perturbation forces they cause on the other particles. The Kirkwood-Riseman theory uses simplified and averaged expressions for the velocity and force perturbations (Kirkwood and Riseman 1948; Bloomfield et al. 1967; Hess et al. 1986), and provides relatively simple expressions for the hydrodynamic aggregate size (Chen et al. 1984; Lattuada et al. 2004). Stokesian dynamics simulation, which originally aims at the dynamic behaviour of suspensions (Brady and Bossis 1988), provides a further mean for calculating the hydrodynamic forces on aggregates (Bossis et al. 1991; Sierou and Brady 2001; Binder et al. 2006). There, the hydrodynamic forces on the particle surfaces are expanded with low-order multipole functions (Durlofsky et al. 1987). Similar with regard to the basic principle is the multipole expansion of the perturbed velocity fields (Filippov 2000). This approach allows not only the calculation of hydrodynamic forces but the reconstruction of the entire velocity field as well. A further way of modelling the forces on the particles is the method of reflections, as recently proposed (Gastaldi and Vanni 2011).

4.4.2.1 Porous Sphere Approach

The porous sphere approach treats an aggregate as a porous particle without resolving the local structure. Rather, a constant or radially varying porosity, and thus permeability k, is assumed. Consequently, an averaged flow field for the internal region of the aggregate ($r \leq R_C$, R_C being the radius of the envelopping sphere) can be calculated by solving the Brinkman equation:

$$\nabla p = \eta \nabla^2 \mathbf{u} - \frac{\eta}{k} \mathbf{u}, \tag{4.78}$$

while the Stokes equation ($\nabla p = \eta \nabla^2 \mathbf{u}$) prevails outside the aggregate ($r > R_C$). The total drag results from the hydrodynamic forces at the inner and outer surface and is a function of the permeability k.

$$\mathbf{F}_D = -6\pi\eta R_C f(k) \cdot v_P. \tag{4.79}$$

For isolated porous spheres with uniform permeability, Neale et al. (1973) derived the following correction term:

$$f(k) = \left[\frac{3}{2} k/R_C^2 + \left(1 - \sqrt{k/R_C^2} \tanh \sqrt{R_C^2/k}\right)^{-1}\right]^{-1}. \tag{4.80}$$

The permeability is inversely proportional to the viscous friction in the pores or—in the case of aggregates—to the viscous drag on the individual particles. Thus, permeability depends on particle size and packing density φ (Veerapaneni and Wiesner 1996; Vanni 2000):

$$k = \frac{2a^2}{9\varphi} \cdot \frac{L(\varphi)}{S(\varphi)}, \tag{4.81}$$

where the functions L and S (L, $S \leq 1$) correct for the long-range and short-range hydrodynamic interactions between the particles in the aggregate (Vanni 2000).

For fractal aggregates, the local variation in porosity, and thus in permeability, cannot be neglected. Veerapaneni and Wiesner (1996) assumed spherically shaped aggregates consisting of discrete shells with radially decreasing porosity according to a fractal aggregate growth. For a continuous version of this model, Vanni (2000) solved the governing equations numerically. He showed that the details of the model (in particular, the precise formulation of L and S) do not dramatically affect the calculated drag. Moreover, the decisive parameter of this model is the surface porosity $1 - \varphi_S$. That means that Neale's solution (Eq. (4.80)) is a fairly good approximation of the numerical results when the surface permeability k_S is employed.

The major concern about the porous-sphere approach is that it implicitly assumes a geometric, mostly a spherical, boundary between the continuous phase and the porous particle and that it ignores the heterogeneous spatial distribution of primary particles. This approach should, therefore, be best applicable for fairly dense aggregates ($d_f > 2$) that are much larger than the primary particles.

It should be mentioned that, since the 1990s, Gmachowski (1990, 1995, 1998, 2005) has presented a series of papers which combined functional relationships derived from porous-sphere-models with experimental data. He deduced relations for the fractal prefactor k_C and the hydrodynamic diameter R_h, which are frequently cited in the literature (e.g. Vanni 2000; Tang et al. 2002; Gruy and Cugniet 2004; Lee and Kramer 2004). However, his experiments mainly concern coarse flocculated polystyrene latices (floc size > 100 μm), which are probably not comparable with the aggregate types discussed here (e.g. in Gmachowski's papers the fractal prefactors increase monotonically with fractal dimension, which is in striking contrast to the behaviour found in Fig. 4.8).

4.4.2.2 Kirkwood-Riseman Theory

The Kirkwood-Riseman theory allows for an approximate calculation of the hydrodynamic drag of arbitrary clusters of small objects (molecules, particles) with undefined shape. It is based on the fact that, for low Reynolds-numbers, the velocity perturbations originating from the primary particles superpose linearly. The hydrodynamic force on a primary particle j results from the total perturbation field around j ($\mathbf{v}_j$) and from the undisturbed flow $\mathbf{u}^\infty$. To first approximation, it can be assumed that the presence of j does not affect $\mathbf{v}_j$, i.e. the spatial extension of particle j is neglected. Moreover, the perturbation of the velocity field at the place of j can be related to the perturbation forces acting on the other particles:

$$\mathbf{v}_j = -\sum_{j \neq m} \mathrm{J}_{jm} \mathbf{F}_m, \tag{4.82}$$

where **J** is the Oseen tensor given by

$$J_{ij} = \frac{1}{8\pi\eta r_{ij}}\left(\mathbf{I} + \frac{\mathbf{r}_{ij}\mathbf{r}_{ij}}{r_{ij}^2}\right) \text{ or } J_{ij} = \frac{1}{8\pi\eta a}\left(\delta_{ij} + \frac{3}{8}\frac{a}{r_{ij}}\left(1 + \hat{\mathbf{r}}_{ij}\hat{\mathbf{r}}_{ij}\right)\left(1 - \delta_{ij}\right)\right). \quad (4.83)$$

Kirkwood and Riseman (1948) derived expressions for the orientation averaged perturbation by averaging of the tensor J_{ij} and the forces $\mathbf{F}_m$ separately. Assuming monodisperse particles, the following expression for the hydrodynamic diameter (of translation) is obtained (Bloomfield et al. 1967):

$$R_{\mathrm{h,KR}}^{-1} = (Na)^{-1} + \frac{N-1}{N}\left\langle r_{ij}^{-1}\right\rangle_{i \neq j} \quad (4.84)$$

or for continuous formulation:

$$R_{\mathrm{h,KR}}^{-1} = (Na)^{-1} + \frac{1}{N}\int r^{-1} g(r)\mathrm{d}V. \quad (4.85)$$

For large aggregates ($N \gg 1$, $a \ll R_{\mathrm{h}}$), Eq. (4.84) states that the hydrodynamic radius of an aggregate is the harmonic mean of the interparticle distances. Moreover, Eq. (4.85) reveals that in the large size limit the hydrodynamic radius R_{h} of fractal aggregates with scale-invariant structure (i.e. $g(r)$ = funct. (r/R_{g})) is proportional to the geometric dimension of the aggregate. The hydrodynamic size, thus, obeys the same fractal law (Eq. (4.9)) as the radius of gyration. Hess et al. (1986) showed that for spherically symmetric aggregates the subsequent relationship would be derived

$$\frac{R_{\mathrm{h}}}{R_{\mathrm{C}}} = 2 - \frac{2}{d_{\mathrm{f}}}. \quad (4.86)$$

Due to its simplicity, the Kirkwood-Riseman theory has been widely used in the literature for estimation of the hydrodynamic diameter of colloidal aggregates (Chen et al. 1984; Hess et al. 1986; Wiltzius 1987; Naumann and Bunz 1991; Lattuada et al. 2003; Sandkühler et al. 2005a). However, this is an approximate approach which considers the hydrodynamic interaction only within a first-order correction of the unperturbed force and which neglects the finite size and the shape of the primary particles. Hence, the Kirkwood-Riseman theory is best applicable for very porous aggregates and is expected to fail for very compact ones (de la Torre and Bloomfield 1977; Binder et al. 2006). Attempts to consider the size and shape of the primary particles within this framework lead to comparably cumbersome expressions of the hydrodynamic drag (de la Torre and Bloomfield 1977) and were obviously not able to compete with other theories or numerical methods.

4.4.2.3 Stokesian Dynamics

Stokesian dynamics is a numerical technique for simulating the dynamic behaviour of colloidal suspensions (sedimentation, rheology), where the motions of the individual particles is driven by Brownian and volume forces (including particle interactions) and coupled by hydrodynamic interaction. In a more general approach than in Eq. (4.69), the hydrodynamic forces are traced back to the generalised particle velocities $\mathbf{v}_\mathrm{p}$ and the velocity gradients $\mathbf{E}$:

$$\begin{pmatrix} \mathbf{F} \\ \mathbf{S} \end{pmatrix} = -\Re \cdot \begin{pmatrix} \mathbf{v}_\mathrm{p} - \mathbf{u}^\infty \\ -\mathbf{E}^\infty \end{pmatrix}, \tag{4.87}$$

where $\mathbf{F}$ represents the hydrodynamic force and torque and $\mathbf{S}$ denotes the hydrodynamic stresses acting on the particles. Note that Eq. (4.87) is a very general representation, which comprises the hydrodynamics of translational and rotational motion. The matrix $\mathfrak{R}$ is called resistance matrix and its reciprocal is the mobility matrix $\mathbf{M}$. The calculation of $\mathfrak{R}$ is based on a separate treatment of many-body far-field interactions and the strong lubrication effects between two approaching particles.

$$\Re = (\mathbf{M}^\infty)^{-1} + \Re_{2\mathrm{b}} - \Re_{2\mathrm{b}}^\infty. \tag{4.88}$$

Both far-field and lubrication interaction are computed by a multipole expansion in forces and torques (*multipole force expansion*—MFE). A low order expansion is used for the far-field interaction ($\mathbf{M}^\infty$), whereas for lubrication, multipoles of high order (>2) are required.

The principles of Stokesian dynamics were developed by Durlofsky, Brady, and Bossis in the 1980s (Durlofsky et al. 1987; Brady and Bossis 1988). The computational costs increase with the third power of particle number (N^3). In 2001, Sierou and Bossis introduced a modified numerical procedure for solving the matrix equations, called the *accelerated Stokesian dynamics*, which considerably reduced the computational effort ($N^{1.25}\log N$).

Although originally developed for the hydrodynamics of suspensions, Stokesian dynamics can be employed for the calculation of hydrodynamic drag of aggregates. However, there is a qualitative difference between the hydrodynamics of aggregates and that of clusters of freely moving particles. When exposed to a velocity field, the spatial configurations of the latter change, i.e. the particles move relative to each other. In contrast, the particle positions within an aggregate are fixed; there is no relative motion and lubrication does not need to be considered. Bossis et al. (1991) showed that the total hydrodynamic stress $\mathbf{S}$ on a fixed aggregate increases with the third power of aggregate size, whereas the stress on a free cluster grows with the number of particles. Hence, for aggregates, Eq. (4.87) simplifies to:

$$\begin{pmatrix} \mathbf{F} \\ \mathbf{S} \end{pmatrix} = -(\mathbf{M}^\infty)^{-1} \cdot \begin{pmatrix} \mathbf{v}_\mathrm{p} - \mathbf{u}^\infty \\ -\mathbf{E}^\infty \end{pmatrix}. \tag{4.89}$$

Stokesian dynamics was recently used for the calculation of drag forces of fractal aggregates as well as for studying their settling behaviour (Binder et al. 2006, 2009). It could be shown that the computational costs are lower than for pure numerical methods such as the Lattice-Boltzmann method. It was applied to non-sintered aggregates of spheres, but can principally be used for other regular particle shapes as well—as long as the particle surface can be "easily" described when formulating the hydrodynamic boundary conditions. For that reason, irregular particle shapes and sintered structures defy a calculation by Stokesian dynamics.

4.4.2.4 Multipole Expansion of the Velocity Field

In 2000, Filippov introduced the *multipole velocity expansion* (MVE) for the calculation of the velocity field $\mathbf{u}$ in and around an aggregate (or cluster of particles) which moves with defined velocity through a viscous medium. Once this velocity field is known, the hydrodynamic forces can be calculated straightforwardly. The approach is based on the following expansion of the velocity perturbation $\mathbf{v}_j$ around a particle j (cf. Happel and Brenner 1983, p. 260):

$$\mathbf{v}_j = \sum_n \left(\nabla\Phi_n^j + \nabla \times \left(\mathbf{r}_j \chi_n^j\right) - \frac{n-2}{2n(2n-1)} \mathbf{r}_j^2 \frac{1}{\eta} \nabla p_n^j + \frac{2(n+1)}{2n(2n-1)} \mathbf{r}_j \frac{1}{\eta} p_n^j \right), \tag{4.90}$$

where the Φ_n^j, χ_n^j, and p_n^j are the harmonic functions of nth order ($n = 1\ldots\infty$). They are expanded in irregular solid spherical harmonics u_{mn}^j, which are defined for spherical coordinates $\mathbf{r}_j$ centred in particle j:

$$\Phi_n^j = \sum a_{mn}^j u_{mn}^j, \quad \chi_n^j = \sum c_{mn}^j u_{mn}^j, \quad \frac{1}{\eta} p_n^j = \sum b_{mn}^j u_{mn}^j, \quad |m| \leq n. \tag{4.91}$$

The perturbations decay at large distances ($\mathbf{v}_j = 0$ for $|\mathbf{r}_j| \to \infty$) and superpose linearly in and around the aggregate:

$$\mathbf{u} = \mathbf{u}^\infty + \sum \mathbf{v}_j, \quad j = 1\ldots N \tag{4.92}$$

The boundary conditions established for the particle surface are:

$$\mathbf{u} - \mathbf{u}_{\text{slip}} = \mathbf{v}_{\text{p},i}, \quad \text{for} \left|\mathbf{r} - \mathbf{r}_{\text{c},i}\right| = a_i \tag{4.93}$$

The slip velocity $\mathbf{u}_{\text{slip}}$ is zero for Newtonian fluids in the hydrodynamic regime, but takes finite values for the aerodynamic slip regime and can be related to the Knudsen-number (Filippov 2000). Within the scope of this work (colloidal suspensions), slip can be ignored. The analytical formulation of the boundary conditions requires a step which transforms the spherical harmonics u_{mn}^j centred at the particle j to spherical harmonics u^i_{mn} centred at the particle i. That involves some lengthy calculations and eventually leads to a linear system of equations for the unknown expansion coefficients a^j_{mn}, b^j_{mn}, and c^j_{mn}, which can be solved by appropriate numerical tools. After having determined the expansion coefficients, one can calculate the hydrodynamic force and torque acting on the particles (Eqs. (4.68) and (4.70)):

$$\mathbf{F}_j = -4\pi\eta\left[-\left(b^j_{11} - \frac{1}{2}b^j_{-11}\right)\mathbf{e}_x - \mathrm{i}\left(b^j_{11} + \frac{1}{2}b^j_{-11}\right)\mathbf{e}_y + b^j_{01}\mathbf{e}_z\right], \tag{4.94}$$

$$\mathbf{M}_j = -8\pi\eta\left[-\left(c^j_{11} - \frac{1}{2}c^j_{-11}\right)\mathbf{e}_x - \mathrm{i}\left(c^j_{11} + \frac{1}{2}c^j_{-11}\right)\mathbf{e}_y + c^j_{01}\mathbf{e}_z\right]. \tag{4.95}$$

Details of the mathematical procedure are explained by Filippov (2000) and Kätzel (2007).

The accuracy of the solution increases monotonically with the truncation level L of the multipole expansion. However, as the number of expansion coefficients is $3 \times N \times L \times (L + 2)$, the size of the linear systems of equations scales with N^2 and L^4. Hence, for practical reasons, the truncation level L is usually kept small. In Fig. 4.15 (left), the relative deviation from the exact solution of the hydrodynamic radius is shown for different types of aggregates (straight chains, hcp, and DLCA aggregates) of varying aggregation number. It can be seen that even a field expansion in first-order multipoles (L = 1) will usually result in a reasonable approximation of the hydrodynamic drag. For L = 2, the relative error is less than 1 %, and for L = 3, less than 0.1 %. Moreover, for DLCA aggregates the error decreases with aggregate size (Fig. 4.15, right), which justifies an approximation of the total velocity with low order multipole expansions (L = 2–3), as has already been proposed by Filippov (2000).

In conclusion, the multipole velocity expansion (MVE) is a fairly rigorous calculation scheme for the hydrodynamic behaviour of arbitrary clusters or

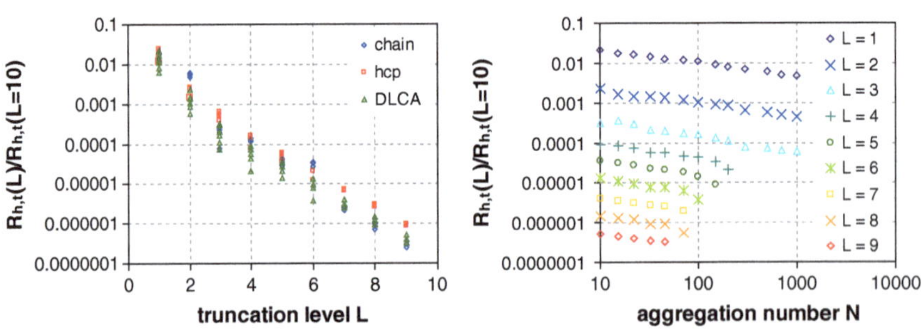

Fig. 4.15 Relative deviation of hydrodynamic radii $R_{h,t}$ calculated with MVE for varying truncation levels L of multipole expansion from the "correct" value (L = 10); *left* dependency on L for straight chains, hcp&DLCA aggregates (N = 10–500), *right* dependency on aggregation number N for DLCA aggregates

aggregates of spherical particles. It is similar to the multipole force expansion (MFE) employed by Stokesian dynamics but, in contrast to that method, it yields a solution for the velocity field, which may help to model other physical phenomena (like thermo- and electrophoresis, Filippov 2001).

The MVE has been elaborated only for spherically shaped primary particles and requires considerably more computational costs than approximate approaches like the Kirkwood-Riseman theory. In return, it accurately computes the velocity field and the hydrodynamic drag of aggregates, irrespective of the aggregate structure. Moreover, the MVE allows for the inclusion of a velocity slip at the particle surface and can, thus, be applied to the aerodynamic slip regime (i.e. for small and moderate Knudsen numbers). Filippov (2000) gave evidence that (for small aggregates, $N \leq 30$) the results of MVE and MFE are essentially identical and agreed with experimental data. Kätzel et al. (2008a) used the MVE in order to study the hydrodynamic behaviour of fractal aggregates within a wide range of fractal dimension ($1 \leq d_f \leq 3$). They found a small but distinct difference between the geometrically defined fractal dimensions (i.e. d_g) and the hydrodynamic dimension (describing the relationship between aggregate mass N and its hydrodynamic diameter x_h).

4.4.2.5 Numerical Techniques

The multipole expansions of the velocity field (MVE) and hydrodynamic forces (MFE) as employed for Stokesian dynamics allow for a fairly accurate computation of hydrodynamic behaviour of aggregates, yet they impose restriction on shape and overlap of the primary particles. These restrictions can be lifted, when the flow field around an aggregate is approximated by numerical techniques.

One of these techniques is the *finite differences method* (FDM), which solves the flow equations at the knots of a (rectangular) grid that is covering the whole fluid domain. It was used by Coelho et al. (1997) for the computation of the hydrodynamic and electrophoretic mobility of various types of fractal aggregates with constant mass ($d_f = 1.6$–2.5, $N = 256$). The aggregates were placed in a periodic box of varying size in order to study the influence of particle concentration ($\varphi_V = 0.001$–0.02). The results on the hydrodynamic radius could be fitted by the following equation:

$$R_h/R_g = \alpha(d_f)\Big/\left(1 - \varphi_V^{1/3} \cdot \beta(d_f)\right), \tag{4.96}$$

where $\alpha(d_f) = 0.3 \cdot d_f + 0.4$ and $\beta/\alpha = -1.6 \cdot d_f + 6.1$. That means that for isolated DLCA aggregates ($d_f = 1.8$) the ratio R_h/R_g becomes 0.94, while for RLCA aggregates ($d_f = 2.1$) a value of 1.03 is obtained.

Coelho et al. (1997) avoided an explicit treatment of isolated aggregates by the variation of box size and subsequent data regression because this case requires a discretisation of large fluid domains or artificial boundary conditions when the

FDM or similar methods—like finite elements or finite volume—are used. This kind of problem is circumvented by the *boundary element method* (BEM), which is based on an integral formulation of the fundamental flow equations at the particles' surface. The BEM considerably reduces computation costs because the flow field in and around the aggregate is not resolved; instead, the hydrodynamic forces are directly calculated. Geller et al. (1993) and Mondy et al. (1996) applied this method to study the drag force on linear chains of spheres and on flake-like particles, respectively. Geller et al. (1993) observed "very good agreement" between their BEM results and the data of other experimental and numerical studies. Similarly, Filippov (2000) observed that BEM results for linear chains are in excellent agreement with the results of his method of multipole velocity expansion. A further numerical technique is the lattice-Boltzmann method (LBM), which replaces the fluid continuum mechanics by the dynamic interaction of fictitious fluid particles. Binder et al. (2006) showed that this technique facilitates the calculation of hydrodynamic drag on partially sintered aggregates.

4.4.2.6 Comparison

The *porous sphere approach* and the *Kirkwood-Riseman theory* facilitate a numerically simple and fast calculation of the hydrodynamic behaviour of aggregates. They do not impose any restriction on particle shape and overlap, but provide only approximate solutions, which are accurate to first order. Conversely, the multipole velocity expansion (MVE) and the multipole force expansion (MFE) allow for a highly accurate calculation of the hydrodynamic drag of aggregates. However, both methods are restricted to mainly spherical particles that do not overlap. In order to conduct precise hydrodynamic calculations for aggregates of irregularly shaped and/or sintered particles, several numerical techniques can be employed. It could be shown that these techniques reproduce experimental data (Geller et al. 1993) and that they are in very good agreement with the multipole expansion techniques (Filippov 2000; Binder et al. 2006). Consequently, there are several well-studied algorithms for calculating the hydrodynamic drag of aggregates consistently and in agreement with experiment. These provide a sound platform for studying the impact of aggregate size and structure on the hydrodynamic behaviour in detail.

4.4.3 Hydrodynamic Drag on Fractal Aggregates

Results of this section are based on MVE method—mainly because of its high accuracy and because it allows for insight into the velocity field of aggregates. The truncation level was set to $L = 2$ in order to study aggregates with up to 1000 primary particles. According to Fig. 4.15, that means an error of less than 0.1 %.

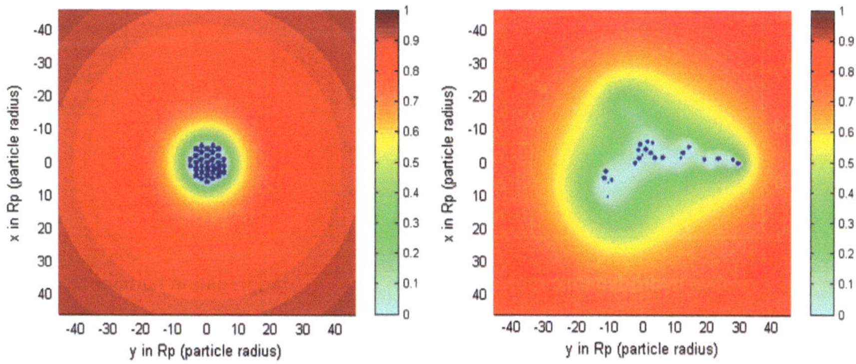

Fig. 4.16 Velocity field around an hcp aggregate (*left*) and a DLCA aggregate (*right*) with N = 150 for creeping flow into the projection area; colour plots show the normal velocity component normalised with velocity of the undisturbed flow field $\mathbf{u}^{\infty}$

In Fig. 4.16, comparison is made between a hexagonal close-packed (hcp) and a DLCA aggregate ($N = 150$). The direction of flow is normal to the projection area, while particles are at rest. Velocities are normalised by the velocity $|\mathbf{u}^{\infty}|$ of the undisturbed flow. The colour maps show the velocity component in the direction of flow; the undisturbed flow is shown as dark red, zero values are indicated with light green, and the particles are depicted in deep blue.

At first, the flow field of the hcp aggregate reveals that the velocity perturbations propagate into space at the order of some multiples of the aggregate size ($O(R_C)$), even though these perturbations are only strong in a small zone around the aggregate. The flow in the pores of the hcp-aggregate is virtually zero. Similar statements apply to the DLCA aggregate, which is approx. 3–4 times larger the hcp aggregate. Again, there are long ranging weak perturbations into the "outer space", rather strong perturbations in the vicinity of the aggregate and almost no flow in the inner region. The shape of the flow field is obviously determined by the asymmetric shape of the convex hull, although there is no real aggregate surface where a boundary condition could apply to. The lack of a significant internal flow agrees principally with calculated flow patterns for permeable spheres with radially varying porosity (Veerapaneni and Wiesner 1996; Vanni 2000). This is due to the screening of the external velocity field by the primary particles, which has a range in the order of the primary particle size ($O(a)$).

Based on the solution of the velocity field, the force and torque acting on the aggregates can be calculated and the hydrodynamic equivalent diameters for translation ($x_{h,t}$) and rotation ($x_{h,r}$) can be derived. They scale with the aggregate mass via a power law that reflects the structural properties of the aggregate. That means, while for hcp aggregates the aggregate mass (N) is proportional to the third power of x_h, there is a fractal-like relationship for DLCA aggregates with a hydrodynamic dimension d_h close to the fractal dimension (Fig. 4.17, left; cf. discussion to Kirkwood-Riseman theory on pp. 164). This fractal relationship

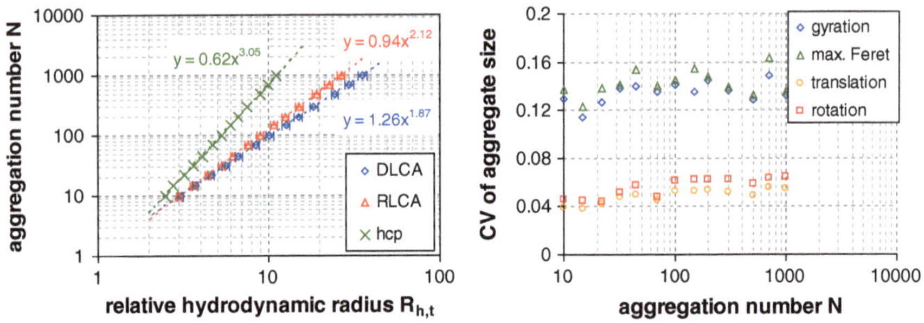

Fig. 4.17 *left* Power law relationship between the aggregation number N and the hydrodynamic radius of translation $R_{h,t}$ for DLCA, RLCA, and hcp aggregates; *right* coefficient of variation (CV) for differently defined aggregate sizes within populations of DLCA aggregates of uniform aggregation number N (for 100 aggregates per N)

represents a population averaged correlation and does not (necessarily) predict the behaviour of a specific aggregate. In fact, similar to morphological size parameters, there is a significant variation in x_h for a given aggregate mass (N). However, this variation is notably weaker than that of morphology, i.e. for x_g and $x_{Feret,max}$ (Fig. 4.17, right). The reason is that the values of x_g and $x_{Feret,max}$ are dominated or defined, respectively, by the largest interparticle distances r_{ij}, which are few in number and show significant stochastic variation among aggregates of a given aggregation process. In contrast, the hydrodynamic behaviour of aggregates can be related to the harmonic mean of the interparticle distances (Eq. (4.84)); it is, thus, primarily determined by the small interparticle distances, which reflect the aggregate's self-similarity. Hence, the observed variation in the hydrodynamic diameter is solely due to the variation in structure (cf. Sect. 4.2.3.4), while the variation of x_g and $x_{Feret,max}$ is additionally affected by configurational diversity.

The power law correlation between the hydrodynamic diameters and the aggregate mass N yield hydrodynamic dimensions ($d_{h,t}$ and $d_{h,r}$) which are close to the geometrically defined fractal dimension (d_g); they are, thus, characteristic for a certain type of aggregate. For small and moderate aggregation numbers N, a slight dependency of d_h from N can be observed (Fig. 4.18), which corresponds to an analogous effect in d_g and which coincides with a clear deviation between the various fractal dimensions. Their correlation plot (Fig. 4.19) reveals, that these differences diminish with increasing aggregation number N and tend to a uniform value. However, convergence obviously occurs only at very large aggregation numbers ($N \gg 1000$). Since experimental results of aggregate systems frequently indicate small and moderate aggregation numbers ($N \leq 1000$) (e.g. Sorensen 2001; Kammler et al. 2005; Boldridge 2010; Babick et al. 2012b), the difference in scaling laws may matter.

Once the functional relationship between the hydrodynamic diameters and the aggregate mass is known, one can establish simple models for the average dynamic properties (e.g. for the settling velocity) of suspensions containing aggregates. However, while the aggregate mass is well defined for numeric aggregates, it is difficult to measure in real-life experiments. It is more likely that the radius of

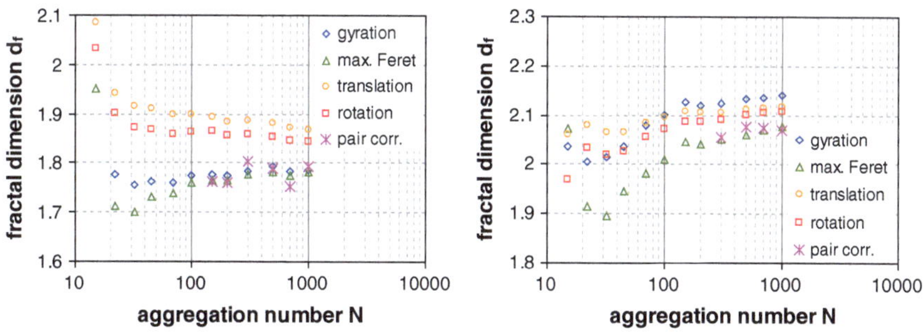

Fig. 4.18 Fractal dimension with regard to the hydrodynamic diameters of translation and rotation and with regard to the diameter of gyration, the maximum Feret-diameter, and the pair correlation function; for DLCA (*left*) and RLCA (*right*) aggregates

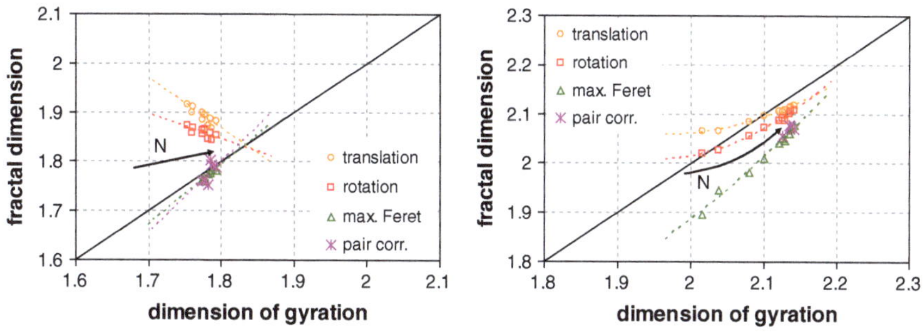

Fig. 4.19 Correlation of the various fractal dimensions with the fractal dimension of gyration; *left* DLCA aggregates, *right* RLCA aggregates (cf. Babick et al. 2012a)

gyration is known from scattering experiments. For that reason, many authors present values for the ratio of the hydrodynamic diameter to that of gyration. Figure 4.20 depicts computed values of this ratio for DLCA and RLCA aggregates. For this purpose, the various approaches discussed in the previous section were employed. Apart from the MVE solution, the graphs show curves for the porous-sphere approach (Gmachowski 1996; Vanni 2000), the Kirkwood-Riseman theory (Eq. (4.84)) and its explicit solution for spherical symmetry (Eq. (4.86)), as well as for Eq. (4.96), which is based on FDM calculations. The shift of the fractal scaling laws for small aggregates was accounted for.

It can be seen that the Kirkwood-Riseman theory considerably underestimates the hydrodynamic size (by approx. 10 %) even for large clusters. This probably results from ignoring the hydrodynamic many-body interactions within the aggregate, which are more pronounced the more compact the aggregate structure is. On the other hand, the simple correlation by Hess et al. considerably overestimates the hydrodynamic radius. This is because they assume a smooth radial density distribution ($\rho(r) \propto r^{d-3}$), which ignores the anisotropy and local density variations of real CCA-formed aggregates.

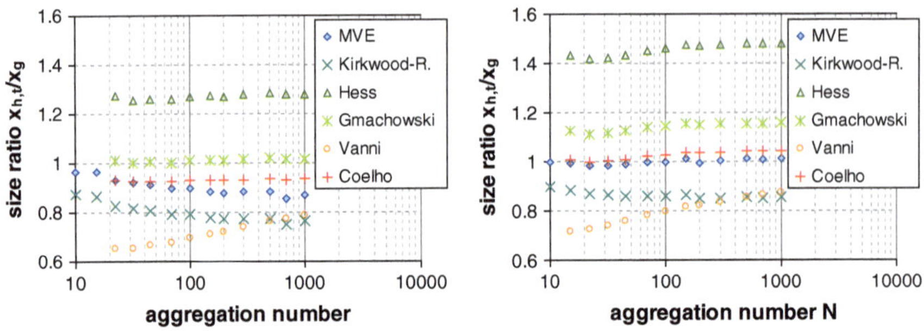

Fig. 4.20 Ratio of $x_{h,t}$ to x_g for DLCA (*left*) and RLCA (*right*) aggregates calculated with multipole velocity expansion (MVE), Kirkwood-Riseman theory (Eq. (4.84)) and its solution by Hess (Eq. (4.86)), as well as with porous sphere approaches from Gmachowski (1996) and Vanni (2000), and based on FDM results by Coelho et al. (1997; Eq. (4.96))

The porous sphere approach by Vanni (2000) adopts essentially the same concept for the internal aggregates structure. Yet, this approach yields smaller hydrodynamic diameters than the multipole expansion method because the local porosity at the convex hull, i.e. the decisive parameter of the model, is systematically underestimated when the aggregate is considered spherical.

The best agreement is found with the approximate expression of Coelho et al. (Eq. (4.96)) that was derived for aggregates with tuneable fractal dimension. This outcome confirms the above given statement that the results of numerical techniques and of the multipole methods are consistent (when appropriately designing the numerical algorithms).

4.5 Characterisation of Suspensions of Pyrogenic Powders

Pyrogenic powders are particle systems produced in a gas phase reaction. These powders are composed of relatively rigid, submicron aggregates of nanosized primary particles with a fractal-like morphology. The aggregates themselves may further agglomerate to large, micrometre-sized clusters. Pyrogenic materials like carbon black and fumed oxides experience manifold applications (e.g. as thickeners or anticaking agents); they are produced in considerable quantities (e.g. carbon black ≥ 10,000,000 tons/a, IARC 2010) and have a global market of several billions of US dollars.

A typical example of such materials is fumed (or: pyrogenic) silica, which is produced by a vapour phase hydrolysis of silicon tetrachloride $SiCl_4$ in an oxygen-hydrogen flame (Brinkmann et al. 2006):

$$2\,H_2 + O_2 + SiCl_4 \rightarrow SiO_2 + 4\,HCl$$

Global production rates of fumed silica rose from 100,000 tons/year in 1991 (Ettlinger 2002) to 300,000 tons/year in 2009 (Ying 2010). It has found application as a polishing agent in CMP slurries for wafer production, as a thixotropic additive in paints, or for rubber reinforcement (Barthel et al. 2002). Schaefer and Hurd (1990) described the synthesis of fumed silica as a multiscale process. Ballistic aggregation (polymerisation) of silica molecules leads to dense clusters that grow in size and solidify by sintering. These primary particles start to aggregate in a diffusion limited cluster-cluster aggregation process, where at first small chain-like aggregates and eventually large porous aggregates are formed. The resulting morphology (size and surface roughness of primary particles, aggregation number) can be controlled via flame temperature and production rate. Schaefer and Hurd measured fractal dimensions of 1.9 for the aggregate structure, which is close to the value of an ideal DLCA process (cf. Sect. 4.1.2). Other authors reported on fractal dimensions in the range from 1.8 to 2.0 for commercial fumed silica of varying grade and from different manufacturers (Kammler et al. 2004; Kätzel et al. 2008b; Ibaseta and Biscans 2010; Boldridge 2010) (Fig. 4.21).

Note that the fractal DLCA-like aggregates which are generated in the flame frequently agglomerate to large, micrometre-sized clusters. This may either occur during the combustion process or later when the pyrogenic powders are stored in silos or big packs (Kammler et al. 2004; Tsantilis and Pratsinis 2004; Ibaseta and Biscans 2010). These agglomerates are relatively compact with fractal dimensions of 2.5.

Aggregates of pyrogenic powders are polydisperse with regard to aggregate mass and primary particle size. However, qualitative and quantitative analyses of TEM and SEM images show, that, within an aggregate, the size distribution of primary particles is rather narrow (Kätzel et al. 2008b; Boldridge 2010), whereas the aggregate-averaged size of primary particles $x_{p,av}$ (or a_{av}) is broadly distributed. Moreover, the results of different authors suggest that the mean size of the primary particles (e.g. their Sauter diameter) anticorrelates with the aggregation number N (Kammler et al. 2004; Kätzel et al. 2008b). This correlation is not very strict, yet it may be roughly approximated as inversely proportional.

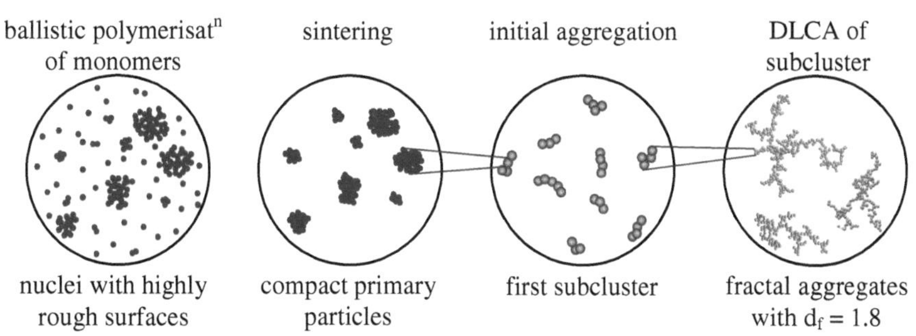

Fig. 4.21 Schematic model for aggregate formation during flame synthesis of SiO_2 (acc. to Schaefer and Hurd 1990)

In order to quantify the morphological properties of fractal pyrogenic aggregates, a variety of characterisation techniques can be used. Size distribution of primary particles and aggregate structure can be studied directly from electron microscope images or with scattering techniques (SLS, SAXS, SANS). The latter facilitate an easy determination of the fractal dimension and can even be used to measure the aggregate size. Aggregate size and structure can alternatively be determined via the dynamic aggregate properties (e.g. diffusion coefficient, settling velocity). That means for suspensions that such properties are, e.g., accessible with dynamic light scattering and centrifugation analysis (cf. Chap. 2).

The intention of this section is to discuss principles, data analysis and performance of different measurement techniques with regard to the characterisation of suspensions of pyrogenic aggregates.

4.5.1 Characterisation of Pyrogenic Aggregates with Static Scattering Techniques

Scattering techniques measure the intensity of a scattered signal versus the value q of scattering vector (cf. Sect. 2.3.4). The principal shape of such a scattering curve is shown in Fig. 4.22.

The curve was calculated for q-values from 0.3 μm^{-1} to 0.3 Å^{-1} and, thus, reveals properties of the whole aggregates as well as of the surface of the primary particles. However, such a broad q-range cannot usually be covered by a single scattering instrument. Typically, static light scattering (SLS) is operated in the range from 0.001 to 0.03 nm^{-1}, while laboratory SAXS instrumentation covers

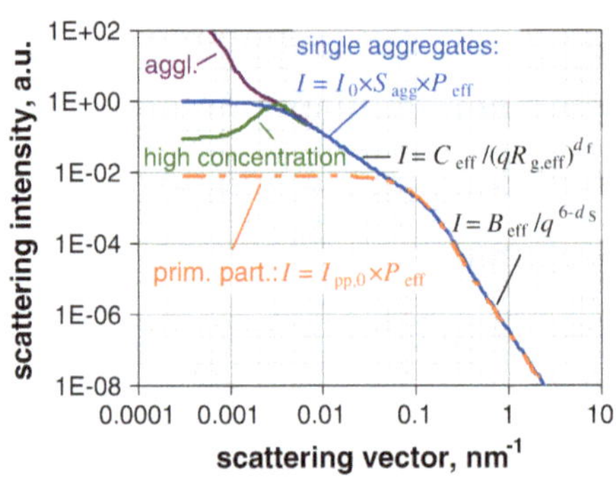

Fig. 4.22 Shape of the global scattering curve for a population of DLCA aggregates consisting of polydisperse primary particles (log normal distribution with $x_{p,av,50}$ = 20 nm and σ_{ln} = 0.4) and obeying the self-preserving cluster-mass distribution (Eq. (4.10), $\overline{N}$ = 200, $R_{g,eff}$ = 311 nm); the blue solid line is the calculated curve for sufficiently separated individual aggregates, while the two other solid lines at low q-values indicate the principal shape of the scattering curve for agglomerated aggregates (e.g. when measured in powders) and for highly concentrated suspensions of aggregates, where ordering effects cannot be ignored; additionally, the contribution of the primary particles to the scattering curve is indicated

larger q-values (0.03–4 nm^{-1}). This means for the pyrogenic aggregates (made up of nanoscaled particles) that SLS allows for a determination of aggregate parameters (radius of gyration, fractal dimension), while SAXS can be applied to study the primary particles (shape, size distribution) and to determine the fractal dimension. A global view on an aggregate system with one scattering experiment can be obtained from USAXS instrumentation ($q = 0.001$–4 nm^{-1}, cf. Sect. 2.3.4.4) which, however, is bound to a synchrotron X-ray source.

A log-log plot of the scattering curve as in Fig. 4.22 immediately reveals important morphological properties of the aggregate system. There are (at least) two linear regions that correspond to the structure dominated scattering and the surface scattering, respectively. The latter occurs at large q-values ($qx_p \gg 1$) where the scattering signal originates from the interaction of wave with the particle surface and where the slope in the log-log plot is a measure of the surface fractal dimension d_S ($I = B_{eff}/q^{6-d_S}$). For smooth particles, d_S is 2 and the intensity declines with the 4th power of q. A second linear region that appears at moderate q-values ($x_{agg}^{-1} \ll q \ll x_p^{-1}$) reflects the interparticles distances r_{ij} within the aggregate. In this range, the scattering curve declines with q^{-d_f}. At low q-values ($qx_{agg} < 1$), the scattering is due to the spatial configuration of the aggregates to each other. In this range, scattering signals are expected to be independent from q if the aggregates are randomly distributed in space (i.e. for highly diluted suspensions). However, if a dense suspension of stable aggregates is probed, the scattering curve will exhibit a local maximum (Bugnicourt et al. 2007). In contrast, the presence of meta-structures (agglomerates) is indicated by a further declining region in the scattering curve (Kammler et al. 2004; Kammler et al. 2005).

The size of the aggregates and of the primary particles can be roughly estimated by the values q_{cr} at which the transition between the different scattering regions occur (Eq. (4.63)). The common procedure is, however, to plot the logarithm of the scattering intensity versus the squared scattering vector (Guinier plot) and to determine the slope of that curve for vanishing q. According to Eq. (4.42), this slope is proportional to the squared radius or diameter of gyration. Prerequisite for this procedure is the absence of meta-structures and of suspension structuring. Moreover, since scattering at $q = 0$ cannot be measured, the required slope (at $q \to 0$) has to be extrapolated from experimental data, which inevitably increases the measurement uncertainty of this parameter (Babick et al. 2012c).

An alternative data analysis consists of fitting experimental scattering curves with model equations. Such equations can be established by combining analytical expressions for the form factor $P(q)$ (e.g. Eq. (B.54)) with those for the aggregate structure (e.g. Eq. (4.26)). Hurd et al. (1987)[6] proposed a four-parametric equation that is based on the Fisher-Burford approximation of the aggregate structure factor (Eq. (4.24)):

[6]The equations in that paper contain misprints. A corrected version is given by Schaefer and Hurd (1990).

$$I_{\mathrm{sca}}(q) = I_{\mathrm{sca},0} \cdot \left(1 + \frac{2}{3d_{\mathrm{f}}} R_{\mathrm{g,eff}}^2 q^2\right)^{-d_{\mathrm{f}}/2} \cdot \left(1 + \frac{2}{3(6 - d_{\mathrm{S}} - d_{\mathrm{f}})} a_{\mathrm{eff}}^2 q^2\right)^{-(6-d_{\mathrm{S}}-d_{\mathrm{f}})/2}. \tag{4.97}$$

A further scattering equation was developed by Beaucage (1995, 1996), who distinguished between the incoherent scattering of all primary particles (I_{pp}) at large q-values and the aggregate scattering (I_{agg}) for small and moderate scattering vectors $\mathbf{q}$:

$$\begin{aligned} I_{\mathrm{sca}} = I_{\mathrm{agg}} + I_{\mathrm{pp}} = &\left[I_{\mathrm{sca},0} \exp\left(-\frac{q^2 R_{\mathrm{g,eff}}^2}{3}\right) + \frac{C_{\mathrm{eff}}}{(qR_{\mathrm{g,eff}})^{d_{\mathrm{f}}}} \operatorname{erf}\left(\frac{qR_{\mathrm{g,eff}}}{\sqrt{6}}\right)^{3 \cdot d_{\mathrm{f}}} \cdot \exp\left(-\frac{q^2 a_{\mathrm{eff}}^2}{5}\right)\right] \\ &+ \cdots \left[I_{\mathrm{pp},0} \cdot \exp\left(-\frac{1}{5} q^2 a_{\mathrm{eff}}^2\right) + \frac{B_{\mathrm{eff}}}{q^{6-d_S}} \cdot \operatorname{erf}\left(\frac{q a_{\mathrm{eff}}}{\sqrt{10}}\right)^{18-3d_{\mathrm{S}}}\right]. \end{aligned} \tag{4.98}$$

The mathematical structure of this equation (six adjustable parameters) is flexible enough to fit a large variety of experimental data and, thus, facilitates their analysis with respect to morphological parameters. Both equations predict a smooth shape of the scattering curve in the surface regime ($qx_{\mathrm{p}} \gg 1$), which is reasonable for polydisperse primary particles, as in the case of pyrogenic powders. On the other hand, both equations restrict the functional space for the scattering function and, thus, cover only a limited range of aggregate morphologies. Consequently, the fit parameters do not necessarily reflect the true structure of the aggregate. This warning applies in particular to the size parameters a_{eff} and $R_{\mathrm{g,eff}}$ and to the primary particle contribution $I_{\mathrm{pp},0}$, while the influences of the fractal dimensions d_{S} and d_{f} are correctly expressed.

Once having obtained an analytical description of the scattering curve, it is possible to calculate the Porod invariant Q (cf. Eq. (2.25)) from the primary particle contributions

$$Q = \int I_{\mathrm{pp}}(q) \cdot q^2 \mathrm{d}q, \tag{4.99}$$

and to use this value for the determination of the Sauter diameter x_{ST} of the primary particles:

$$x_{\mathrm{ST}} = \frac{6}{\pi} \cdot \frac{Q}{B_{\mathrm{eff}}}. \tag{4.100}$$

Moreover, since the scattering regions are sensitive to different morphological properties (e.g. to the total volume at low q and to the total surface at large q), it is possible to derive parameters for the aggregate polydispersity. Assuming a log-normal size distribution for the primary particles (with median $x_{\mathrm{p},50}$ and logarithmic standard deviation $\sigma_{\ln}$), one may define a polydispersity index *PD*, which is related to $\sigma_{\ln}$ (cf. Kammler et al. 2005):

$$PD \underset{\text{def}}{=} \frac{B_{\text{eff}} x_{\text{ST}}^4}{72 I_{\text{pp},0}} = \mathrm{e}^{-6\sigma_{\ln}^2}, \tag{4.101}$$

The average aggregation number can be estimated from:

$$I_0 / I_{\text{pp},0} = \overline{N^2} \Big/ \overline{N} = \overline{N} \cdot \left(1 + \sigma_N^2 / \overline{N}^2\right). \tag{4.102}$$

Kammler et al. (2005) showed further that even the total number concentration of the primary particle $c_{N,\text{p}}$ can be derived:

$$Q^2 / I_{\text{pp},0} = \frac{9}{4}\pi^2 K' \cdot c_{N,\text{p}} \cdot L \cdot \mathrm{e}^{-9\sigma_{\ln}^2}, \tag{4.103}$$

where K' is a material constant and L the length of measurement zone. However, the last equations rely on the knowledge of $I_{\text{pp},0}$, which is rather difficult to extract unambiguously from experimental scattering curves.

Scattering techniques have been widely used for the characterisation of pyrogenic powders. One focus of such investigations was the understanding of the particle and aggregate formation in a flame reactor. Schaefer and Hurd (1990) combined scattering curves from SANS and SAXS with those measured by SLS in order to analyse different grades of a commercial pyrogenic silica product (BET ≥ 200 m^2/g). All powders showed fractal dimensions of the aggregate structure of approx. 1.9, but the fractal dimension of the surface was decreasing with the fineness of the primary particles, i.e. the surface roughness decreases with the growth of the primary particles. Based on their scattering data, the authors located any porosity to the surface and excluded the idea that the primary particles may consist of a compact core with a porous shell. Moreover, contrast matching SANS experiments showed that there is no occluded porosity in the particles.

Kammler et al. (2004) employed USAXS (0.001–4 nm^{-1}) for the characterisation of pyrogenic powders and looked at the impact of process parameters (e.g. hydrogen flow rate, production rate) on the aggregate morphology. They found a significant variation of the fractal dimension in dependence of the process parameters (d_f = 1.6–2.2 in vapour-fed and d_f = 1.9–2.0 in liquid-fed flame reactors). Analysis of the scattering curves by means of the unified scattering Eq. (4.98) delivered the mean aggregation number $\overline{N}$ and the Sauter diameter x_{ST} of the powders. The latter agreed fairly well with the Sauter diameter calculated from BET specific surface and correlated negatively with the mean aggregation number $\overline{N}$. In a further paper, Kammler et al. (2005) reported on in situ USAXS measurements at a flame reactor, which gave insight into the evolution of the morphological parameters along the flame axis.

While Kammler was focussing on the original aggregates formed during the production process, Kätzel et al. (2008b) investigated suspensions of commercial pyrogenic silica, which had undergone a uniform dispersion procedure (high power ultrasonication). Scattering experiments were accomplished with SLS at $q = 0.0035$–$0.025\ \text{nm}^{-1}$) and with SAXS ($q = 0.04$–$1\ \text{nm}^{-1}$) and did not provide a complete scattering curve for the full q-range. A distinct difference for the scaling laws in the fractal region was observed: While SLS measurements yielded fractal dimensions around 1.8, the SAXS measurements indicated values in the range of 2.3–2.5. The same powders, when analysed with SANS ($q = 0.02$–$2\ \text{nm}^{-1}$, Bugnicourt et al. 2007), showed fractal dimensions around 2.3. Kätzel et al. could not resolve this discrepancy, which is not observable in the USAXS data from Kammler et al. (2004). It is probably related to the fact that the fractal dimension values were derived from SLS data at the upper limit of the detected q-range and from SAXS/SANS data at the lower q-limit. The derived d_f-values are, thus, influenced by polydispersity because it broadens the transition between the different scattering regimes (cf. Sect. 4.3.3). Moreover, for SANS and laboratory SAXS instrumentation, the quality of the scattering curves at small q-values is derogated by smearing (cf. Sect. 2.3.4.4) because the necessary de-smearing procedures put an inevitable bias on the measurement data. Hence, such instruments should preferably be used for the characterisation of the primary particles, whereas synchrotron SAXS and USAXS equipment with vanishing smearing behaviour (cf. Sect. 2.3.4.4) allows for a reliable quantification of the aggregate structure.

4.5.2 *Characterisation of Pyrogenic Aggregates with Multi-angle DLS*

4.5.2.1 Effect of Aggregate Rotation and Polydispersity on DLS Experiments

The basic idea behind DLS measurements is that the detected scattering signal is caused by the interference of the scattering fields generated at all illuminated particles and, thus, depends on the particle positions. Each variation in the spatial arrangement of particles, therefore, coincides with a signal change. For individual spheres in a quiescent liquid, such a variation results from the particles' Brownian translation (in the absence of external fields). In the case of non-spherical particles or particle aggregates, i.e., when the scattering object cannot be described by just one scattering centre, the interference field is also affected by Brownian rotation. For soft agglomerates with elastic bonds and gelling structures, the signal fluctuation might even be attributed to vibrational motions. However, for rigid aggregates of pyrogenic powders it is fairly reasonable to neglect the latter and to solely consider translation and rotation.

The translation of individual aggregates is detected by DLS when the distance between the aggregates is larger than the resolution scale of the scattering experiment, i.e. the inverse scattering vector q^{-1}. This is achieved for sufficiently dilute suspension; the corresponding self-diffusion coefficient of the aggregate is given by

Eq. (4.72). Similarly, aggregate rotation becomes significant when the inverse scattering vector q^{-1} is sufficiently small to resolve the interparticle distances r_{ij} within the aggregates ($qr_{ij} > 1$ or $qR_g > 1$). A low scattering resolution, i.e., $qR_g \ll 1$, means that the aggregates contribute to the scattering signal as whole objects (volume scattering). In this case, only the aggregate translation is monitored. Hence, the meaning of the relaxation constant Γ of a DLS experiment and that of the corresponding diffusion coefficient D ($\Gamma = q^2D$) changes with q. The translational diffusion coefficient D_t is only measurable in the limit of perfect forward scattering ($q = 0$), while for finite scattering angles an apparent diffusion coefficient D_{app} is obtained ($D_{app} > D_t$).

The impact of the rotational diffusion on the DLS of aggregates was studied by Lindsay et al. (1988). Their mathematical treatment is based on a multipole expansion of the scattered fields in the limit of RDG scattering, a method previously developed to compute DLS data for anisotropic particles (Aragón and Pecora 1977). Lindsay et al. assumed, that the rotational and translational movements are decoupled—a reasonable approach when the aggregates show only moderate anisotropy (true for DLCA aggregates, cf. p. 160)—and considered vertical polarisation only. They derived analytical expressions for the field and intensity autocorrelation function and proved that they obey a Siegert-type relation (Eq. (2.30)), provided that there is a sufficiently large number (>100) of aggregates in the illuminated zone. For the apparent diffusion coefficient D_{app} of a population of monodisperse aggregates, they eventually obtained:

$$\frac{D_{app}}{D_t} = 1 + \frac{\Theta}{D_t q^2} \cdot \frac{\sum l(l+1)S_l}{\Sigma S} = 1 + \frac{3}{4}\frac{R_{h,t}}{R_{h,r}^3 q^2} \cdot \frac{\sum l(l+1)S_l}{\Sigma S}, \tag{4.104}$$

where S_l are multipole functions which expand the scattered fields in terms of spherical harmonics. The sum over all multipole function S_l is proportional to the static structure factor $S_{agg}(q)$:

$$\sum_l S_l(q) = \frac{1}{4\pi}\sum_{l,j}\frac{\sin(qr_{ij})}{qr_{ij}} = \frac{N^2}{4\pi}S_{agg}(q). \tag{4.105}$$

The multipole functions depend on the specific aggregate configuration and cannot be described universally. However, in the Guinier limit ($qR_g \ll 1$) they reduce to:

$$\lim_{qR_g \to 0} S_l = \frac{N^2}{4\pi} \cdot K_l \cdot (qR_g)^{2l+4\delta_{1l}}. \tag{4.106}$$

Hence, the dominant multipoles for the Guinier regime are S_0 and S_2 and the effective diffusion coefficient can be approximated by:

$$\lim_{qR_g \to 0} \frac{D_{app}}{D_t} = 1 + \frac{3}{4}\frac{R_{h,t}}{R_{h,r}^3 q^2} \cdot \frac{6S_2}{\Sigma S_l} = 1 + \frac{3}{4}\frac{R_{h,t}}{R_{h,r}^3 q^2} \cdot \frac{6K_2 q^4 R_g^4}{S_{agg}}. \tag{4.107}$$

Considering the general solution for static structure factor in the Guinier regime $(1-\frac{1}{3}(qR_g)^2)$ and approximately assuming proportionality between R_g and the hydrodynamic radii $R_{h,t/r}$, the following expression is obtained:

$$\lim_{qR_g \to 0} \frac{D_{app}}{D_t} = 1 + \text{const.} \cdot (qR_g)^2. \tag{4.108}$$

This is an important finding with regard to the analysis of experimental DLS data because the scattering at $q = 0$ is not accessible by experiment so that the parameter $D_{app}(0) = D_t$ has to be derived from extrapolation.

Figure 4.23 compares the apparent diffusion coefficients for populations of DLCA aggregates (relevant for pyrogenic powders) and RLCA aggregates containing 20–1000 primary particles. Multipole functions, hydrodynamic radii, and apparent diffusion coefficients were individually computed for 100 aggregates in each mass fraction (N) and subsequently averaged. The graph shows the normalised diffusion coefficients D_{app}/D_t. The q-axis is scaled with the aggregate radii of gyration R_g. It can be seen that the data points for a given type of aggregate lie on master curves. The differences between the two master curves of DLCA and RLCA are partly due to structure related differences in the multipole functions S_l. However, they primarily result from the different relations between the hydrodynamic radii ($R_{h,t}$ and $R_{h,r}$) and the radius of gyration. Both master curves attain maximum values at around $qR_g = 8$, which are approx. 1.54 for DLCA and 1.45 for RLCA aggregate. These values are slightly lower than those given by other authors (Lindsay et al. 1988, 1989; Lin et al. 1990a), which assumed identity for $R_{h,t}$ and $R_{h,r}$ and a constant ratio $R_{h,t}$ to R_g. For large arguments, a slight decline is observed, which is not seen for the approximate solutions as used e.g. by Lin et al. (1989) or Lattuada et al. (2004).

Real aggregate systems, whether formed by aggregation process or prepared by dispersing pyrogenic powders, will be polydisperse in aggregation number N and in

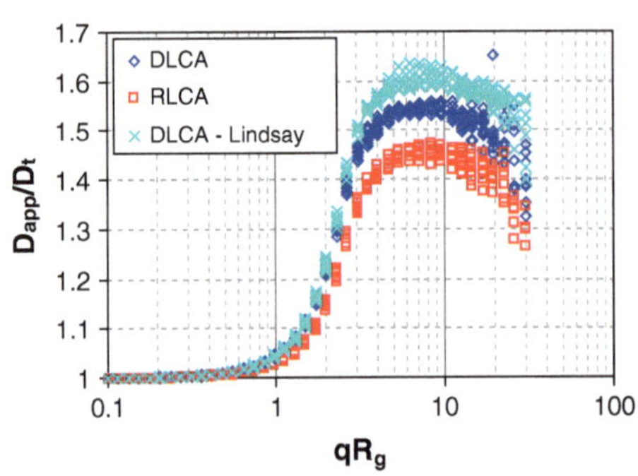

Fig. 4.23 Computed relationships between the normalised apparent diffusion coefficient $D_{app}/D_{app,0}$ and the product qR_g for monodisperse DLCA and RLCA aggregate populations (N = 10–1000); using individual values of R_g, $x_{h,t}$ and $x_{h,r}$ for aggregates (*curves* "DLCA" and "RLCA") and assuming a constant relationship between R_g and x_h ($\beta = 0.87$, acc. to Lindsay et al. 1988)

primary particle size x_p. Data analysis with the method of cumulants then yields the effective, i.e., intensity weighted, diffusion coefficient D_{eff}:

$$D_{eff}(q) = \frac{\iint D_{app}(q) N^2 a^6 \cdot S_{agg}(q) \cdot P(q) \cdot p(N|a) p(a) \mathrm{d}N \mathrm{d}a}{\iint N^2 a^6 \cdot S_{agg}(q) \cdot P(q) \cdot p(N|a) p(a) \mathrm{d}N \mathrm{d}a}. \tag{4.109}$$

Polydispersity enhances the impact of the scattering vector q on the measured diffusion coefficient. That is because the scattering intensity of large aggregates decreases more severely with growing q than that of small aggregates (cf. global scattering curve, Fig. 4.22). Consequently, the signal fluctuation at $q > 0$ is more affected by small particles than at $q = 0$. This holds true for any type of particle shape.

The influence of polydispersity is exemplarily shown in Fig. 4.24 for DLCA aggregates. Variation in primary particle size (monodisperse or log-normal distributed with $\sigma_{\ln,x} = 0.4$) as well as in aggregation number (log-normal distributed and a self-preserving distribution function, Eq. (4.10)) were considered. As in the previous section, a reciprocal correlation between $\overline{N}$ and $x_{p,av}$ was assumed for the calculation. The normalised diffusion coefficients are plotted versus a q-axis that is scaled with the effective radius of gyration $R_{g,eff}$ (cf. Eq. (4.59)):

$$R_{g,eff}^2 = \iint R_g^2 N^2 a^6 p(N|a) p(a) \mathrm{d}N \mathrm{d}a \Big/ \iint N^2 a^6 p(N|a) p(a) \mathrm{d}N \mathrm{d}a. \tag{4.110}$$

Due to the presumed anticorrelation between $\overline{N}$ and $x_{p,av}$, there is only a weak influence of the distribution in x_p. In contrast, the distribution of N has a clear effect.

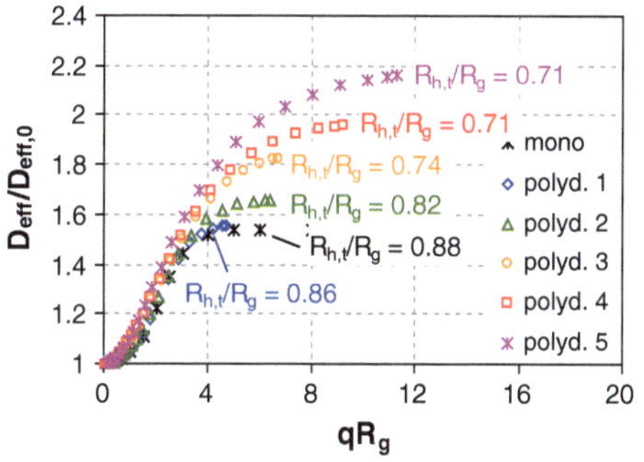

Fig. 4.24 Computed relationships between the normalised effective diffusion coefficient $D_{eff}/D_{eff,0}$ and the product qR_g for mono- and polydisperse DLCA aggregate populations ($x_{p,av,50}$ = 20 nm, $\overline{N} = 200$), where "polyd. 1" means uniform x_p and log-normal N ($\sigma_{\ln} = 0.2$), "polyd. 2": log-normal x_p ($\sigma_{\ln} = 0.4$) and log-normal N ($\sigma_{\ln} = 0.2$), "polyd. 3": uniform x_p and log-normal N ($\sigma_{\ln} = 0.6$), "polyd. 4": log-normal x_p ($\sigma_{\ln} = 0.4$) and log-normal N ($\sigma_{\ln} = 0.6$), "polyd. 5": log-normal x_p ($\sigma_{\ln} = 0.4$) and self-preserving N (Eq. (4.10)); additionally, a reciprocal correlation between $\overline{N}$ and $x_{p,av}$ is assumed; calculation for 475 nm wavelength and scattering angles: 0.1°–180°; indication is given for the ratio between the averaged hydrodynamic radius of translation to the effective radius of gyration

For a self-preserving distribution (curve "polyd. 5" in Fig. 4.24), the effective diffusion coefficient can grow to more than twice the value of the averaged translational diffusion coefficient (i.e. D_{eff} at $q = 0$). That means that the results of DLS instruments operating at scattering vectors $\mathbf{q}$ (e.g. different scattering angles) may considerably differ due aggregate rotation and polydispersity.

Consider a suspension of DLCA aggregates made up of polydisperse primary particles with a global average diameter $x_{p,av,50}$ of 20 nm, yet with almost uniform size $x_{p,av}$ within the aggregates. Each aggregate fraction with $x_{p,av}$ shows a self-preserving distribution of the aggregation number with the average value $\overline{N}$, which is reciprocal to $x_{p,av}$. The average aggregation is 200. For such a system, one obtains a radius of gyration of 426 nm and a hydrodynamic radius of translation of 301 nm. In that case, backscattering DLS experiments (i.e. at 180°) would yield an effective hydrodynamic radius $R_{h,eff}$ of 139 nm, while DLS measurements at a scattering angle of 30° would lead to $R_{h,eff} = 193$ nm; still an underestimation by more than a third.

Polydispersity does not only affect the q-dependency of the effective diffusion coefficient D_{eff}, but also influences the ratio of the hydrodynamic diameter to that of gyration ($R_{h,t}/R_g$). In Fig. 4.24, these ratios are given for each graph corresponding to a certain degree of polydispersity. Obviously, this ratio is lowered when the variation in aggregation number and primary particle size is increased. While for monodisperse DLCA population $R_{h,t}/R_g$ amounts to 0.88 (cf. Fig. 4.20), a ratio of 0.71 is expected for a self-preserving distribution cluster mass distribution with superimposed polydispersity in primary particle size. This kind of correlation results from the fact that the effective radius of gyration $R_{g,eff}$ is a rms-average of the intensity weighted distribution (i.e. a disproportionately high impact of large aggregates), whereas the effective hydrodynamic diameter of translation is a harmonic mean of the intensity weighted distribution (i.e. disproportionately highly influenced by small aggregates). The ratio between $R_{h,t}$ and R_g may, thus, support the evaluation of the polydispersity of an aggregate system.

4.5.2.2 Characterisation of Pyrogenic Powders with Light Scattering Techniques

The theoretical result that the effective diffusion coefficients of fractal aggregates form a master curve when plotted in a dimensionless manner as in Fig. 4.23 was first experimentally confirmed by Lin et al. (1989), who monitored the aggregation of monodisperse colloidal gold, silica, and polystyrene particles under DLCA and RLCA conditions. The shape of the master curves was used to quantify the polydispersity of the cluster mass distribution (for RLCA). Similar results are provided by Sandkühler et al. (2005b), who discussed i.a. the effect of gravitational settling and internal dynamics on the shape of the master curve. For pyrogenic materials, such an analysis of DLS data was first presented by Kätzel et al. (2008b). These authors examined fumed silica powders of different grade, which is defined by the BET and related to the fineness of the primary particles and the polydispersity of the aggregate mass. Their results were re-evaluated and supplemented with regard to the impact of the dispersion procedure by Babick et al. (2012c).

Both papers refer to aqueous suspensions of commercial fumed silica products (WACKER Chemie AG), which were vigorously dispersed by ultrasonication in order to destroy all non-fractal agglomerates. The light scattering experiments (wavelength: 632 nm) were performed with SLS instrumentation at scattering angles from 15° to 145° and with multi-angle DLS equipment in the range of 30°–150°. Since there are no measurement data for perfect forward scattering (0°) the effective radius of gyration $R_{g,eff}$ and the effective translational diffusion coefficient $D_{eff,0}$ were derived by extrapolation, which induces an uncertainty in the dimensionless presentation of the DLS results (Babick et al. 2012c). In order to check the universality of DLS data for pyrogenic aggregates, it is, therefore, more appropriate to normalise the effective diffusion coefficients with measured diffusion coefficients. For this reason, the reference point was chosen to be $qR_{g,eff} = 2$. The resulting dimensionless plot $D_{eff}/D_{eff,2}$ versus $qR_{g,eff}$ for all grades of fumed silica and for different states of dispersion is shown in Fig. 4.25.

Obviously, all data points in the range of $qR_{g,eff} = 1–4$ lie essentially on one curve. That means that all aggregates of the examined fumed silica product belong to the same class of aggregates and obey approximately a universal scattering behaviour. The experimental correlation of Fig. 4.25 can be approximated for $qR_{g,eff} \leq 4$ with the following equation:

$$\frac{D_{eff}}{D_{eff}\left(qR_{g,eff} = 2\right)} = a - \frac{b}{1 + c \cdot \left(qR_{g,eff}\right)^2}, \tag{4.111}$$

which yields a quadratic influence of $qR_{g,eff}$ on D_{eff} for small arguments as predicted by theory (Eq. (4.107)). The equation can further be used to extrapolate the effective translational diffusion coefficient $D_{eff,0}$ and the effective translational diameter $x_{h,t}$, respectively. Based on these values, Babick et al. (2012c) obtained values for the ratio $x_{h,t}/x_g$ in the range of 0.63–0.83. These values are lower than for ideal monodisperse DLCA and RLCA aggregates (approx. 0.88 and 1.01, respectively), yet they could apply to ideal fractal aggregates when polydispersity is accounted for

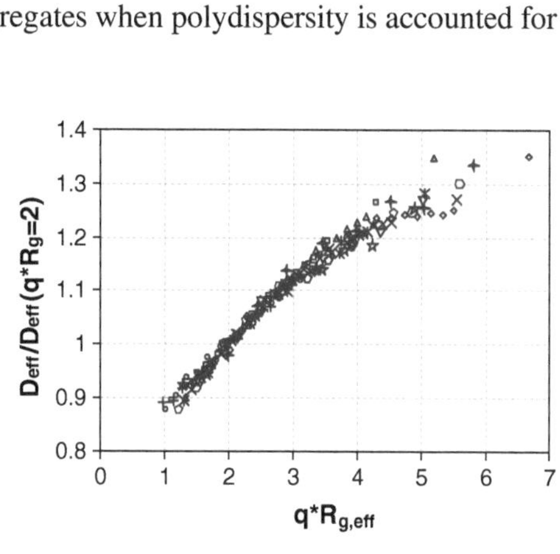

Fig. 4.25 Effective diffusion coefficients D_{eff} from MA-DLS normalised with the $D_{eff}(qR_{g,eff} = 2)$ versus the scattering vector q scaled with the effective radii of gyration $R_{g,eff}$ as determined by SLS, for 7 grades of pyrogenic silica and different dispersion procedures (Babick et al. 2012c)

(cf. Fig. 4.24). Slight deviations in the data curves of Fig. 4.25 at $qR_{g,eff} \geq 4$ indicate differences in the shape of the cluster mass distribution, but due to the limited amount of experimental data for large $qR_{g,eff}$-values, it remains unclear to which degree the polydispersity of the aggregate systems differ.

The empiric master curve for the DLS data of pyrogenic powders offers an opportunity to calculate the effective translational diffusion coefficient $D_{eff,0}$ or the effective hydrodynamic diameter from one DLS experiment, i.e. for DLS at one scattering angle. Babick et al. (2012c) gave the following implicit formulation:

$$\frac{x_{h.t}}{x_{h,eff}} = 1.754 - \frac{0.754}{1 + 0.0170 \cdot (q \cdot x_P)^2 \cdot \left(\frac{x_{h.t}}{x_p}\right)^{2.36}}, \tag{4.112}$$

where $x_{h,eff}$ is the hydrodynamic diameter calculated from the effective diffusion coefficient via the Stokes-Einstein-equation and x_p is the Sauter diameter of the primary particles.

When Eq. (4.112) is applied to the experimental DLS data, i.e. without any information from SLS, $x_{h,t}$ can be accurately computed for any scattering angle. Figure 4.26 shows the performance of such a data processing for the experiments of Fig. 4.25. Approximate constancy of the calculated $x_{h,t}$ values is observed for all suspensions. The total deviation is not larger than 9 % and the relative standard deviation is less than 2.5 %. Employing Eq. (4.112) to standard DLS measurements of pyrogenic silica powders may, therefore, significantly enhance the comparison of DLS results from different research laboratories.

Eventually, the experimental data from Fig. 4.25 can be compared with the theoretical predictions for monodisperse and polydisperse DLCA aggregates (cf. Fig. 4.24). There is a principal agreement between experiment and theory, even though the quantitative deviations cannot be neglected. To some extent, the deviations can be attributed to the uncertainty of the parameters $D_{eff,0}$ and $R_{g,eff}$, which are not directly measured, but derived from extrapolation. A systematic overestimation of $R_{g,eff}$ would, for instance, cause a stretching of the experimental master curve along the abscissa. However, the primary reason for the discrepancy between the experimental data and the DLCA curves is certainly the deviation of the real

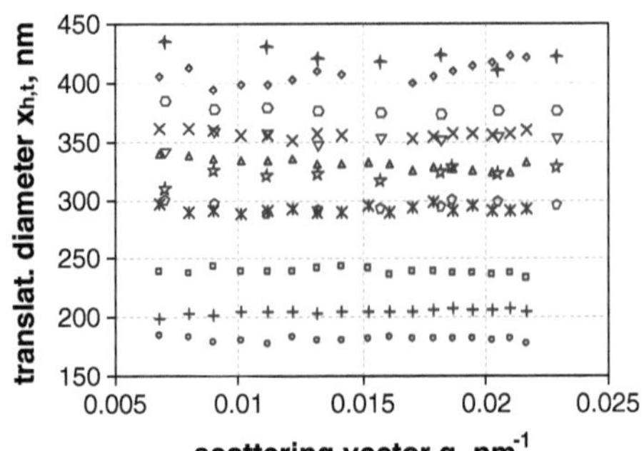

Fig. 4.26 Translational hydrodynamic diameter calculated from the data in Fig. 4.25 by means of Eq. (4.112)

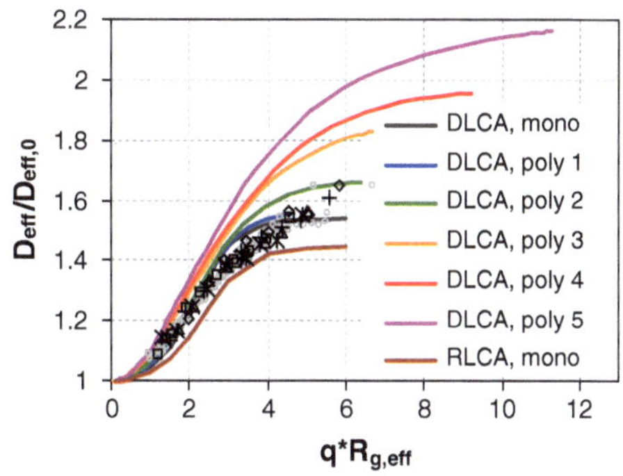

Fig. 4.27 Comparison of theoretical master curves ($D_{\text{eff}}/D_{\text{eff},0}$ vs. $qR_{g,\text{eff}}$) for DLCA and RLCA aggregate populations with the experimental values obtained for pyrogenic silica (Babick et al. 2012c); the legend of the polydisperse DLCA curves is as in Fig. 4.24

morphology of pyrogenic aggregates from that of ideal DLCA aggregates. That concerns the shape of the primary particles, the degree of sintering, the polydispersity, or the fractal dimension. In order to illustrate the effect of the latter factor, the theoretical master curve of monodisperse RLCA aggregates ($d_f = 2.1$) is included in Fig. 4.27 as well. According to it, the actual shape of the experimental master curve could indicate a fractal dimension slightly larger than that of ideal DLCA aggregates (i.e. $d_{f,\text{pyrogenic}} > 1.8$).

Summarising, it can be stated that the DLS behaviour of pyrogenic aggregates is principally understood and that, even for DLS experiments at only one (finite) scattering angle, the determination of the mean translational hydrodynamic diameter $x_{h,t}$ is possible. Nevertheless, there still remain open questions on the real structure (e.g. fractal dimension d_f or distribution type for the aggregate mass).

4.5.3 Sizing of Pyrogenic Aggregates with Optical Centrifugation and DLS

The two previous sections have shown that static scattering techniques and multi-angle dynamic light scattering (MA-DLS) can be used to gain insight into the morphology of colloidal aggregates. These techniques, however, require careful sample preparation and are typically time-consuming. Furthermore, they have to be operated by experienced staff with scientific background and the costs for acquisition and maintenance are usually high. For that reason, they are mainly found in specialised research facilities. Application laboratories in industry or academia, where particle systems are investigated for the purpose of basic material characterisation, quality assurance, or process control, typically employ conventional particle sizing techniques such as optical centrifugation analysis (OCA) or DLS at a fixed scattering angle (mostly in sideward and backscattering mode). Such instruments yield distributions of an equivalent diameter that is specific to the method or even instrument. Moreover, these distributions are weighted by a specific type of quantity. Most frequently, the intensity of a scattering signal, the extinction of a transmitted beam, or some other optical quantity is used. Any comparison of such different measurement results is meaningless if the structure-property-relationships are not properly accounted for. For fractal aggregates, such relationships have been developed in the course of this chapter. This section first discusses the performance of optical centrifugation and DLS when employed for aggregated particle systems. This is accomplished by means of calculated size distributions for DLCA aggregates. Subsequently, the experimental data of pyrogenic materials are evaluated with regard to the aggregation number and the translational hydrodynamic diameter. For that purpose, a priori knowledge (with regard to primary particle size or fractal dimension) is used.

4.5.3.1 Prediction of OCA and DLS Results for Systems of Colloidal Aggregates

Consider a population of DLCA aggregates of uniform primary particles (x_p = 20 nm) and a log normal distribution of the aggregation number N weighted by the aggregate number concentration ($N_{50,0}$ = 200 and σ_{ln} = 0.4). The solid material is silica and the continuous phase is water. By means of the fractal relationships for the diameter of gyration x_g and the translational hydrodynamic diameter $x_{h,t}$—cf. Sects. 4.2.3 and 4.4.3—it is possible to convert this distribution in number weighted size distributions. Corresponding distribution functions for the geometric aggregate size (i.e. the maximum Feret diameter $x_{Feret,max}$) and for the Stokes-diameter x_{Stokes} are obtained by employing Fig. 4.3 and Eq. (4.77), respectively. The comparison of these distribution functions reveals remarkable differences (Fig. 4.28, left). In particular, the Stokes diameter is much smaller than the other diameters since the employed definition of x_{Stokes} (Eq. (4.76)) assumes

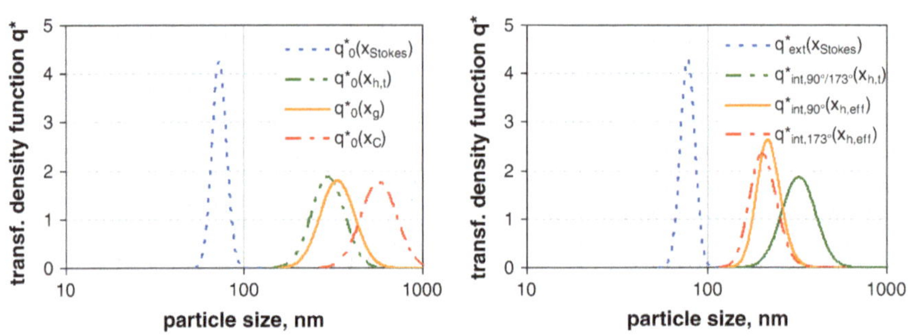

Fig. 4.28 *Left* Transformed number weighted density function of the Stokes diameter, the translational hydrodynamic diameter, the maximum Feret diameter (x_{Stokes}, $x_{h,t}$, $x_{Feret,max}$), and the diameter of gyration x_g, calculated for DLCA aggregates with primary particles of 20 nm diameter and a lognormal distribution of N ($N_{50,0}$ = 200 and σ_{ln} = 0.4); *right* corresponding density functions of the Stokes, the translational, and the effective hydrodynamic diameter (x_{Stokes}, $x_{h,t}$, $x_{h,eff}$) weighted by light extinction (q_{ext}) and scattering intensity at 90° and 173° ($q_{int,90°/173°}$) and number (q_0)

identity of particle and solid density. It is further worth mentioning that the hydrodynamically defined equivalent diameter is considerably smaller than the geometric size. This is different to compact, densely packed agglomerates, where the flow inside the pores is practically zero (cf. Fig. 4.16).

The next step is the conversion of these number weighted distributions in size distributions weighted by scattering intensity or light extinction, which requires a model for the scattering behaviour. For pyrogenic aggregates, which are composed of nanosized primary particles (x_p < 100 nm), one benefits from the fact that the intracluster multiple scattering can be neglected. Hence, their light scattering is well approximated by the Rayleigh-Debye-Gans (RDG) approach (cf. Sect. 4.3.2), thus simply related to the—known—static structure function S_{agg} (Eq. (4.26)).

Extinction weighted distributions are measured by OCA. Here, illumination with NIR radiation (λ = 880 nm) is assumed and the aperture of optical detection system is neglected. Light scattering is considered for two scattering angles (90° and 173°), for a He–Ne laser (λ = 632 nm), and vertical polarisation, which are frequent settings in DLS experiments. Additionally, the influence of aggregate rotation on the effectively measured hydrodynamic size has to be considered (Eq. (4.112)). That causes a systematic difference between the two DLS setups. The resulting distribution functions are shown in Fig. 4.28(right).

The main outcome is the striking difference between the OCA and DLS size distributions ($q_{ext}(x_{Stokes})$ and $q_{int}(x_{h,eff})$, respectively). According to the former, the aggregate systems would be clearly classified as nanomaterial (x < 100 nm), whereas the latter would indicate the absence of nanoparticles, even though the effective hydrodynamic diameters $x_{h,eff}$ measured by DLS are smaller than the translational hydrodynamic diameters $x_{h,t}$.

> The effective hydrodynamic diameter $x_{h,eff}$, i.e. the immediate result of a DLS measurement, is frequently taken to describe the aggregate size. It should be noted that for ideal DLCA aggregates this parameter may be just half of the diameter of gyration x_g and only a third of the geometric aggregate size $x_{Feret,max}$.

The conversion from the distributions in the left diagram of Fig. 4.28 to those in the right one is twofold because it concerns the particle property as well as the type of quantity. For OCA and DLS, the distributions are *extinction weighted* and *intensity weighted*, respectively. However, in practice it is typically much more important to know the *volume* or *number* weighted distribution functions. For the given DLCA population, the differently weighted distribution functions of $x_{h,t}$ are compared in Fig. 4.29. This comparison includes the intensity weighted distribution of compact spheres with diameter $x = x_{h,t}$ obeying Rayleigh scattering ($I_{sca} \propto x^6$) as well as Mie scattering (assuming silica particles in water). The intensity weighted distributions q_{int} for the DLCA aggregates are very close to the volume weighted distribution q_3 and hardly show any influence of the scattering angle. In contrast, fine compact particles would either be extremely sensitive to coarse particle fractions (below <100 nm, Rayleigh scattering) or may become strongly dependent on the scattering angle and insensitive for certain size fractions (above 100 nm, Mie scattering).

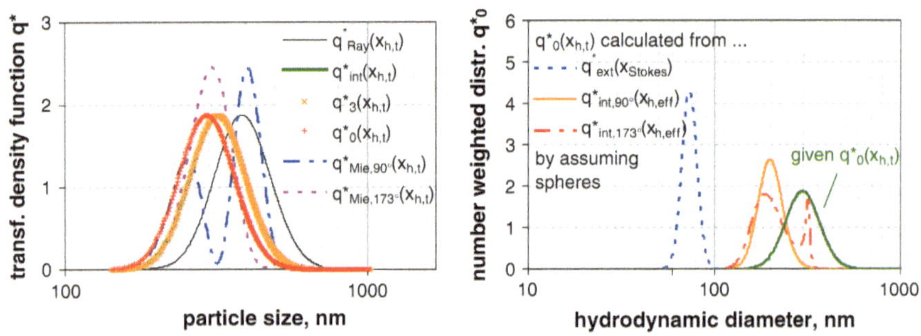

Fig. 4.29 *Left* Density function of the translational hydrodynamic diameter $x_{h,t}$ weighted by number (q_0), volume (q_3), and light scattering intensity (q_{int}), assuming Rayleigh scattering for the whole aggregate ($I_{sca} \propto x^6$) and Rayleigh-Debye-Gans scattering for scattering angles at 173° and 90°; calculated for the DLCA aggregate population of Fig. 4.28; *right* number weighted distributions of translational hydrodynamic diameter $x_{h,t}$ when the distribution functions on the r.h.s. of Fig. 4.28 are converted under the assumption of spherical SiO_2 particles experiencing Mie scattering

The close similarity of q_3 and q_{int} is characteristic for the fractal scattering regime, i.e. for $x_{agg}^{-1} < q < x_p^{-1}$, in which the scattering intensity scales with $\overline{N} \cdot x_p^{6-d_f}$, while the aggregate volume scales with $\overline{N} \cdot x_p^3$ (cf. Eqs. (4.45) and (4.61)). Differences between q_3 and q_{int}, thus, mainly reflect the polydispersity of the primary particles x_p (Babick et al. 2012a).

Eventually the exemplary DLCA population is used to illustrate the effect of ignoring the aggregate structure when analysing DLS and OCA data. For that purpose the virtual measurement distributions in Fig. 4.28—$q_{ext}(x_{Stokes})$, $q_{int,90°}(x_{h,eff})$, and $q_{int,173°}(x_{h,eff})$—are converted to number weighted distribution functions by assuming spherical particles (i.e. $x_{h,eff} = x_{Stokes} = x_{h,t}$) and Mie scattering (for silica; Fig. 4.29, right). The latter coincides with a non-monotonic, oscillating influence of size on the scattering intensity and leads to an artificial multimodality (in particular for DLS in backscattering mode), which might be misunderstood as the presence of different particle populations. Without appropriate consideration of the aggregate morphology, it is, therefore, not possible to derive meaningful results on aggregate size distribution from conventional sizing techniques.

4.5.3.2 Deriving Aggregate Morphology from OCA and DLS Measurements

The situation that the principal morphology of particles is completely unknown before characterisation may occur during the development of new routes for particle synthesis or when taking samples from the environment. However, when particle characterisation is part of quality assurance and process monitoring in industrial practice or for research applications, there is always prior knowledge on the particle shape or the state of aggregation. This also applies to pyrogenic materials, which are known to consist of fractal aggregates with structures similar to that of ideal DLCA aggregates (p. 174). Moreover, when using such materials, such as polishing or thickening agents, products of defined grade are employed. Hence, the average size of primary particles is known as well. What has to be measured is the aggregate size distribution after dispersing the powders or during product application under hydrodynamic stress. *Size* may refer to the translational hydrodynamic size (e.g. relevant for diffusion) or to the aggregation number (e.g. relevant for the total number of primary particles) and is preferentially weighted by number or volume. Such distributions can in principle be gained from OCA or DLS measurements by means of the mathematical framework presented in this chapter. The following pages discuss such a way of analysis for a series of measurements conducted with different grades of fumed silica (Babick et al. 2012b).

Experiments were conducted with commercial fumed silica products (WACKER Chemie AG) distinguished by the BET surface (from 50 to 300 m^2/g). The products are essentially the same as those used for the MA-DLS experiments in Sect. 4.5.2. However, samples were taken from different lots than before and different dispersion procedures were employed. Aqueous suspensions were prepared at 1 wt% and diluted to 0.1 wt% for DLS and centrifugation experiments. The optical centrifugation was accomplished with a cuvette centrifuge (LUMiSizer, L.U.M. GmbH) using a NIR detector (880 nm), while DLS experiments were performed with a cross-correlation technique (NANOPHOX, Sympatec) at 90° and an auto-correlation backscattering technique (HPPS, Malvern; 173°, both with laser wavelength of 632 nm).

The characterisation equipment corresponds to the parameters assumed for the theoretical considerations above. The measurements were analysed with regard to the extinction weighted distribution of the Stokes diameter $q_{ext}(x_{Stokes})$ and with regard to the intensity weighted distribution of the effective hydrodynamic diameter $q_{int,\theta}(x_{h,eff})$. While for centrifugation the size distribution could directly be derived from the extinction profiles (cf. Sect. 2.3.3.1), instrument specific algorithms were used for inverting the correlation spectra in the case of DLS (cf. Sect. 2.3.5.1). These algorithms were operated with standard parameters.

A typical result of this comprehensive characterisation is exemplarily shown for one material in Fig. 4.30. In this example, as for all other powders, the measured size distributions complied with the expectations from theory. That is, the backscattering DLS instrument measures slightly finer distributions than the 90°-DLS instrument and both yield considerably coarser distribution functions than the optical centrifuge. Additionally, there is a systematic difference between the distribution widths of the DLS instruments, which can be assigned to the specific inversion algorithms. Indeed, polydispersity values obtained by the method of cumulants (Eq. 2.34) perfectly agreed.[7]

[7]A more detailed analysis in Babick et al. (2012b) revealed that the polydispersity of the computed distribution q_{int} contradicts the "physical" polydispersity obtained by the method of cumulants in the case of the 90° DLS instrument, while they agree well for the backscattering DLS instrument.

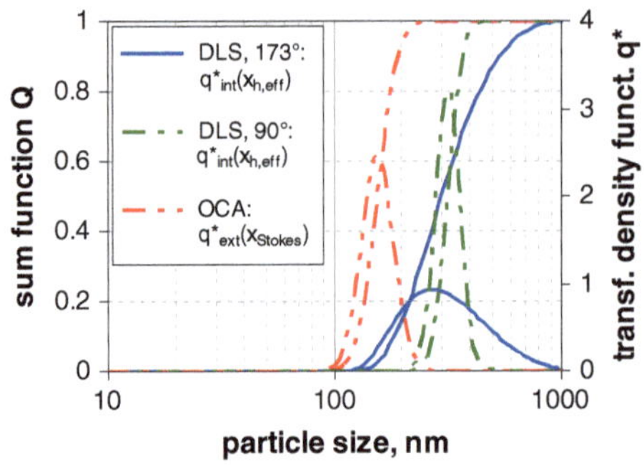

Fig. 4.30 Measured size distributions from optical centrifugation analysis (OCA) and dynamic light scattering (90° and 173°), i.e. distribution of the effective hydrodynamic diameter $x_{h,eff,90°/173°}$ weighted by scattering intensity as well as of the Stokes diameter x_{Stokes} weighted by optical extinction; pyrogenic silica (BET = 50 m^2/g)

Conversion from these distribution functions to number weighted distributions of $x_{h,t}$ and N were run by assuming:

- the validity of Eq. (4.112) for the effective diffusion coefficient (based on experiment),
- monodisperse and spherical primary particles,
- whose diameter is given by $6/S_V$ (with S_V based on the manufacturer's BET values),
- the fractal dimension is 1.8 for all aggregates (based on SLS measurements),
- the diameter of gyration obeys a fractal law with the fractal dimensions and prefactors given in Sect. 4.2.3.5 (based on DLCA simulation),
- the translational hydrodynamic diameter obeys a fractal law with the hydrodynamic dimensions and prefactors given in Sect. 4.4.3 (based on DLCA simulation and hydrodynamic computations),
- the static structure factor is described by Eq. (4.26) (based on DLCA and RLCA simulation),
- insignificance of intracluster multiple scattering (since $x_p < 100$ nm, Sect. 4.3.2.2).

Since these assumptions are only partly verified by experiment, the conversion produces uncertainty in the final results that is superposed or even magnified by the measurement uncertainty. For that reason, the transformation from $x_{h,eff}$ in $x_{h,t}$ is first considered for the DLS results. The left diagram in Fig. 4.31 shows that the difference between the two DLS results is minimised when the results are plotted as $q_{int}(x_{h,t})$, while the right diagram confirms that this holds true for all examined materials. However, when the intensity weighted distributions are converted to number weighted distributions, this agreement gets lost as a result of the very different distribution widths.

On the other hand, a remarkably good agreement between the number weighted distributions $q_0(x_{h,t})$ and $q_0(N)$ of the centrifugation and of the backscattering DLS is observed for almost all silica grades (Figs. 4.32 and 4.33). Severe deviations are

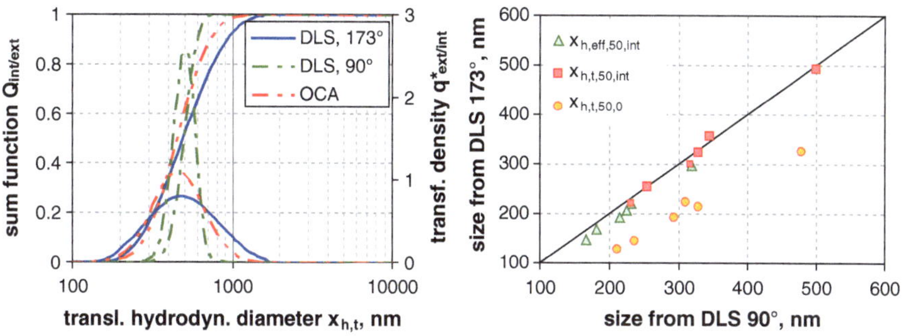

Fig. 4.31 *Left* Intensity weighted distribution of the translational hydrodynamic diameter $q_{ext}(x_{h,t})$ and $q_{int}(x_{h,t})$ calculated from the measured size distributions shown in Fig. 4.30; *right* correlation between the median values of the intensity and number weighted distributions measured with 90° and 173° DLS instrumentation (original data in Fig. 4.30)

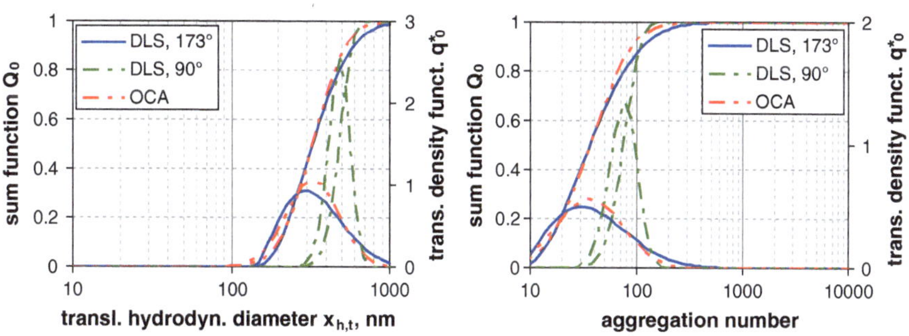

Fig. 4.32 Number weighted distribution of the translational hydrodynamic diameter $q_0(x_{h,t})$ and number weighted distribution of the aggregation number $q_0(N)$ calculated from the measured size distributions shown in Fig. 4.30

found only for the finest silica grades, where the centrifugation experiments were difficult to analyse due to very small turbidity. The origin of these deviations could not be clarified definitely; a failure of the assumptions underlying the conversion step cannot be excluded.

Nevertheless, the conversion of experimental distribution functions to number weighted distributions of the translational hydrodynamic diameter and the aggregation number has yielded similar results for centrifugation and (backscattering) DLS. This is remarkable with respect to the long list of required assumptions and the uncertainty in measurement data (in particular with regard to polydispersity). That implies the principal correctness of the DLCA approximation for well-dispersed pyrogenic materials and of the corresponding structure-property-relationships employed. Therefore, the conversion results can be—carefully—used for further discussion.

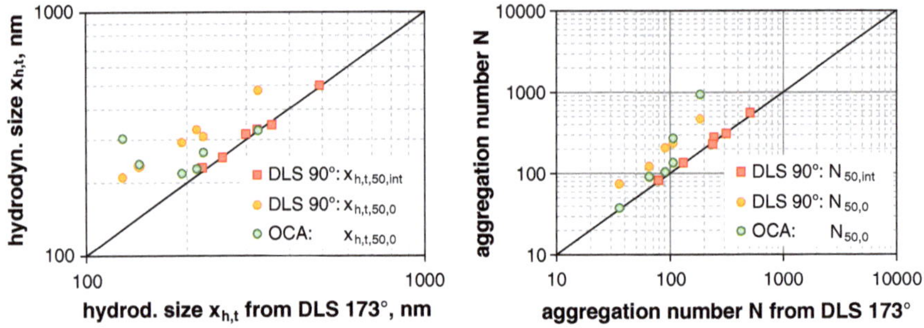

Fig. 4.33 Parity plots for the number and intensity weighted medians of the distribution in $x_{h,t}$ and N derived from OCA and DLS measurements on pyrogenic silica suspensions (six different grades with BET = 50–300 m^2/g) including the measurements shown in Fig. 4.30

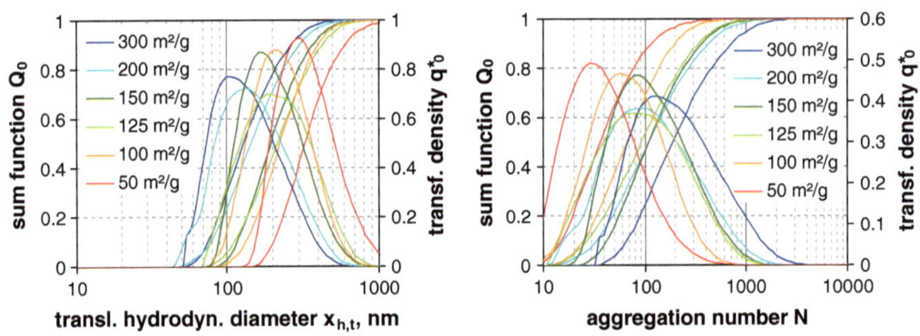

Fig. 4.34 Number weighted distribution of the translational hydrodynamic diameter $q_0(x_{h,t})$ and number weighted distribution of the aggregation number $q_0(N)$ calculated from size distributions measured with OCA and DLS on pyrogenic silica suspensions (six different grades with BET = 50–300 m^2/g)

Figure 4.34 compares the $q_0(x_{h,t})$ and $q_0(N)$ for all analysed materials. All samples had undergone a uniform, vigorous dispersion procedure before measurement. The resulting aggregates had hydrodynamic diameters in the submicrometre range. Significant numbers of nanosized aggregates ($x_{h,t} < 100$ nm) were only detected for the finest silica grades.[8] The aggregation numbers varied between tens to more than a thousand particles, which is the range typically reported for aggregates of pyrogenic materials (e.g. Sorensen 2001; Kammler et al. 2004; Kätzel et al. 2008b).

The aggregation data can be further analysed with regard to the correlation between the primary particle size x_p and the aggregation number N. As already mentioned above (p. 175), there is a negative correlation between the reported

[8]Since the detection is based on optical signals, the methods can be blind to very small aggregates. The minimum aggregate size can never be reliably detected via scattering or extinction.

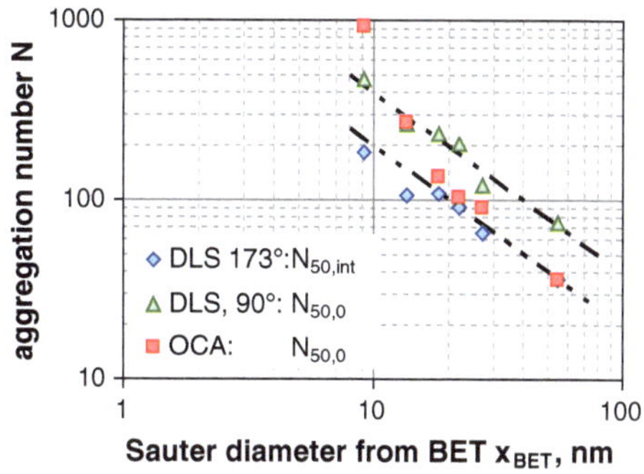

Fig. 4.35 Correlation between the medians of the number weighted distribution of the aggregation number $q_0(N)$ with the Sauter diameter as derived from the BET surface; data from the right chart of Fig. 4.34

values of the aggregation number and the averaged primary particle size. These findings referred to aggregate formation in a flame reactor (Kammler et al. 2004) as well as to the aggregate dispersion in aqueous suspensions (Kätzel 2008b). Obviously, the existence of such an anticorrelation can be confirmed (Fig. 4.35); it is approximately quantified by a reciprocal relationship.

4.5.3.3 Conclusion

The perception that pyrogenic materials consist of fractal aggregates offers an opportunity to characterise suspensions of such materials that goes beyond a conventional data analysis of particle sizing instruments. First of all, it helps to explain differences between the results of different particle sizing instruments, like the fact that centrifugation measurements always lead to much finer distribution functions than static and dynamic light scattering experiments. Moreover, by consequent applying of analytical relationships derived for DLCA aggregates, it is possible to convert experimental distribution functions which are instrument-specific into number or volume weighted distributions of morphological aggregate properties (e.g. aggregation number, radius of gyration) and important physical properties (e.g. the translational hydrodynamic diameter). Even though such a conversion—as any conversion—introduces new uncertainty (because measurement errors at the borders of the distribution function are disproportionately magnified), it provides a powerful approach for understanding the morphology of colloidal aggregates (e.g. by revealing intrinsic correlations between the morphological parameters).

4.6 Interaction Forces Between Aggregates

As already stated in Chap. 3, there is a significant influence of particle interaction on the macroscopic properties of colloidal suspensions (e.g. viscosity, sedimentation, light scattering, sound propagation). Moreover, the attractive part of the particle interaction ultimately causes aggregation (or coagulation) in colloidal systems.

The kinetic of aggregation determines the morphology of the evolving aggregates and is a decisive factor for the formation and structure of particle networks (gelling systems). Last but not least, the aggregation kinetics is used as a measure for the stability of colloidal suspensions (cf. Sect. 5.2.4).

On the other hand, the strength and range of particle interactions are considerably affected by the size and structure of particles and aggregates. With ongoing aggregation, i.e. growing size of interacting objects, the relevance of interaction is being reduced. As a result, a transition from reaction-limited to diffusion-limited aggregation can be observed (di Biasio et al. 1994; Sandkühler et al. 2005b; cf. Sect. 4.1.2). Nonetheless, the effect of size and structure of aggregates on the interaction forces is not fully understood yet. That is mainly due to the absence of simple analytical expressions for the interaction between aggregates and the diversity of aggregate configurations. Simulations have shown a significant impact of the range of interaction on the resulting aggregate morphology (Fernández-Toledano et al. 2007; Videcoq et al. 2007).

The total interaction between two particles or aggregates results from the superposition of attractive and repulsive forces. The main attractive component is the van-der-Waals interaction, whereas the repulsion (at least in aqueous systems) is primarily due to the interaction between the electric double layers surrounding any charged particle in a liquid medium. For spherical particles, the total effect of van-der-Waals and double layer interaction is described by the DLVO theory (cf. Sect. 3.2). A common approach for the interaction of aggregates consists in "summing all the intercluster monomer-monomer pair interactions" (e.g. di Biasio et al. 1994). However, in particular the computation of the double layer interaction between aggregates requires the consideration of the double layer overlap of the primary particles and the corresponding charge regulation. Thus, double layer interaction is intrinsically a many-body phenomenon, the calculation of which requires high computational costs. Yet, the pairwise-interaction approach or even more simple approaches can be sufficiently accurate for certain situations.

4.6.1 Van-der-Waals Interaction Between Two Aggregates

The van-der-Waals interaction is the most prominent cause for the adhesion of particles and, thus, for aggregation. It results from the weak electric coupling between neutral atoms and molecules, which can be considered as fluctuating or permanent electric dipoles, and is of attractive nature (cf. Sect. 3.2.1). For macroscopic bodies, these microscopic interactions superpose for all pairs of atoms and/or molecules, which yields a corresponding net attraction. In general, the van-der-Waals interaction decays very rapidly with the distance between the interacting objects. In the case of the interacting aggregates, two limiting situations may be distinguished: At large aggregate separation, the two aggregates act like volume equivalent spheres, whereas at short distances only the van-der-Waals interaction of the closest pair of primary particles should be relevant. However, for intermediate distances, the aggregate structure cannot be neglected.

In order to study how structure influences the van-der-Waals interaction, a simplified calculation scheme will be employed. This is because analytical expressions of Lifshitz's macroscopic approach on the electrodynamic interaction exist only for a few regular geometries (e.g. interacting half-spaces—or Lifshitz 1956, or spheres—Langbein 1971). Furthermore, it is helpful to be reminded that fractal aggregates are usually not closely packed and, therefore, that the mean interparticle distance is considerably larger than the primary particle size (in particular for fractal dimensions below 2.5). It thus seems fairly reasonable to assume that the van-der-Waals interaction between a pair of primary particles is little affected by the presence of the other primary particles (i.e. the many-body aspect can be ignored). Consequently, the van-der-Waals interaction energy between two aggregates can be approximated by summation of the interaction between all pairs of primary particles from aggregate *A* with those of aggregate *B*:

$$V_{\mathrm{vdW},A-B} = \sum\sum V_{\mathrm{vdW},i-j}, \quad (i \in A, j \in B). \tag{4.113}$$

The pair-contribution $V_{\mathrm{vdW},i-j}$ can be further approximated with Eq. (3.23) by accounting for the retardation of the Hamaker coefficient at large distances (Eq. (3.25)).

Arunachalam et al. (1998) examined the performance of such *discrete pairwise sum* (i.e. Eq. (4.113)) for three spherical molecule clusters (CCl_4, $N = 55$, ∅ 2.82 nm) with two of them in contact. They showed that the deviation from a Lifshitz approach, which accounts for the interaction between the electron fluctuations of the single molecules, is very small. The contribution of the intermolecular coupling to the total interaction energy was in the order of 1 %. Similarly, Kim et al. (2007) compared several van-der-Waals calculation schemes for a pair of nanoclusters (2340 molecules, ∅ 5.88 nm) with the coupled dipole method (CDM), which is claimed to be exact in the non-retarding region. For the discrete pairwise sum, they observed significant differences only for very small gaps (approx. 15 % for 1 nm), while for growing distances the discrete pairwise sum converges relatively fast with the CDM and Langbein's solution.

The discrete pairwise approach was already employed by Naumann and Bunz (1993) for fractal-like aggregates. They presented an analytical expression for the orientation-averaged interaction energy of two aggregates with a fractal dimension of 2. However, as they assumed a spherical shape and a fractal law for the radial mass density ($\rho(r) \propto r^{df-3}$), and because their approach ignores the retardation of the dispersion interaction, their equations provide rather rough estimates for a given set of aggregates.

Babick et al. (2011) computed the van-der-Waals interaction between two aggregates for a linear arrangement as well as for an orientation-averaged arrangement. The former refers to a situation where the aggregate centres of gravity and their outermost primary particles lay on a straight line with the latter forming

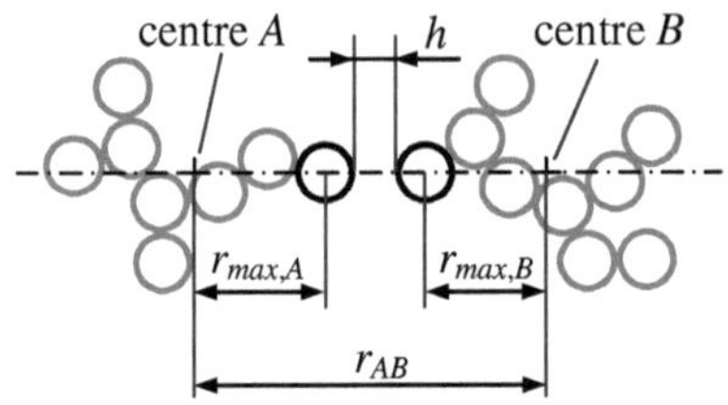

Fig. 4.36 Linear arrangement of aggregates for calculating the influence of the minimum surface distance, i.e. the distance between the closest pair of primary particles

the closest pair of primary particles (Fig. 4.36). This configuration allows for a controlled variation of the minimum surface distance h. In contrast, the orientation averaged alignment was used to study the dependency on the centre distance.

Orientation averaging was achieved by randomly rotating (100×) both aggregates, which remained separated by a fixed centre distance. For DLCA aggregates, the structural variation has to additionally be considered. For this reason, the orientation averaging was repeated for 55 different aggregate pairs (arbitrarily chosen from a set of 100 aggregates). In order to avoid overlap of primary particles, the centre distance r_{AB} was restricted to values larger than the sum of the maximum radial extensions of the two aggregates.

The authors focussed on the influence of the aggregate structure on the van-der-Waals interaction. For that purpose, they first considered deterministic aggregates with tuneable fractal dimensions (cf. Sect. 4.1.3). Besides this, they included hexagonal close-packed (hcp) aggregates and straight-chain aggregates in their study. The results for a linear arrangement according to Fig. 4.36 are reproduced in Fig. 4.37(left). This graph contains additional curves which are the solutions for a pair of primary particles:

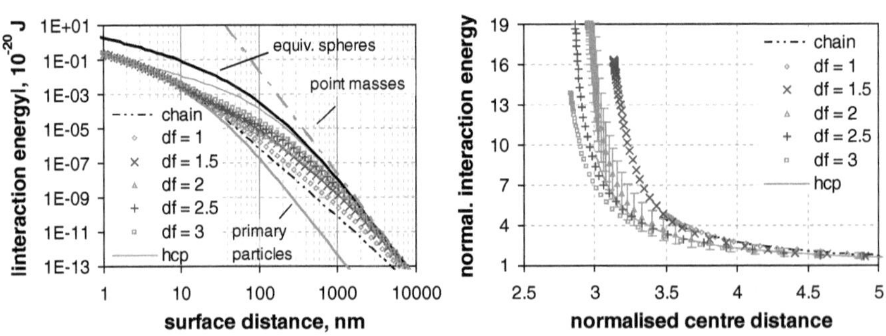

Fig. 4.37 *Left* Value of the interaction energy versus surface distance (linear arrangement, cf. Fig. 4.36), *right* value of the orientation averaged interaction energy versus centre distance; for deterministic aggregates (N = 200, x_p = 20 nm) including an ideal chain and a hexagonal close-packing (hcp); asymptotes for a pair of primary particles, a pair of volume equivalent spheres, and a pair of point masses are shown as well; material properties for silica (Table B.1)

$$V_{\mathrm{vdW}}(h) = -\frac{A_{132}(h)}{12} \cdot \left(\frac{x_{\mathrm{p}}^2}{(x_{\mathrm{p}}+h)h} + \left(\frac{x_{\mathrm{p}}}{x_{\mathrm{p}}+h}\right)^2 + 2 \cdot \ln\left(1 - \left(\frac{x_{\mathrm{p}}}{x_{\mathrm{p}}+h}\right)^2\right)\right), \tag{4.114}$$

a pair of volume equivalent spheres:

$$V_{\mathrm{vdW}}(h) = -\frac{A_{132}(h)}{12} \cdot \left(\frac{x_V^2}{r_{AB}^2 - x_V^2} + \left(\frac{x_V}{r_{AB}}\right)^2 + 2 \cdot \ln\left(1 - \left(\frac{x_V}{r_{AB}}\right)^2\right)\right), \tag{4.115}$$

and a pair of two point masses:

$$V_{\mathrm{vdW}}(h = r_{AB}) = -\frac{A_{132}(r_{AB})}{36} \cdot \left(\frac{x_V}{r_{AB}}\right)^6. \tag{4.116}$$

For small surface distances ($h \ll x_{\mathrm{p}}$) all curves converge with the solution for a pair of primary particles, whereas at very large distances ($h \gg x_{\mathrm{agg}}$) all curves merge with the solution of volume equivalent spheres and point mass arrangement. The impact of structure is most pronounced for moderate distances ($x_{\mathrm{p}} < h < x_{\mathrm{agg}}$). In general, the higher the interaction energy is, the less porous the aggregate is, i.e. the more pairs of primary particles are spaced in a relatively close distance (with regard to the aggregate surface distance). However, for medium fractal dimensions ($1.5 \leq d_{\mathrm{f}} \leq 2.5$), the curves are fairly close and do not exhibit a characteristic dependency on d_{f}. This is probably due to the negative correlation between the fractal prefactor and the fractal dimension, which can be observed for any kind of aggregation mechanism (cf. Fig. 4.8) and was employed for designing the deterministic aggregates. The fractal curves deviate considerably from those of the linear particle chains and of the hexagonal close-packed (hcp) aggregates. At moderate surface distances, the interaction of the latter is up to 10 times stronger than that of the former.

According to Fig. 4.37, it is possible to locate the transition from the primary particle impact to the structure impact at $h = 0.5 \times x_{\mathrm{p}}$. This surface distance corresponds to an interaction energy of $0.0057 \times A_{132}$ (Eq. (4.114)). Even for the relatively strong interacting system of titania particles in water ($A_{132,0} = 5.35 \times 10^{-20}$ J), this is not more than $0.075 \times \mathrm{k_B}T$ at room temperature. Hence, the primary particle approximation can be applied to all situations because the van-der-Waals energy has no physical relevance, i.e. it is exceeded by the thermal energy, when this approximation fails.

Regarding two approaching fractal aggregates, it is quite conceivable that there are only a few (maybe just one) pairs of primary particles that really contribute to the van-der-Waals interaction between the aggregates. That means that the attraction between such aggregates is almost independent of their size (for a given x_{p}). However, the motion of aggregates is best described for the aggregate centres and their mutual alignment is steadily changing due to Brownian rotation and/or hydrodynamic shear. For this reason, the interaction energy has to be studied as a function of the aggregate centre distance r_{AB} and for random alignment as well.

The right diagram of Fig. 4.37 plots the orientation averaged interaction energies of the aggregates from the left diagram versus the centre distance, which is scaled with the radii of gyration of the respective aggregates. Besides this, the interaction energy is normalised with that of the corresponding point masses (Eq. (4.116)). Additionally, an indication is given for $d_f = 2$, to which degree the aggregate orientation affects the interaction energy. Regarding this impact, the variation among the different aggregate structures is not very large, yet it is significant. This is because the aggregate structure, in particular the radial density distributions, is controlling the minimum surface distance between two aggregates at a given centre distance. Consequently, the average surface distance between the few interacting pairs of primary particles is determined by aggregate size and structure. That means that aggregate structure affects the van-der-Waals interaction between aggregates simply because of steric effects.

As a result, it can be shown that the relationship between V_{vdW} and centre distance r_{AB} of a given type of aggregate and given material is independent of the aggregation number. Figure 4.38 shows such curves for silica and titania aggregates of varying mass (N). It can be seen that the variations caused by N are much smaller than those caused by orientation and aggregate configuration. It is very likely that the existing deviations are mainly due to the relatively small set of employed DLCA aggregates. The comparison between silica and titania reveals a strong influence of the material on the normalised interaction curves. This is due to the retarding dispersion interaction, which introduces a material influence (cf. Table B.1). Figure 4.38, therefore, suggests the existence of a material-specific master curve for the average interaction energy, which may considerably enhance the computation of aggregate interaction, structuring, and motion in dense suspensions.

In summary, the van-der-Waals interaction energy can be fairly well approximated by computing the interaction energy of the closest pair(s) of primary

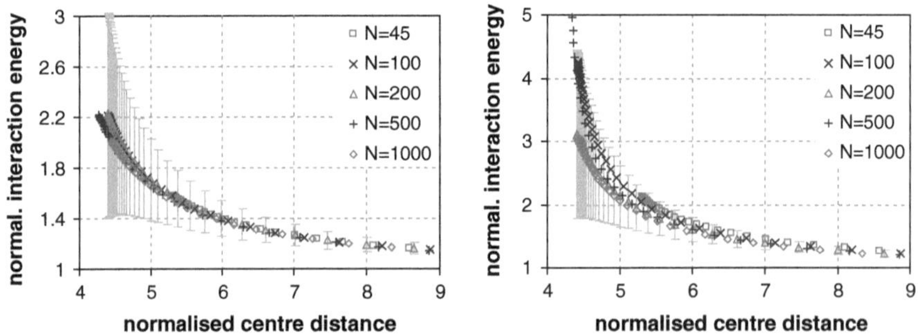

Fig. 4.38 Averaged normalised interaction energy versus normalised centre distance for randomly arranged DLCA aggregates of varying aggregation number N ($x_p = 20$ nm); error bars for $N = 200$ represent variation due to changing orientation and aggregate configuration; *left* result for silica; *right* result for titania (rutile) (cf. Table B.1)

particles. Strictly speaking, this approximation is applicable only for surface distances within the radius of the primary particles ($h \leq 0.5 \times x_p$), but for larger distances, van-der-Waals interactions are negligible. Nonetheless, since the morphology and outer radial density of an aggregate affects the surface distance between two aggregates of given centre distance r_{AB}, there is an apparent impact of the aggregates' structure on their van-der-Waals interaction. The proposed discrete pairwise approach provides a possibility to account for that.

4.6.2 Double Layer Interaction Between Two Aggregates

This section addresses the calculation of the double layer interaction energy between aggregates. A numeric calculation scheme is described in detail and further employed for a qualitative analysis on the factors that determine the strength of the double layer interaction. Finally, the benefit of rigorous numerical methods and the application of approximate solutions are discussed.

4.6.2.1 Modelling the Electric Double Layer for Particle Aggregates

The fundamental equation for calculating the structure of electric double layers is the Poisson-Boltzmann-equation (PBE, Eq. (3.16)), which applies to any particle shape and aggregate structure. Analytical solutions of the PBE can be derived for isolated particles of regular shape (e.g. spheres) and/or small surface potentials (Debye-Hückel approximation). These solutions are based on known values for either the electric charge density σ_0 or the electric potential ψ_0 on the particle surface. Both parameters are locally changed, when the double layers of two particles overlap (cf. Sect. 3.2.2). That inevitably occurs inside a particle aggregate. As a result, the surface potential and/or surface charge in the aggregate deviate from those of isolated primary particles and vary across the aggregate surface. In order to account for this internal overlap a charge regulation model is required, which quantifies the mutual influence between these two parameters.

Due to the complex morphology and the internal double layer overlap with charge regulation, there is no general analytical expression for the double layer of particle aggregates. Instead, the spatial distribution of the electric potential ψ and the double layer interaction between aggregates have to be computed numerically. There are several methods that seem appropriate for solving the PBE in the case of aggregates. Some methods are based on the discretisation of space or interfaces, e.g. the finite difference method (FDM; Coelho et al. 1996), the finite element method (FEM; Reiner and Radke 1991; Ring 1982), and the boundary element method (BEM; Kwon et al. 1998; Sun and Walz 2001). A different, quasi analytical approach is the multipole expansion (ME), which approximates the potential distribution $\psi(\mathbf{r})$ by a series of spherical harmonics (Glendinning and Russel 1983; Krozel and Saville 1992; Carnie and Chan 1993; Carnie et al. 1994). Similar to the

ME, but of less mathematical rigour, is the singularity method (SM), which reconstructs the potential field by a superposition of singularity fields from point charges (Phillips 1995). Additionally, it is possible to approximate compact, isometric aggregates as porous spheres (PS), for which analytical expressions of the double layer structure are available (Ohshima 2008). Some of the mentioned methods assume a certain shape of the aggregates or primary particles, some of them are based on the linearised PBE (BEM, ME, PS). Only a few have already been employed for particle clusters (Coelho et al. 1996; Kwon et al. 1998; Schießl et al. 2012).

The computation costs of the numerical methods are rather high and increase disproportionately rapidly with the aggregate size. Therefore, approximate solutions may be justified for limiting situations such as very thin double layers. A common approximation for the double layer interaction of aggregates is to ignore any interaction term except for that of the closest pair of primary particles (e.g. Sandkühler et al. 2005b; Runkana et al. 2005; Ahmad et al. 2008; Gauer et al. 2009; Jia and Iwata 2010), which is assumed to have a double layer as in the isolated case (*approximation of primary particle interaction*—APPI). However, its validity has not been sufficiently examined for aggregates of nanoparticles, i.e. particles whose size is in the typical range of the double layer thickness (e.g. the Debye length $1/\kappa$ of an aqueous 0.001 M NaCl solution is 10 nm). There is no quantitative criterion that can be employed. Another approximation consists in adding up all pairwise interactions between the primary particles (*discrete pairwise sum*). As in APPI, the individual double layers are considered unchanged compared to the isolated case (e.g. Fernández-Toledano et al. 2007; Peng et al. 2010).

4.6.2.2 Calculation of Double Layer Interaction with the Singularity Method

Studies on the double layer interaction between aggregates, including the effect of aggregate morphology and double layer thickness, are rare. In a recent paper, Schießl et al. (2012) presented results from a numerical analysis for DLCA and hexagonal close-packed (hcp) aggregates and discussed the applicability of approximate solutions. The numerical computations were conducted with the singularity method, which was introduced by Phillips (1995). This method considers charge regulation and is in principle applicable to any particle shape. The primary result is the distribution of electric potential and ion concentration around the particle's surface, which can be straightforwardly used to calculate the interaction force. However, for the general case there is no fast route for deriving the interaction energy. By restricting to spherical primary particles, Schießl et al. (2012) derived an elegant procedure to compute the interaction energy once the potential distribution has been obtained. The main ideas of the singularity method will be presented in the following.

The singularity method is based on the linearised PBE (Eq. (3.17)) and on the simple analytical expression of the radially-symmetric potential field $\psi(r)$ around a point charge:

$$\psi^* = q^* \frac{e^{-\kappa a r^*}}{r^*}, \quad (4.117)$$

where ψ^*, r^*, and q^* are normalised expressions of the electric potential, the radial coordinate, and the strength of the singularity (cf. Appendix C.2).

Any solution to the linearised PBE can be expressed as a linear combination of singularity fields (Eq. (4.117)). The problem is, thus, reduced to (i) properly locating the singularities (inside the particles) and to (ii) adjusting the singularity strengths q^* in such a way that the boundary condition (3.28) is fulfilled. That is verified at a number of control points on the surface of the primary particles.

The singularities may be classified in two groups. The first ones are located in the centre of each primary particle. They represent the surface charge of the isolated primary particles and do not change their strength with double layer overlap. For uniform primary particles we get:

$$q^*_{c,k} = q^*_c = e^{\kappa a}. \quad (4.118)$$

The second group of singularities is located off-centre, e.g. on concentric shells inside the primary particles. Their strength depends on the specific structure of the aggregates and the double layer overlap. The total potential field is obtained by:

$$\psi^* = \sum_{k=1}^{N} \psi^*_{c,k} + \sum_{l=1}^{\alpha N} \psi^*_{oc,l} \quad (4.119)$$

and has to obey the boundary condition:

$$K^*_{reg}\psi^* + \mathbf{n} \bullet \nabla\psi^* = S^*_{reg}, \quad \text{or} \quad (4.120)$$

$$\sum_{k=1}^{N} K^*_{reg}\psi^*_{c,kj} + \sum_{l=1}^{\alpha N} K^*_{reg}\psi^*_{oc,lj} + \sum_{k=1}^{N} \mathbf{n}_j \bullet \nabla\psi^*_{c,kj} + \sum_{k=1}^{\alpha N} \mathbf{n}_j \bullet \nabla\psi^*_{oc,lj} = S^*_{reg} \quad (4.121)$$

at each control point j.

Due to the simple analytical expression of the singularity field, the gradient in Eq. (4.121) can be easily deduced:

$$\nabla\psi^*_{oc,lj} = \frac{\partial\psi^*_{oc,lj}}{\partial r^*_{lj}} \cdot \mathbf{e}_{lj} = -q^*_{oc,l} \frac{\kappa a \cdot r^*_{lj} + 1}{r^{*2}_{lj}} \cdot e^{-\kappa a \cdot r^*_{lj}} \cdot \mathbf{e}_{lj}. \quad (4.122)$$

Finally, one obtains a set of linear equations that define the singularity weights:

$$\mathbf{A} \cdot \mathbf{q}_{oc}^* = \mathbf{b}, \tag{4.123}$$

$$\text{with } A_{jl} = \frac{e^{-\kappa a \cdot r_{lj}^*}}{r_{lj}^*} \left(K_{\text{reg}}^* - \mathbf{n}_j \bullet \mathbf{e}_{lj} \frac{\kappa a \cdot r_{lj}^* + 1}{r_{lj}^*} \right), \tag{4.124}$$

$$\text{and } b_j = S_{\text{reg}}^* - \sum_{k=1}^{N} \frac{e^{-\kappa a \cdot r_{kj}^*}}{r_{kj}^*} \left(K_{\text{reg}}^* - \mathbf{n}_j \bullet \mathbf{e}_{kj} \frac{\kappa a \cdot r_{kj}^* + 1}{r_{kj}^*} \right) q_c^*. \tag{4.125}$$

The size of the coefficient matrix **A** depends on the number of control points per primary particle n_{cp}, the number of off-centre singularities per primary particle n_{oc}, and the number of primary particles N_{p} involved (the sum of N from all interacting aggregates).

$$\text{size}(\mathbf{A}) = N_{\text{p}} \cdot n_{\text{oc}} \times N_{\text{p}} \cdot n_{\text{cp}} = N_{\text{p}}^2 \cdot n_{\text{oc}} \cdot n_{\text{cp}}. \tag{4.126}$$

The solution of the matrix Eq. (4.123) facilitates the reconstruction of the total potential field at any location in the continuous phase. That is prerequisite for the calculation of interaction energies, which involves the knowledge of the charge and potential at each surface point of the aggregates. According to Eq. (3.31), the interaction energy is obtained from:

$$\begin{aligned} V_{\text{DL}}(h) = &\frac{1}{2} S_{\text{reg},1} \sum_{m=1}^{N_1} \int \left(\sum_{k_2=1}^{N_2} \psi_{\text{c},k_2} + \sum_{l=1}^{\alpha N} \psi_{\text{oc},l} - \sum_{l_1=1}^{\alpha N_1} \psi_{\text{oc},l_1}^{\text{iso}} \right) \text{d}S_m \\ &+ \frac{1}{2} S_{\text{reg},2} \sum_{n=1}^{N_2} \int \ldots \text{d}S_n. \end{aligned} \tag{4.127}$$

The elementary step of this computation is to solve the integral of the potential field of singularity k or l over the surface of the primary particle m or n. For spherical particles of radius a, this yields:

$$\int \psi \, \text{d}S = \begin{cases} 4\pi q a \frac{e^{-\kappa R}}{\kappa R} \sinh(\kappa a), & \text{external source: } R > a \\ 4\pi q a \frac{e^{-\kappa a}}{\kappa R} \sinh(\kappa R), & \text{internal source: } R < a \end{cases}, \tag{4.128}$$

where R is the distance between singularity and the centre of the surface.

In principle, Eqs. (4.123)–(4.128) allow for the calculation of the interaction energy between an arbitrary number of aggregates with individual structure and size. The main assumptions are the Debye-Hückel approximation, the sphericity of the primary particles, their uniformity in size, and surface regulation properties. Note that Eqs. (4.120)–(4.128) have to be replaced when charge regulation with constant potential is assumed (Appendix C.2).

Schießl et al. (2012) discussed the performance of the singularity method with regard to numeric stability and agreement to other numerical methods. There are apparently two limits:

(i) strong double layer overlap—because the Debye-Hückel approximation is then questionable (in particular if the potential is not maintained constant), and
(ii) very thin double layers ($\kappa a > 20$)—which means a strong decay of the electric potential in the vicinity of the particle surface; in this case, the singularity method performs rather weakly and eventually leads to a considerable underestimation of the interaction energy (in particular for constant charge regulation). However, for very thin double layers, the interaction can be ascribed to the double layer overlap of just one pair of primary particles and numerical methods are no longer required.

In any other case, excellent agreement between all numerical methods was found.

4.6.2.3 Double Layer Interaction Between Particle Aggregates

The calculation of double layer interaction between aggregates is rather costly (with regard to computation), even if it is based on the Debye-Hückel approximation, as in BEM or in the singularity method. This probably explains the lack of systematic studies which are based on the degree to which the double layer interaction of aggregates depends on aggregate structure and double layer thickness. In their paper, Schießl et al. (2012) presented first results for DLCA and hcp aggregates with varying double layer thickness, which illustrate the main effects and can even be used for quantitative evaluation. The paper employs the singularity method and a linear charge regulation model (including the limiting cases *constant charge* and *constant potential*). The aggregates in that study are rather small, with N from 20 to 150; however, these values are still in the range experimentally found for aggregates of fumed silica (N = 90–1000; Kätzel et al. 2008b; Fig. 4.34) and for soot aggregates (N = 10–120, Sorensen 2001). The results being presented in this section have been mainly obtained in the context of the aforementioned publication by Schießl et al. (i.a. Schießl 2010). The conclusions are identical.

Before starting any double layer calculation (i.e. solving the PBE), one has to define an appropriate charge regulation model for the interface (cf. Sect. 3.2.2). Charge regulation becomes important when double layers overlap, i.e., when the concentration of counter-ions locally increases, as for colliding particles or inside an aggregate. It depends on the electrochemical properties of the particle phase as well as on the interaction between solutes and interface. When the surface charge keeps unaffected by the overlap (*constant charge* model), the regulation capacity K_{reg} is virtually zero. In contrast, when the regulation capacity is large ($K_{reg} \gg \varepsilon_m \varepsilon_0 \kappa$), the *constant potential* model applies. Most interfaces do not fit to either of these cases, but show both charge and potential regulation. One way to approximately describe this intermediate charge regulation is Carnie's "canonical intermediate" state

(Carnie and Chan 1993; Carnie et al. 1994), which is defined as $K_{reg} = \varepsilon_m\varepsilon_0\ \kappa$. Schießl et al. 2012 showed for DLCA aggregates that the charge regulation model has a strong impact on the potential distribution inside the aggregates and on the strength of the double interaction. For aggregates with thick double layers ($\kappa a < 1$), the interaction energies may differ by a factor of 10 (cf. Schießl et al. 2012, Fig. 3a). For this reason, the effects of structure and double layer thickness have been examined for the intermediate charge regulation only.

Figures 4.39 and 4.40 show distributions of the electric potential in the central plane of isolated aggregates with N = 150. Two types of aggregates—hcp and DLCA aggregates—are considered for thick and thin double layers with regard to the primary particle size (κa = 0.1 and κa = 1, respectively). These figures illustrate fairly well that the structure, i.e., packing density, primarily affects the internal

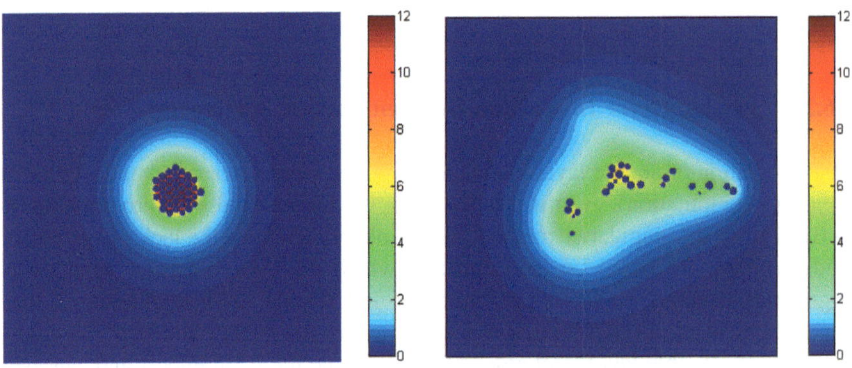

Fig. 4.39 Normalised electric potential in the central plane through the double layers ($\kappa a = 0.1$) of an hcp and a DLCA aggregate (N = 150); normalisation with the surface potential of an isolated primary particle, intermediate charge regulation

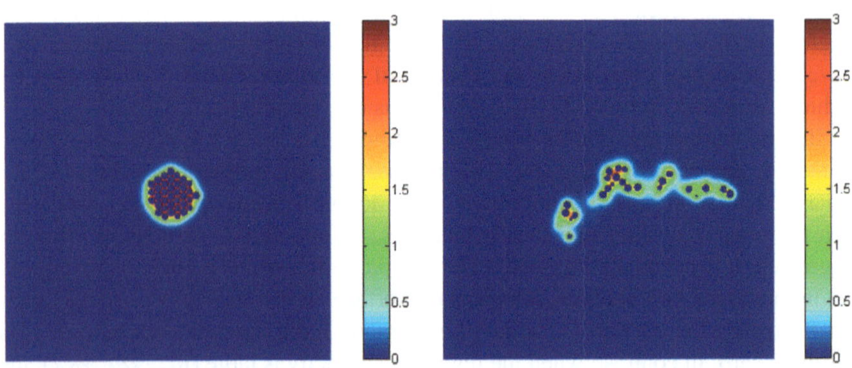

Fig. 4.40 Normalised electric potential in the central plane through the double layers (κa = 1) of an hcp and a DLCA aggregate (N = 150); normalisation with the surface potential of an isolated primary particle, intermediate charge regulation

double layer overlap. Hence, the double layer overlap is much more pronounced in hcp aggregates than in DLCA ones.

For the DLCA aggregate employed ($N = 150$), the average coordination number, i.e., the number of touching neighbours, is 2.0 and its maximum is 4. In contrast, the average and maximum values for the hcp aggregate are 9.0 and 12, respectively.

Moreover, the differences in packing density may even lead to qualitative differences with regard to overlap: For moderate and thin double layers ($\kappa a \geq 1$), the overlap is approximately pairwise in DLCA aggregates, but it remains a many-body interaction within the compact, highly coordinated hcp aggregates. Nevertheless, since for thin double layers the outermost particles of the aggregates are less affected by the internal overlap, one can assume that the interaction between two aggregates is then mainly an interaction between the closest pair(s) of primary particles.

In the case of thick double layers, internal overlap occurs independently from the packing density. The double layer's shape does not reflect the position of individual primary particles, but rather is fairly smooth. For example, the potential distribution of the DLCA aggregate has a concave and anisometric shape for $\kappa a = 1$, but it resembles a spheroidal form when κa is reduced to 0.1. Thus, the aggregates interact with each other similarly to compact spheroids, but usually with a surface potential higher than for the isolated sphere (apart from the constant potential regime).

The dependency of the potential distributions on the aggregate structure and double layer thickness coincides with a corresponding dependency for the interaction energy. Between two aggregates with similar interface, the repulsion increases with the packing density and the thickness of the double layer. This is confirmed by Fig. 4.41, which plots the interaction energy versus the minimum surface distance for DLCA and for hcp aggregates ($N = 70$). It can be clearly seen that for moderate double layers ($\kappa a = 1$) the repulsion between two hcp aggregates is considerably larger than between the fractal ones, whereas no difference is observed for thin double layers ($\kappa a = 5$).

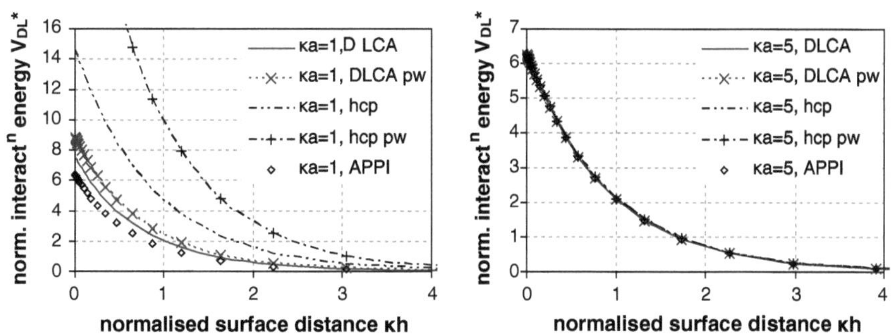

Fig. 4.41 Normalised interaction energy for intermediate charge regulation, comparison with pairwise summation approach (pw), and interaction between one pair of primary particles (APPI); DLCA and hcp aggregates ($N = 70$); *left* $\kappa a = 1$; *right* $\kappa a = 5$ (all on one curve)

In Fig. 4.41, the results of the singularity method are compared with two approximate solutions—the APPI and pairwise summation approximation. For $\kappa a = 5$, there is obviously no deviation between the approximations and the numerical calculation—independent from the aggregate type. In the case of DLCA aggregates, both approximations perform well even for $\kappa a = 1$. However, both approaches fail as soon as the double layer thickness is large compared to the primary particle size ($\kappa a < 1$). The APPI systematically underestimates the interaction energy, whereas the pairwise summation approximation overvalues it. Regarding the deviation from the numerical solution, none of the two approaches can be preferred. However, the APPI requires less computation steps than the pairwise summation approach.

One must thus ask, in which kind of situation a numerical calculation scheme is required. Assuming an aqueous solution, we can estimate the Debye length $1/\kappa$ with 10 nm for an ionic strength of 0.001 M (usually less than tap water) and with 3 nm for 0.01 M (approximate value for artificial fresh water). Hence, the condition $\kappa a < 1$, which requires numerical calculation, is fulfilled for primary particles at the lower nanoscale. Examples of such products may be pyrogenic oxides, where the radii of the primary particles lie in a range below 50 nm (cf. Fig. 4.35). In non-aqueous solvents, the range of the electrostatic interaction (the Bjerrum length) is typically increased and the salt concentrations are relatively low. Then, the need for a numerical double layer calculation may be extended to aggregates of primary particles up to the micrometre region.

The examples presented in this section consider the interaction energy of aggregates with a fixed orientation to each other. Clearly, the orientation influences the interaction energy, especially for moderate and thick double layers, and may, thus, affect the aggregation of particles and particle clusters. In an early paper, Usher (1929) showed qualitatively that the double layer overlap around the joint of a particle doublet locally increases the interparticle repulsion and causes any potential aggregation to take place at the poles of the doublet. In such a way, chain-like aggregates are formed (cf. as well Hoffmann 1943; Rees 1951; Thomas and McCorkle 1971). Hence, the local variation of the double layer interaction between two particle clusters affects the resulting aggregate structure. In a 2-dimensional diffusion driven aggregation simulation, Fernández-Toledano et al. (2007) observed that the classic distinction between DLCA and RLCA has to be supplemented by a third fundamental aggregation regime, which occurs for long-ranging, repulsive interaction, and leads to a linear growth of the aggregates (Table 4.2). This regime prevails only for the initial aggregation state.

> We have observed that the range of the interaction modifies not only the kinetics of the coagulation process at short times but also the structure of the small aggregates formed in this process. Moreover, although the effect of the interaction range appears only in the first stages of the coagulation, the consequences of this affect the complete development of the process.
>
> (Fernández-Toledano et al. 2007)

Table 4.2 Fundamental aggregation regime for colloidal particles in a quiescent liquid (based on Fernández-Toledano et al. 2007)

Oriented RLCA	(classic) RLCA	DLCA
Long-ranging repulsion, e.g. for $\kappa R_{agg} < 1$	Short-ranging repulsion, e.g. for $1 < \kappa R_{agg}$	Negligible repulsion, e.g. for $1 \ll \kappa a$
Interaction energy grows with cluster size, oriented aggregation (anisometric sub-clusters)	Interaction energy independent from cluster size, no preference for adhesion of sub-clusters	Interaction energy independent from cluster size, adhesion of sub-clusters at first contact
Initial aggregation	Initial and medium aggregation	Final aggregation regime

Hitherto, the oriented RLCA was not given much attention. This is probably because most experimental studies use aqueous solutions with background electrolyte, which leads to thin double layers ($1 < \kappa a$). Additionally, the computational costs for aggregation simulation increase when the orientation dependency of aggregate interaction is to be accounted for. Finally, there is no simple model that describes the double layers aggregates when $\kappa a < 1$. Numerical methods, like the singularity method, facilitate the calculation of orientation dependent double layer interactions and may, thus, be employed to enhance our understanding of reaction-limited cluster aggregation of fine particles in organic solvents or salt-free aqueous solutions.

A further application of the numerical methods is the calculation of aggregate interactions in dense suspension, where the effective suspension properties depend on the structuring of aggregates. This is of particular importance for thick double layers $\kappa a < 1$. However, tackling this problem will considerably amplify the computational effort. It seems recommendable to develop approximate solutions, e.g. by replacing individual aggregates with spheroids or by orientation averaging.

4.6.2.4 Conclusion

This section has discussed the calculation of double layer interaction between two aggregates and, thus, adds to the previous section on the van-der-Waals interaction between aggregates. The calculation requires the solution of the Poisson-Boltzmann equation (PBE) for not very regular physical boundaries and, therefore, has to be conducted by numerical methods. Results for the double layer structure and of interaction energy were computed for DLCA and hcp aggregates, with the double layer thickness or its inverse (κa) as adjustable parameters. The most useful outcome is probably that the interaction of aggregates can be traced back to the interaction of the closest pair of the primary particles (APPI) when the double layer is relatively thin ($\kappa a \geq 5$). For aggregates with a small fractal dimension (like DLCA aggregates), the limit may even be lowered to $\kappa a = 1$. In both cases, the pairwise summation approach may be used as well, yet without additional benefit. For thick double layers ($\kappa a \leq 1$), the double layer overlap inside the aggregate dramatically affects the surface charge and potential of the primary particles and results in a smooth, spheroidal shape of the aggregate's double layer. In this case, the aggregate structure clearly influences the interaction potential. Even more, the aggregation of particles with thick double layers is affected by the potential distribution around the interacting aggregates and leads to an oriented RLCA regime.

References

Aggregation

A.L. Ahmad, M.F. Chong, S. Bhatia, Population balance model (PBM) for flocculation process: Simulation and experimental studies of palm oil mill effluent (POME) pretreatment. Chem. Eng. J. **140**(1–3), 86–100 (2008). doi:10.1016/j.cej.2007.09.014

B. Al Zaitone, H.-J. Schmid, W. Peukert, Simulation of structure and mobility of aggregates formed by simultaneous coagulation, sintering and surface growth. Aerosol Sci. **40**(11), 950–964, 2009. doi:10.1016/j.jaerosci.2009.08.007

D. Asnaghi, M. Carpineti, M. Giglio, M. Sozzi, Coagulation kinetics and aggregate morphology in the intermediate regimes between diffusion-limited and reaction-limited cluster aggregation. Phys. Rev. A **45**(2), 1018–1023 (1992). doi:10.1103/PhysRevA.45.1018

W.D. Brown, R.C. Ball, Computer simulation of chemically limited aggregation. J. Phys. A: Math. Gen. **18**(9), L517–L521 (1985). doi:10.1088/0305-4470/18/9/006

A. di Biasio, G. Bolle, C. Cametti, P. Codastefano, F. Sciortino, P. Tartaglia, Crossover region in the aggregation of colloids. Phys. Rev. E **50**(2), 1649–1652 (1994). doi:10.1103/PhysRevE.50.1649

A.V. Filippov, M. Zurita, D.E. Rosner, Fractal-like aggregates: relation between morphology and physical properties. J. Colloid Interface Sci. **229**(1), 261–273 (2000). doi:10.1006/jcis.2000.7027

N. Fuchs, Über die Stabilität und Aufladung der Aerosole. Z. Phys. **89**(11–12), 736–743 (1934). doi:10.1007/BF01341386

C. Gauer, H. Wu, M. Morbidelli, Effect of surface properties of elastomer colloids on their coalescence and aggregation kinetics. Langmuir **25**(20), 12073–12083 (2009). doi:10.1021/la901702s

F. Gruy, Population balance for aggregation coupled with morphology changes. Colloids Surf. A **374**(1–3), 69–76 (2011). doi:10.1016/j.colsurfa.2010.11.010

R. Hogg, Agglomeration models for process design and control. Powder Technol. **69**(1), 69–76 (1992). doi:10.1016/0032-5910(92)85009-K

E.P. Honig, G.J. Roebersen, P.H. Wiersema, Effect of hydrodynamic interaction on the coagulation rate of hydrophobic colloids. J. Colloid Interface Sci. **36**(1), 97–109 (1971). doi:10.1016/0021-9797(71)90245-1

R.K. Iler, *The chemistry of silica* (Wiley, New York, 1979). ISBN 0-471-02404-X

R. Jullien, Aggregation phenomena and fractal aggregates. Contemp. Phys. **28**(5), 477–493 (1987). doi:10.1080/00107518708213736

R. Jullien, M. Kolb, Hierarchical model for chemically limited cluster-cluster aggregation. J. Phys. A Math. Gen. **17**(12), L639–L643 (1984). doi:10.1088/0305-4470/17/12/003

U. Kätzel, R. Bedrich, M. Stintz, R. Ketzmerick, T. Gottschalk-Gaudig, H. Barthel, Dynamic light scattering for the characterization of polydisperse fractal systems: I. Simulation of the diffusional behavior. Part. Part. Syst. Charact. **25**(1), 9–18 (2008a). doi:10.1002/ppsc.200700005

R. Klein, D.A. Weitz, M.Y. Lin, H.M. Lindsay, R.C. Ball, P. Meakin, Theory of scattering from colloidal aggregates, in *Trends in Colloid and Interface Science IV*, eds. by M. Zulauf, P. Lindner, P. Terech. Progress in Colloid & Polymer Science, vol. 81 (Steinkopff Verlag, 1990), pp. 161–168. doi:10.1007/BFb0115545

M. Kolb, R. Botet, R. Jullien, Scaling of kinetically growing clusters. Phys. Rev. Lett. **51**(13), 1123–1126 (1983). doi:10.1103/PhysRevLett.51.1123

Y. Kusaka, T. Fukasawa, Y. Adachi, Cluster-cluster aggregation simulation in a concentrated suspension. J. Colloid Interface Sci. **363**(1), 34–41 (2011). doi:10.1016/j.jcis.2011.07.024

M. Lattuada, H. Wu, M. Morbidelli, Hydrodynamic radius of fractal clusters. J. Colloid Interface Sci. **268**(1), 96–105 (2003). doi:10.1016/j.jcis.2003.07.028

M. Lattuada, H. Wu, P. Sandkühler, J. Sefcik, M. Morbidelli, Modelling of aggregation kinetics of colloidal systems and its validation by light scattering measurements. Chem. Eng. Sci. **59**(8–9), 1783–1798 (2004). doi:10.1016/j.ces.2004.01.033

M.Y. Lin, H.M. Lindsay, D.A. Weitz, R. Klein, R.C. Ball, P. Meakin, Universal diffusion-limited colloid aggregation. J. Phys. Condens. Matter **2**(13), 3093–3113 (1990a). doi:10.1088/0953-8984/2/13/019

W. Lin, M. Kobayashi, M. Skarba, C. Mu, P. Galletto, M. Borkovec, Heteroaggregation in binary mixtures of oppositely charged colloidal particles. Langmuir **22**(3), 1038–1047 (2006). doi:10.1021/la0522808

C. Lu, R. Pelton, PEO flocculation of polystyrene-core poly(vinylphenol)-shell latex: an example of ideal bridging. Langmuir **17**(25), 7770–7776 (2001). doi:10.1021/la010893o

D.W. Mackowski, Electrostatics analysis of radiative absorption by sphere clusters in the Rayleigh limit: application to soot particles. Appl. Opt. **34**(18), 3535–3545 (1995). doi:10.1364/AO.34.003535

P. Meakin, Formation of fractal clusters and networks by irreversible diffusion-limited aggregation. Phys. Rev. Lett. **51**(13), 1119–1122 (1983). doi:10.1103/PhysRevLett.51.1119

P. Meakin, The effects of rotational diffusion on the fractal dimensionality of structures formed by cluster-cluster aggregation. J. Chem. Phys. **81**(10), 4637–4639 (1984a). doi:10.1063/1.447398

P. Meakin, Diffusion-limited aggregation in three dimensions: results from a new cluster-cluster aggregation model. J. Colloid Interface Sci. **102**(2), 491–504 (1984b). doi:10.1016/0021-9797(84)90252-2

P. Meakin, F. Family, Structure and dynamics of reaction-limited aggregation. Phys. Rev. A **36**(11), 5498–5501 (1987). doi:10.1103/PhysRevA.36.5498

V. Runkana, P. Somasundaran, P.C. Kapur, Reaction-limited aggregation in presence of short-range structural forces. AIChE J. **51**(4), 1233–1245 (2005). doi:10.1002/aic.10375

P.G. Saffman, J.S. Turner, On the collision of drops in turbulent clouds. J. Fluid Mech. **1**(1), 16–30 (1956). doi:10.1017/S0022112056000020

P. Sandkühler, J. Sefcik, M. Morbidelli, Scaling of the kinetics of slow aggregation and gel formation for a fluorinated polymer colloid. Langmuir **21**(5), 2062–2077 (2005a). doi:10.1021/la048055s

P. Sandkühler, M. Lattuada, H. Wu, J. Sefcik, M. Morbidelli, Further insights into the universality of colloidal aggregation. Adv. Colloid Interface Sci. **113**(2–3), 65–83 (2005b). doi:10.1016/j.cis.2004.12.001

D.W. Schaefer, A.J. Hurd, Growth and structure of combustion aerosols: fumed silica. Aerosol Sci. Technol. **12**(4), 876–890 (1990). doi:10.1080/02786829008959400

J. Schlomach, M. Kind, Investigations on the semi-batch precipitation of silica. J. Colloid Interface Sci. **277**(2), 316–326 (2004). doi:10.1016/j.jcis.2004.04.051

R.H. Smellie Jr, V.K. La Mer, Flocculation, subsidence and filtration of phosphate slimes. VI. A quantitative theory of filtration of flocculated suspensions. J. Colloid Sci. **23**(6), 589–599 (1958). doi:10.1016/0095-8522(58)90071-0

M.V. Smoluchowski, Drei Vorträge über Diffusion, Brownsche Molekularbewegung und Koagulation von Kolloidteilchen. Phys. Z. **17**, 557–571, 585–599 (1916)

R. Thouy, R. Jullien, A cluster-cluster aggregation model with tuneable fractal dimension. J. Phys. A: Math. Gen. **27**(9), 2953–2963 (1994). doi:10.1088/0305-4470/27/9/012

V.A. Tolpekin, M.H.G. Duits, D.V.D. Ende, J. Mellema, Aggregation and breakup of colloidal particle aggregates in shear flow, studied with video microscopy. Langmuir **20**(7), 2614–2627 (2004). doi:10.1021/la0357581
T.A. Witten Jr, L.M. Sander, Diffusion-limited aggregation, a kinetic critical phenomenon. Phys. Rev. Lett. **47**(19), 1400–1403 (1981). doi:10.1103/PhysRevLett.47.1400

Structure of (Fractal) Aggregates

F. Babick, K. Schießl, M. Stintz, Characterization of pyrogenic powders with conventional particle sizing technique: I. Prediction of measured size distributions. Part. Part. Syst. Charact. **29**(2), 104–115 (2012). doi:10.1002/ppsc.201000024
R.C. Ball, D.A. Weitz, T.A. Witten, F. Leyvraz, Universal kinetics in reaction-limited aggregation. Phys. Rev. Lett. **58**(3), 274–277 (1987). doi:10.1103/PhysRevLett.58.274
R. Bedrich, *Geometrische und hydrodynamische Eigenschaften fraktaler Aggregate*. Diploma thesis, Technische Universität Dresden, Institut für theoretische Physik (2006)
D. Boldridge, Morphological characterization of fumed silica aggregates. Aerosol Sci. Technol. **44** (3), 188–192 (2010). doi:10.1080/02786820903499462
E. Bugnicourt, J. Galy, J.-F. Gérarda, F. Bouéb, H. Barthel, Structural investigations of pyrogenic silica–epoxy composites: combining small-angle neutron scattering and transmission electron microscopy. Polymer **48**(4), 949–958 (2007). doi:10.1016/j.polymer.2006.12.012
J.L. Burns, Y.-D. Yan, G.J. Jameson, S. Biggs, A light scattering study of the fractal aggregation behavior of a model colloidal system. Langmuir **13**(24), 6413–6420 (1997). doi:10.1021/la970303f
P. Dimon, S.K. Sinha, D.A. Weitz, C.R. Safinya, G.S. Smith, W.A. Varady, H.M. Lindsay, Structure of aggregated gold colloids. Phys. Rev. Lett. **57**(5), 595–598 (1986). doi:10.1103/PhysRevLett.57.595
S.R. Forrest, T.A. Witten, Long-range correlations in smoke-particle aggregates. J. Phys. A Math. Gen. **12**(5), L109–L117 (1979). doi:10.1088/0305-4470/12/5/008
L. Gmachowski, Flow drag in polydisperse systems. Wat. Res. **32**(9), 2655–2659 (1998). doi:10.1016/S0043-1354(98)00039-6
L. Gmachowski, Calculation of the fractal dimension of aggregates. Colloids. Surf. A **211**(2–3), 197–203 (2002). doi:10.1016/S0927-7757(02)00278-9
F. Gruy, P. Cugniet, Experimental study of small aggregate settling. J. Colloid Interface Sci. **272** (2), 465–471 (2004). doi:10.1016/j.jcis.2003.11.026
F. Gruy, Light-scattering cross section as a function of pair distribution density. J. Quant. Spectrosc. Radiat. Transfer **10**(3), 240–246 (2009). doi:10.1016/j.jqsrt.2008.10.003
A. Guinier, La diffusion des rayons X sous les très faibles angles appliquèe á l'etude de fines particules et de suspensions colloidales. CR Hebd. Seance Acad. Sci. **206**(19), 1374–1376 (1938)
W. Hess, H.L. Frisch, R. Klein, On the hydrodynamic behavior of colloidal aggregates. Z. Phys. B Condens. Matter **64**(1), 65–67 (1986). doi:10.1007/BF01313690
N. Ibaseta, B. Biscans, Fractal dimension of fumed silica: comparison of light scattering and electron microscope methods. Powder Technol. **203**(2), 206–210 (2010). doi:10.1016/j.powtec.2010.05.010
R. Jullien, Les phénomènes d'agrégation et les agrégats fractals. Ann. Télécommun. **41**(7–8), 343–372 (1986). doi:10.1007/BF02997881
N.G. Khlebtsov, Spectroturbidimetry of fractal clusters: test of density correlation function cutoff. Appl. Opt. **35**(21), 4261–4270 (1996). doi:10.1364/AO.35.004261
A.A. Lall, M. Seipenbusch, W. Rong, S.K. Friedlander, On-line measurement of ultrafine aggregate surface area and volume distributions by electrical mobility analysis: II. Comparison of measurements and theory. J. Aerosol Sci. **37**(3), 272–282 (2006). doi:10.1016/j.jaerosci.2006.01.006

M. Lattuada, *Aggregation kinetics and structure of gels and aggregates in colloidal systems* (Diss. ETH Nr. 15187). PhD thesis, ETH Zurich (2003)

M. Lattuada, H. Wu, M. Morbidelli, Hydrodynamic radius of fractal clusters. J. Colloid Interface Sci. **268**(1), 96–105 (2003). doi:10.1016/j.jcis.2003.07.028

M. Lattuada, H. Wu, P. Sandkühler, J. Sefcik, M. Morbidelli, Modelling of aggregation kinetics of colloidal systems and its validation by light scattering measurements. Chem. Eng. Sci. **59**(8–9), 1783–1798 (2004). doi:10.1016/j.ces.2004.01.033

M. Lattuada, H. Wu, J. Sefcik, M. Morbidelli, Detailed model of the aggregation event between two fractal clusters. J. Phys. Chem. B **110**(13), 6574–6586 (2006). doi:10.1021/jp056538e

C. Lee, T.A. Kramer, Prediction of three-dimensional fractal dimensions using the two-dimensional properties of fractal aggregates. Adv. Colloid Interfaces Sci. **112**(1–3), 49–57 (2004). doi:10.1016/j.cis.2004.07.001

M.Y. Lin, H.M. Lindsay, D.A. Weitz, R.C. Ball, R. Klein, P. Meakin, Universality of fractal aggregates as probed by light scattering. Proc. R. Soc. London, A **423**(1864), 71–87 (1989). doi:10.1098/rspa.1989.0042

M.Y. Lin, H.M. Lindsay, D.A. Weitz, R. Klein, R.C. Ball, P. Meakin, Universal diffusion-limited colloid aggregation. J. Phys. Condens. Matter **2**(13), 3093–3113 (1990a). doi:10.1088/0953-8984/2/13/019

M.Y. Lin, H.M. Lindsay, D.A. Weitz, R.C. Klein, R. Ball, P. Meakin, Universal reaction-limited aggregation. Phys. Rev. A **41**(4), 2005–2020 (1990b). doi:10.1103/PhysRevA.41.2005

H.M. Lindsay, R. Klein, D.A. Weitz, M.Y. Lin, P. Meakin, Structure and anisotropy of colloid aggregates. Phys. Rev. A **39**(6), 3112–3119 (1989). doi:10.1103/PhysRevA.39.3112

B. Mandelbrot, How long is the coast of Britain? Statistical self-similarity and fractional dimension. Science **156**(3775), 636–638 (1967). doi:10.1126/science.156.3775.636

B. Mandelbrot, *Fractals, form, chance, and dimension*, 1st edn. (Freeman & Co, San Fransisco, 1977). ISBN 0-716-70473-0

P. Meakin, Effects of cluster trajectories on cluster-cluster aggregation: a comparison of linear and Brownian trajectories in two-dimensional and 3-dimensional simulations. Phys. Rev. A **29**(2), 997–999 (1984c). doi:10.1103/PhysRevA.29.997

K.-H. Naumann, H. Bunz, Van der Waals interactions between fractal particles. J. Aerosol Sci. **24**(S1), S181–S182 (1993). doi:10.1016/0021-8502(93)90183-A

J.S. Olfert, The effective density and fractal dimension of particles emitted from a light-duty diesel vehicle with a diesel oxidation catalyst. J. Aerosol Sci. **38**(1), 69–82 (2007). doi:10.1016/j.jaerosci.2006.10.002

K. Park, D.B. Kittelson, P.H. McMurry, Structural properties of diesel exhaust particles measured by transmission electron microscopy (TEM): Relationships to particle mass and mobility. Aerosol Sci. Technol. 38(9), 881–889 (2004). doi:10.1080/027868290505189

L.F. Richardson, The problem of contiguity: an appendix of statistics of deadly quarrels. Gen. Syst. Yearbook **6**(139), 139–187 (1961)

P. Sandkühler, M. Lattuada, H. Wu, J. Sefcik, M. Morbidelli, Further insights into the universality of colloidal aggregation. Adv. Colloid Interface Sci. **113**(2–3), 65–83 (2005b). doi:10.1016/j.cis.2004.12.001

D.W. Schaefer, J.E. Martin, P. Wiltzius, D.S. Cannell, Fractal geometry of colloidal aggregates. Phys. Rev. Lett. **52**(26), 2371–2374 (1984). doi:10.1103/PhysRevLett.52.2371

D.W. Schaefer, A.J. Hurd, Growth and structure of combustion aerosols—Fumed silica. Aerosol Sci. Technol. **12**(4), 876–890 (1990). doi:10.1080/02786829008959400

B. Schleicher, S. Künzel, H. Burtscher, In sifu measurement of size and density of submicron aerosol particles. J. Appl. Phys. **78**(7), 4416–4422 (1995). doi:10.1063/1.359849

J. Schlomach, M. Kind, Investigations on the semi-batch precipitation of silica. J. Colloid Interface Sci. **277**(2), 316–326 (2004). doi:10.1016/j.jcis.2004.04.051

W.G. Shin, J. Wang, M. Mertler, B. Sachweh, H. Fissan, D.Y.H. Pui, Structural properties of silver nanoparticle agglomerates based on transmission electron microscopy: relationship to particle mobility analysis. J. Nanopart. Res. **11**(1), 163–173 (2009). doi:10.1007/s11051-008-9468-y

W.G. Shin, G.W. Mulholland, D.Y.H. Pui, Determination of volume, scaling exponents, and particle alignment of nanoparticle agglomerates using tandem differential mobility analyzers. J. Aerosol Sci. **41**(7), 665–681 (2010). doi:10.1016/j.jaerosci.2010.04.009
C.M. Sorensen, G.C. Roberts, The prefactor of fractal aggregates. J. Colloid Interface Sci. **186**(2), 447–452 (1997). doi:10.1006/jcis.1996.4664
C.M. Sorensen, Light scattering by fractal aggregates: a review. Aerosol Sci. Technol. **35**(2), 648–687 (2001). doi:10.1080/02786820117868
P. Tang, J. Greenwood, J.A. Raper, A model to describe the settling behavior of fractal aggregates. J. Colloid Interface Sci. **247**(1), 210–219 (2002). doi:10.1006/jcis.2001.8028
R. Thouy, R. Jullien, Geometrical properties of aggregates with tuneable fractal dimension. J. Phys. A Math. Gen. **30**(19), 6725–6735 (1997). doi:10.1088/0305-4470/30/19/013
P.G.J. van Dongen, M.H. Ernst, Dynamic scaling in the kinetics of clustering. Phys. Rev. Lett. **54** (13), 1396–1399 (1985). doi:10.1103/PhysRevLett.54.1396
A. Virtanen, J. Ristimäki, J. Keskinen, Method for measuring effective density and fractal dimension of aerosol agglomerates. Aerosol Sci. Technol. **38**(5), 437–446 (2010). doi:10.1080/02786820490445155
H. von Koch, Une courbe continue sans tangente, obtenue par une construction géometrique élémentaire. Arkiv för Matematik **1**, 681–704 (1904)
A.P. Weber, U. Baltensperger, H.W. Gäggeler, A. Schmidt-Ott, In Situ characterization and structure modification of agglomerated aerosol particles. J. Aerosol Sci. **27**(6), 915–929 (1996). doi:10.1016/0021-8502(96)00013-4
D.A. Weitz, J.S. Huang, M.Y. Lin, J. Sung, Limits of the fractal dimension for irreversible kinetic aggregation of gold colloids. Phys. Rev. Lett. **54**(13), 1416–1419 (1985). doi:10.1103/PhysRevLett.54.1416
P. Wiltzius, Hydrodynamic behaviour of fractal aggregats. Phys. Rev. Lett. **58**(7), 710–713 (1987). doi:10.1103/PhysRevLett.58.710
T.A. Witten Jr, L.M. Sander, Diffusion-limited aggregation, a kinetic critical phenomenon. Phys. Rev. Lett. **47**(19), 1400–1403 (1981). doi:10.1103/PhysRevLett.47.1400

Scattering of Aggregates

F. Babick, K. Schießl, M. Stintz, Characterization of pyrogenic powders with conventional particle sizing technique: I. Prediction of measured size distributions. Part. Part. Syst. Charact. **29**(2), 104–115 (2012a). doi:10.1002/ppsc.201000024
F. Babick, S. Gropp, M. Stintz, Dynamic light scattering of dispersed pyrogenic powders. On the CD-ROM: *PARTEC 2013, International Congress for Particle Technology,* Nuremberg, 23–25 April 2013. ISBN 978-3-00-040578-5 (available from NürnbergMesse GmbH, Messezentrum, 90471 Nürnberg, Germany), p 368
M.V. Berry, I.C. Percival, Optics of fractal clusters such as smoke. Opt. Acta **33**(5), 577–591 (1986). doi:10.1080/713821987
O. Cruzan, Translational addition theorems for spherical vector wave functions. Q. Appl. Math. **20** (1), 33–40 (1962)
P. Debye, Zerstreuung von Röntgenstrahlen. Ann. Phys., IV **46**(6), 809–823 (1915). doi:10.1002/andp.19153510606
B.L. Henke, E.M. Gullikson, J.C. Davis, X-Ray Interactions: Photoabsorption, scattering, transmission, and reflection at E = 50-30,000 eV, Z = 1-92. At. Data Nucl. Data Tables **54**(2), 181–342 (1993). doi:10.1006/adnd.1993.1013
S. Jacquier, F. Gruy, Application of scattering theories to the characterization of precipitation processes, in *Light Scattering Reviews 5. Single Light Scattering and Radiative Transfer*, ed. by A.A. Kokhanovsky, part 1 (Springer Praxis Books, Berlin Heidelberg, 2010), pp. 37–78. ISBN 978-3-642-10335-3. doi:10.1007/978-3-642-10336-0_2

H.K. Kammler, G. Beaucage, R. Mueller, S.E. Pratsinis, Structure of flame-made silica nanoparticles by ultra-small-angle X-ray scattering. Langmuir **20**(5), 1915–1921 (2004). doi:10.1021/la030155v

U. Kätzel, F. Gruy, F. Babick, W. Klöden, Light extinction at agglomerates of spheres—A practical test on the submicroscale. J. Colloid Interface Sci. **289**(1), 116–124 (2005). doi:10.1016/j.jcis.2005.03.041

U. Kätzel, M. Vorbau, M. Stintz, T. Gottschalk-Gaudig, H. Barthel, Dynamic light scattering for the characterization of polydisperse fractal systems: II. Relation between structure and DLS results. Part. Part. Syst. Charact. **25**(1), 19–30 (2008b). doi:10.1002/ppsc.200700005

N.G. Khlebtsov, Spectroturbidimetry of fractal clusters: test of density correlation function cutoff. Appl. Opt. **35**(21), 4261–4270 (1996). doi:10.1364/AO.35.004261

D.W. Mackowski, Analysis of radiative scattering for multiple sphere configurations. Proc. R. Soc., London, A **433**, 599–614 (1991). doi:10.1098/rspa.1991.0066

G. Mie, Beiträge zur Optik trüber Medien, speziell kolloidaler Goldlösungen. Ann. Phys., IV **25** (3), 377–445 (1908). doi:10.1002/andp.19083300302

E.M. Purcell, C.R. Pennypacker, Scattering and absorption of light by nonspherical dielectric grains. Astrophys. J. **186**(2), 705–714 (1973). doi:10.1086/152538

S.B. Singham, C.F. Bohren, Scattering of unpolarized and polarized light by particle aggregates of different size and fractal dimension. Langmuir **9**(5), 1431–1435 (1993). doi:10.1021/la00029a044

C.M. Sorensen, Light scattering by fractal aggregates: a review. Aerosol Sci. Technol. **35**(2), 648–687 (2001). doi:10.1080/02786820117868

S. Stein, Addition theorems for spherical wave functions. Q. Appl. Math. **19**(1), 15–24 (1961)

R.T. Wang, B.Å.S. Gustafson, Angular scattering and polarization by randomly oriented dumbbells and chains of spheres, in *Proceedings of the 1983 Scientific Conference on Obscuration and Aerosol Research*, eds. by J. Farmer, R. Kohl (U.S. Army Aberdeen, Md., 1984), pp. 237–247

W. Wiscombe, Improved Mie scattering algorithms. Appl. Opt. **19**(9), 1505–1509 (1980). doi:10.1364/AO.19.001505

Y.-L. Xu, Electromagnetic scattering by an aggregate of spheres. Appl. Opt. **34**(21), 4573–4588 (1995). doi:10.1364/AO.34.004573

Y.-L. Xu, Electromagnetic scattering by an aggregate of spheres: Far field. Appl. Opt. **36**(36), 9496–9508 (1997). doi:10.1364/AO.36.009496

Y.-L. Xu, Efficient evaluation of vector translation coefficients in multiparticle light-scattering theories. J. Comp. Phys. **139**(1), 137–165 (1998a). doi:10.1006/jcph.1997.5867

Y.-L. Xu, Electromagnetic scattering by an aggregate of spheres: asymmetry parameter. Phys. Lett. A **249**(1–2), 30–36 (1998b). doi:10.1016/S0375-9601(98)00708-7

Y.-L. Xu, B.Å.S. Gustafson, Experimental and theoretical results of light scattering by aggregates of spheres. Appl. Opt. **36**(30), 8026–8030 (1997). doi:10.1364/AO.36.008026

Hydrodynamics of Aggregates

F. Babick, K. Schießl, M. Stintz, Characterization of pyrogenic powders with conventional particle sizing technique: I. Prediction of measured size distributions. Part. Part. Syst. Charact. **29**(2), 104–115 (2012a). doi:10.1002/ppsc.201000024

F. Babick, M. Vorbau, M. Stintz, Characterization of pyrogenic powders with conventional particle sizing technique: II. Experimental data. Part. Part. Syst. Charact. **29**(2), 116–127 (2012b). doi:10.1002/ppsc.201000025

H. Barthel, M. Heinemann, M. Stintz, B. Wessely, Particle sizes of fumed silica. Chem. Eng. Technol. **21**(9), 745–752 (1998). doi:10.1002/(SICI)1521-4125(199809)21:9<745:AID-CEAT745>3.0.CO;2-Q

C. Binder, C. Feichtinger, H.-J. Schmid, N. Thürey, W. Peukert, U. Rüde, Simulation of the hydrodynamic drag of aggregated particles. J. Colloid Interface Sci. **301**(1), 155–167 (2006). doi:10.1016/j.jcis.2006.04.045
C. Binder, M.A.J. Hartig, W. Peukert, Structural dependent drag force and orientation prediction for small fractal aggregates. J. Colloid Interface Sci. **331**(1), 243–250 (2009). doi:10.1016/j.jcis.2008.11.021
V.A. Bloomfield, W.O. Dalton, K.E. van Holde, Frictional coefficients of multisubunit structures. I. Theory. Biopolymers **5**(2), 135–148 (1967). doi:10.1002/bip.1967.360050202
D. Boldridge, Morphological characterization of fumed silica aggregates. Aerosol Sci. Technol. **44** (3), 188–192 (2010). doi:10.1080/02786820903499462
G. Bossis, A. Meunier, J.F. Brady, Hydrodynamic stress on fractal aggregates of spheres. J. Chem. Phys. **94**(7), 5064–5070 (1991). doi:10.1063/1.460543
J.F. Brady, G. Bossis, Stokesian dynamics. Ann. Rev. Fluid Mech. **20**, 111–157 (1988). doi:10.1146/annurev.fl.20.010188.000551
Z.-Y. Chen, J.M. Deutch, P. Meakin, Translational friction coefficient of diffusion limited aggregates. J. Chem. Phys. **80**(6), 2982–2983 (1984). doi:10.1063/1.447012
D. Coelho, J.-F. Thovert, R. Thou, P.M. Adler, Hydrodynamic drag and electrophoresis of suspensions of fractal aggregates. Fractals **5**(3), 507–522 (1997). doi:10.1142/S0218348X97000401
E. Cunningham, On the velocity of steady fall of spherical particles through fluid medium. Proc. R. Soc. London **83**(563), 357–365 (1910). doi:10.1098/rspa.1910.0024
B. Dahneke, Viscous resistance of straight-chain aggregates of uniform spheres. Aerosol Sci. Technol. **1**(2), 179–185 (1982). doi:10.1080/02786828208958586
J.G. de la Torre, V.A. Bloomfield, Hydrodynamic properties of macromolecular complexes. I. Translation. Biopolymers **16**(8), 1747–1763 (1977). doi:10.1002/bip.1977.360160811
L. Durlofsky, J.F. Brady, G. Bossis, Dynamic simulation of hydrodynamically interacting particles. J. Fluid Mech. **180**, 21–49 (1987). doi:10.1017/S002211208700171X
P.S. Epstein, On the resistance experienced by spheres in their motion through gases. Phys. Rev. **23**(6), 710–733 (1924). doi:10.1103/PhysRev.23.710
A.V. Filippov, Drag and torque on clusters of N arbitrary spheres at low Reynolds number. J. Colloid Interface Sci. **229**(1), 184–195 (2000). doi:10.1006/jcis.2000.6981
A.V. Filippov, Phoretic motion of arbitrary clusters of N spheres. J. Colloid Interface Sci. **241**(2), 479–491 (2001). doi:10.1006/jcis.2001.7730
A. Gastaldi, M. Vanni, The distribution of stresses in rigid fractal-like aggregates in a uniform flow field. J. Colloid Interface Sci. **357**(1), 18–30 (2011). doi:10.1016/j.jcis.2011.01.080
A.S. Geller, L.A. Mondy, D.J. Rader, M.S. Ingber, Boundary element method calculations of the mobility of nonspherical particles—1. Linear chains. J. Aerosol Sci. **24**(5), 597–609 (1993). doi:10.1016/0021-8502(93)90017-4
L. Gmachowski, M. Iwata, T. Adachi, T. Murase, Analysis of floc structure based on the extended Brinkman model. J. Chem. Eng. Jpn. **23**(5), 543–549 (1990). doi:10.1252/jcej.23.543
L. Gmachowski, Mechanism of shear aggregation. Wat. Res. **29**(8), 1815–1820 (1995). doi:10.1016/0043-1354(95)00006-7
L. Gmachowski, Hydrodynamics of aggregated media. J. Colloid Interface Sci. **178**(1), 80–86 (1996). doi:10.1006/jcis.1996.0095
L. Gmachowski, Flow drag in polydisperse systems. Wat. Res. **32**(9), 2655–2659 (1998). doi:10.1016/S0043-1354(98)00039-6
L. Gmachowski, Aggregate structure and hydrodynamics of aggregated systems. Colloids. Surf., A **255**(1–3), 105–110 (2005). doi:10.1016/j.colsurfa.2004.12.028
F. Gruy, P. Cugniet, Experimental study of small aggregate settling. J. Colloid Interface Sci. **272** (2), 465–471 (2004). doi:10.1016/j.jcis.2003.11.026
J. Happel, H. Brenner, *Low Reynolds number hydrodynamics with special applications to particulate matter* (Kluwer, Dortrecht, 1983). ISBN 90-247-2877-0

W. Hess, H.L. Frisch, R. Klein, On the hydrodynamic behavior of colloidal aggregates. Z. Phys. B Condens. Matter **64**(1), 65–67 (1986). doi:10.1007/BF01313690

H.K. Kammler, G. Beaucage, D.J. Kohls, N. Agashe, J. Ilavsky, Monitoring simultaneously the growth of nanoparticles and aggregates by in situ ultra-small-angle X-ray scattering. J. Appl. Phys. **97**(5), 054309 (2005). doi:10.1063/1.1855391

U. Kätzel, *Dynamic light scattering for the characterization of polydisperse fractal systems by the example of pyrogenic silica*. Ph.D. thesis, Technische Universität Dresden (2007). urn: nbn:de: swb:14-1197634640783-66357

U. Kätzel, R. Bedrich, M. Stintz, R. Ketzmerick, T. Gottschalk-Gaudig, H. Barthel, Dynamic light scattering for the characterization of polydisperse fractal systems: I. Simulation of the diffusional behavior. Part. Part. Syst. Charact. **25**(1), 9–18 (2008a). doi:10.1002/ppsc. 200700005

J.G. Kirkwood, J. Riseman, The intrinsic viscosities and diffusion constants of flexible macromolecules in solution. J. Chem. Phys. **16**(6), 565–573 (1948). doi:10.1063/1.1746947

A.A. Lall, S.K. Friedlander, On-line measurement of ultrafine aggregate surface area and volume distributions by electrical mobility analysis: I. Theoretical analysis. J. Aerosol Sci. **37**(3), 260–271 (2006). doi:10.1016/j.jaerosci.2006.01.006

M. Lattuada, H. Wu, M. Morbidelli, Hydrodynamic radius of fractal clusters. J. Colloid Interface Sci. **268**(1), 96–105 (2003). doi:10.1016/j.jcis.2003.07.028

M. Lattuada, H. Wu, P. Sandkühler, J. Sefcik, M. Morbidelli, Modelling of aggregation kinetics of colloidal systems and its validation by light scattering measurements. Chem. Eng. Sci. **59**(8–9), 1783–1798 (2004). doi:10.1016/j.ces.2004.01.033

C. Lee, T.A. Kramer, Prediction of three-dimensional fractal dimensions using the two-dimensional properties of fractal aggregates. Adv. Colloid Interfaces Sci. **112**(1–3), 49–57 (2004). doi:10.1016/j.cis.2004.07.001

L.A. Mondy, A.S. Geller, D.J. Rader, M. Ingber, Boundary element method calculations of the mobility of nonspherical particles—II. Branched chains and flakes. J. Aerosol Sci. **27**(4), 537–546 (1996). doi:10.1016/0021-8502(95)00574-9

K.-H. Naumann, H. Bunz, Aerodynamic properties of fractal aerosol particles. J. Aerosol Sci. **22** (S1), S161–S164 (1991). doi:10.1016/S0021-8502(05)80059-X

G. Neale, N. Epstein, W. Nader, Creeping flow relative to permeable spheres. Chem. Eng. Sci. **28** (10), 1865–1874 (1973). doi:10.1016/0009-2509(73)85070-5

P. Sandkühler, J. Sefcik, M. Morbidelli, Scaling of the kinetics of slow aggregation and gel formation for a fluorinated polymer colloid. Langmuir **21**(5), 2062–2077 (2005a). doi:10.1021/ la048055s

A. Sierou, J.F. Brady, Accelerated Stokesian dynamics simulations. J. Fluid Mech. **448**, 115–146 (2001). doi:10.1017/S0022112001005912

C.M. Sorensen, Light scattering by fractal aggregates: a review. Aerosol Sci. Technol. **35**(2), 648–687 (2001). doi:10.1080/02786820117868

M. Staiger, P. Bowen, J. Ketterer, J. Bohonek, Particle size distribution measurement and assessment of agglomeration of commercial nanosized ceramic particles. J. Dispersion Sci. Technol. **23**(5), 619–630 (2002). doi:10.1081/DIS-120015367

P. Tang, J. Greenwood, J.A. Raper, A model to describe the settling behavior of fractal aggregates. J. Colloid Interface Sci. **247**(1), 210–219 (2002). doi:10.1006/jcis.2001.8028

M. Vanni, Creeping flow over spherical permeable aggregates. Chem. Eng. Sci. **55**(3), 685–698 (2000). doi:10.1016/S0009-2509(99)00316-4

S. Veerapaneni, M.R. Wiesner, Hydrodynamics of fractal aggregates with radially varying permeability. J. Colloid Interface Sci. **177**(1), 45–57 (1996). doi:10.1006/jcis.1996.0005

G.M. Wang, C.M. Sorensen, Diffusive mobility of fractal aggregates over the entire knudsen number range. Phys. Rev. E **60**(3), 3036–3044 (1999). doi:10.1103/PhysRevE.60.3036

P. Wiltzius, Hydrodynamic behaviour of fractal aggregats. Phys. Rev. Lett. **58**(7), 710–713 (1987). doi:10.1103/PhysRevLett.58.710

Pyrogenic Powders

S.R. Aragón, R. Pecora, Theory of dynamic light scattering from large anisotropic particles. J. Chem. Phys. **66**(6), 2506–2516 (1977). doi:10.1063/1.434246

F. Babick, K. Schießl, M. Stintz, Characterization of pyrogenic powders with conventional particle sizing technique: I. Prediction of measured size distributions. Part. Part. Syst. Charact. **29**(2), 104–115 (2012a). doi:10.1002/ppsc.201000024

F. Babick, M. Vorbau, M. Stintz, Characterization of pyrogenic powders with conventional particle sizing technique: II. Experimental data. Part. Part. Syst. Charact. **29**(2), 116–127 (2012b). doi:10.1002/ppsc.201000025

F. Babick, S. Gropp, U. Kätzel, M. Vorbau, Dynamic light scattering of dispersed fumed silica aggregates. Powder Technol. **215–216**, 39–45 (2012c). doi:10.1016/j.powtec.2011.10.064

H. Barthel, M. Dreyer, T. Gottschalk-Gaudig, V. Litvinov, E. Nikitina, Fumed silica—Rheological additive for adhesives, resins, and paints. Macromol. Symp. **187**(1), 573–584 (2002). doi:10.1002/1521-3900(200209)187:1<573:AID-MASY573>3.0.CO;2-1

G. Beaucage, Approximations leading to a unified exponential/power-law approach to small-angle scattering. J. Appl. Cryst. **28**(6), 717–728 (1995). doi:10.1107/S0021889895005292

G. Beaucage, Small-angle scattering from polymeric mass fractals of arbitrary mass-fractal dimension. J. Appl. Cryst. **29**(2), 134–146 (1996). doi:10.1107/S0021889895011605

D. Boldridge, Morphological characterization of fumed silica aggregates. Aerosol Sci. Technol. **44** (3), 188–192 (2010). doi:10.1080/02786820903499462

U. Brinkmann, M. Ettlinger, D. Kerner, R. Schmoll, Synthetic amorphous silicas, in *Colloidal silica: Fundamentals and Applications*, eds. by H.E. Bergna, W.O. Roberts, Chap. 43. Surfactant Science Series, vol. 131 (Taylor & Francis, Boca Raton, 2006), pp. 575–588. ISBN 0-8247-0967-5

E. Bugnicourt, J. Galy, J.-F. Gérarda, F. Bouéb, H. Barthel, Structural investigations of pyrogenic silica–epoxy composites: combining small-angle neutron scattering and transmission electron microscopy. Polymer **48**(4), 949–958 (2007). doi:10.1016/j.polymer.2006.12.012

M. Ettlinger, Pyrogenic silica, in *Ullmann's Encyclopedia of Industrial Chemistry*, 7th edn (Wiley-VCH, 2002)

A.J. Hurd, D.W. Schaefer, J.E. Martin, Surface and mass fractals in vapor-phase aggregates. Phys. Rev. A **35**(5), 2361–2364 (1987). doi:10.1103/PhysRevA.35.2361

IARC, Working group on the evaluation of carcinogenic risks to humans, *Carbon black, titanium dioxide, and talc*, in: *IARC Monographs on the Evaluation of Carcinogenic Risks to Humans*, vol. 93 (International Agency for Research on Cancer (IARC), Lyon, 2010), pp. 56–63

N. Ibaseta, B. Biscans, Fractal dimension of fumed silica: comparison of light scattering and electron microscope methods. Powder Technol. **203**(2), 206–210 (2010). doi:10.1016/j.powtec.2010.05.010

H.K. Kammler, G. Beaucage, R. Mueller, S.E. Pratsinis, Structure of flame-made silica nanoparticles by ultra-small-angle X-ray scattering. Langmuir **20**(5), 1915–1921 (2004). doi:10.1021/la030155v

H.K. Kammler, G. Beaucage, D.J. Kohls, N. Agashe, J. Ilavsky, Monitoring simultaneously the growth of nanoparticles and aggregates by in situ ultra-small-angle x-ray scattering. J. Appl. Phys. **97**(5), 054309 (2005). doi:10.1063/1.1855391

U. Kätzel, M. Vorbau, M. Stintz, T. Gottschalk-Gaudig, H. Barthel, Dynamic light scattering for the characterization of polydisperse fractal systems: II. Relation between structure and DLS results. Part. Part. Syst. Charact. **25**(1), 19–30 (2008b). doi:10.1002/ppsc.200700005

M. Lattuada, H. Wu, P. Sandkühler, J. Sefcik, M. Morbidelli, Modelling of aggregation kinetics of colloidal systems and its validation by light scattering measurements. Chem. Eng. Sci. **59**(8–9), 1783–1798 (2004). doi:10.1016/j.ces.2004.01.033

M.Y. Lin, H.M. Lindsay, D.A. Weitz, R.C. Ball, R. Klein, P. Meakin, Universality of fractal aggregates as probed by light scattering. Proc. R. Soc. London, A **423**(1864), 71–87 (1989). doi:10.1098/rspa.1989.0042

M.Y. Lin, H.M. Lindsay, D.A. Weitz, R. Klein, R.C. Ball, P. Meakin, Universal diffusion-limited colloid aggregation. J. Phys. Condens. Matter **2**(13), 3093–3113 (1990a). doi:10.1088/0953-8984/2/13/019

H.M. Lindsay, R. Klein, D.A. Weitz, M.Y. Lin, P. Meakin, Effect of rotational diffusion on quasielastic light scattering from fractal colloid aggregates. Phys. Rev. A **38**(5), 2614–2626 (1988). doi:10.1103/PhysRevA.38.2614

H.M. Lindsay, R. Klein, D.A. Weitz, M.Y. Lin, P. Meakin, Structure and anisotropy of colloid aggregates. Phys. Rev. A **39**(6), 3112–3119 (1989). doi:10.1103/PhysRevA.39.3112

P. Sandkühler, M. Lattuada, H. Wu, J. Sefcik, M. Morbidelli, Further insights into the universality of colloidal aggregation. Adv. Colloid Interface Sci. **113**(2–3), 65–83 (2005b). doi:10.1016/j.cis.2004.12.001

D.W. Schaefer, A.J. Hurd, Growth and structure of combustion aerosols—Fumed silica. Aerosol Sci. Technol. **12**(4), 876–890 (1990). doi:10.1080/02786829008959400

C.M. Sorensen, Light scattering by fractal aggregates: a review. Aerosol Sci. Technol. **35**(2), 648–687 (2001). doi:10.1080/02786820117868

S. Tsantilis, S.E. Pratsinis, Soft- and hard-agglomerate aerosols made at high temperatures. Langmuir **20**(14), 5933–5939 (2004). doi:10.1021/la036389w

L. Ying, China's fumed silica industry urgently needs upgrading. Chin. Chem. Rep. **21**(22), 27–28 (2010)

Interaction Among Aggregates

A.L. Ahmad, M.F. Chong, S. Bhatia, Population balance model (PBM) for flocculation process: simulation and experimental studies of palm oil mill effluent (POME) pretreatment. Chem. Eng. J. **140**(1–3), 86–100 (2008). doi:10.1016/j.cej.2007.09.014

V. Arunachalam, W.H. Marlow, J.X. Lu, Development of a picture of the van der Waals interaction energy between clusters of nanometre-range particles. Phys. Rev. E **58**(3), 3451–3457 (1998). doi:10.1103/PhysRevE.58.3451

F. Babick, K. Schießl, M. Stintz, Van-der-Waals interaction between two fractal aggregates. Adv. Powder Technol. **22**(2), 220–225 (2011). doi:10.1016/j.apt.2010.11.014

S.L. Carnie, D.Y.C. Chan, Interaction free energy between identical spherical colloidal particles: the linearized Poisson-Boltzmann theory. J. Colloid Interface Sci. **155**(2), 297–312 (1993). doi:10.1006/jcis.1993.1039

S.L. Carnie, D.Y.C. Chan, J.S. Gunning, Electrical double-layer interaction between dissimilar spherical colloidal particles and between a sphere and a plate: the linearized Poisson-Boltzmann theory. Langmuir **10**(9), 2993–3009 (1994). doi:10.1021/la00021a024

D. Coelho, M. Shapiro, J.F. Thovert, P.M. Adler, Electroosmotic phenomena in porous media. J. Colloid Interface Sci. **181**(1), 169–190 (1996). doi:10.1006/jcis.1996.0369

A. di Biasio, G. Bolle, C. Cametti, P. Codastefano, F. Sciortino, P. Tartaglia, Crossover region in the aggregation of colloids. Phys. Rev. E **50**(2), 1649–1652 (1994). doi:10.1103/PhysRevE.50.1649

J.C. Fernández-Toledano, A. Moncho-Jordá, F. Martínez-López, A.E. González, R. Hidalgo-Álvarez, Two-dimensional colloidal aggregation mediated by the range of repulsive interactions. Phys. Rev. E **75**(4), 041408 (2007). doi:10.1103/PhysRevE.75.041408

C. Gauer, H. Wu, M. Morbidelli, Effect of surface properties of elastomer colloids on their coalescence and aggregation kinetics. Langmuir **25**(20), 12073–12083 (2009). doi:10.1021/la901702s

A.B. Glendinning, W.B. Russel, The electrostatic repulsion between charged spheres from exact solutions to the linearized Poisson-Boltzmann equation. J. Colloid Interface Sci. **93**(1), 95–104 (1983). doi:10.1016/0021-9797(83)90388-0

K. Hoffmann, Über die Entstehung kettenförmiger Aggregate bei der Koagulation. Kolloid Z. **103**(2), 161–163 (1943). doi:10.1007/BF01502099

J. Jia, S. Iwata, Nonequilibrium structure of primary particles in colloidal bidispersion. Colloid Polym. Sci. **288**(14–15), 1485–1493 (2010). doi:10.1007/s00396-010-2274-y

U. Kätzel, M. Vorbau, M. Stintz, T. Gottschalk-Gaudig, H. Barthel, Dynamic light scattering for the characterization of polydisperse fractal systems: II. Relation between structure and DLS results. Part. Part. Syst. Charact. **25**(1), 19–30 (2008b). doi:10.1002/ppsc.200700005

H.-Y. Kim, J.O. Sofo, D. Velegol, M.W. Cole, A.A. Lucas, Van der Waals dispersion forces between dielectric nanoclusters. Langmuir **23**(4), 1735–1740 (2007). doi:10.1021/la061802w

J.W. Krozel, D.A. Saville, Electrostatic interactions between two spheres: solutions of the Debye-Hückel equation with a charge regulation boundary condition. J. Colloid Interface Sci. **150**(2), 365–373 (1992). doi:10.1016/0021-9797(92)90206-2

G.W. Kwon, Y.S. Won, B.J. Yoon, Electrical double-layer interactions of regular arrays of spheres. J. Colloid Interface Sci. **205**(2), 423–432 (1998). doi:10.1006/jcis.1998.5675

D. Langbein, Non-retarded dispersion energy between macroscopic spheres. J. Phys. Chem. Solids **32**(7), 1657–1667 (1971). doi:10.1016/S0022-3697(71)80059-8

E.M. Lifshitz, The theory of molecular attractive forces between solids. Sov. Phys. JETP **2**(1), 73–83 (1956)

K.-H. Naumann, H. Bunz, Van der Waals interactions between fractal particles. J. Aerosol Sci. **24** (S1), S181–S182 (1993). doi:10.1016/0021-8502(93)90183-A

H. Ohshima, Electrostatic interaction between soft particles. J. Colloid Interface Sci. **328**(1), 3–9 (2008). doi:10.1016/j.jcis.2008.08.009

Z. Peng, E. Doroodchi, G. Evans, DEM simulation of aggregation of suspended nanoparticles. Powder Technol. **204**(1), 91–102 (2010). doi:10.1016/j.powtec.2010.07.023

R.J. Phillips, Calculation of multisphere linearized Poisson-Boltzmann interactions near cylindrical fibers and planar surfaces. J. Colloid Interface Sci. **175**(2), 386–399 (1995). doi:10.1006/jcis.1995.1469

A.L.G. Rees, Directed aggregation in colloidal systems and the formation of protein fibers. J. Phys. Colloid Chem. **55**(8), 1340–1344 (1951). doi:10.1021/j150491a008

E.S. Reiner, C.J. Radke, Electrostatic interactions in colloidal suspensions: tests of pairwise additivity. AIChE J. **37**(6), 805–824 (1991). doi:10.1002/aic.690370604

T.A. Ring, Double-layer interaction energy for two unequal spheres. J. Chem. Soc., Faraday Trans. **2**(78), 1513–1528 (1982). doi:10.1039/f29827801513

V. Runkana, P. Somasundaran, P.C. Kapur, Reaction-limited aggregation in presence of short-range structural forces. AIChE J. **51**(4), 1233–1245 (2005). doi:10.1002/aic.10375

P. Sandkühler, M. Lattuada, H. Wu, J. Sefcik, M. Morbidelli, Further insights into the universality of colloidal aggregation. Adv. Colloid Interface Sci. **113**(2–3), 65–83 (2005b). doi:10.1016/j.cis.2004.12.001

K. Schießl, *Berechnung der Wechselwirkungsenergien zwischen fraktalen Aggregaten*. Diploma thesis, Technische Universität Dresden, Institut für Verfahrenstechnik und Umwelttechnik (2010)

K. Schießl, F. Babick, M. Stintz, Calculation of double layer interaction between colloidal aggregates. Adv. Powder Technol. **23**(2), 139–147 (2012). doi:10.1016/j.apt.2011.01.005

C.M. Sorensen, Light scattering by fractal aggregates: a review. Aerosol Sci. Technol. **35**(2), 648–687 (2001). doi:10.1080/02786820117868

N. Sun, J.Y. Walz, A model for calculating electrostatic interactions between colloidal particles of arbitrary surface topology. J. Colloid Interface Sci. **234**(1), 90–105 (2001). doi:10.1006/jcis.2000.7248

I.L. Thomas, K.H. McCorkle, Theory of oriented flocculation. J. Colloid Interface Sci. **36**(1), 110–118 (1971). doi:10.1016/0021-9797(71)90246-3

F.L. Usher, A mechanism of gelatinisation. Proc. R. Soc. Lond. A **125**, 143–151 (1929). doi:10.1098/rspa.1929.0157

A. Videcoq, M. Han, P. Abélard, C. Pagnoux, F. Rossignol, R. Ferrando, Influence of the potential range on the aggregation of colloidal particles. Phys. A **374**(2), 507–516 (2007). doi:10.1016/j.physa.2006.08.051

Chapter 5
Dispersion of Colloidal Suspensions and Their Stability

Suspensions of colloidal particles frequently show some degree of agglomeration, which clearly affects the macroscopic suspension properties (e.g. viscosity or turbidity). In order to adjust a defined particle size distribution they need a pre-treatment that destroys the agglomerates and aggregates (i.e. *dispersion*). The thus obtained size distribution depends on the employed dispersion techniques, on the microscopic stresses acting on the particles, on the total energy input, as well as on the mechanical strength of agglomerates and aggregates. A profound understanding of these influences facilitates the design of industrial dispersion processes (e.g. for polishing slurries) or the development of SOPs for the off-line characterisation of colloidal systems.

However, the state of dispersion is not a constant property, but rather may change over time, e.g. due to Ostwald ripening and recrystallisation, yet most prominently due to coagulation. Most often, such changes are to be avoided, which requires appropriate measures to stabilise the suspension. This is predominantly achieved by changing the interfacial properties so that, e.g., double layer and steric repulsions is increased. Indeed, dispersion and stabilisation cannot be separated when preparing colloidal suspensions; first stabilisation measures already need to be taken during the dispersion process. The quality of a suspension to keep its state of dispersion is generally called stability. Yet in practice, the term is related to very specific suspension properties and is used as a quantitative parameter. As a result, there is a certain variety of measurands and experimental tests for describing the suspension stability.

This chapter addresses the dispersion behaviour of colloidal suspensions as well as their stability. Since both terms are closely related to the particle size distribution they cannot be discussed without commenting on the employed characterisation techniques. Selected case studies illustrate the general considerations with a clear focus on pyrogenic powders.

F. Babick, *Suspensions of Colloidal Particles and Aggregates*,
Particle Technology Series 20, DOI 10.1007/978-3-319-30663-6_5

5.1 Dispersion of Colloidal Suspensions

The occurrence of coarse, micrometre-sized agglomerates in suspensions of colloidal particles is rather common and can be related to particle synthesis (e.g. precipitation), powder processing (e.g. spray drying), or agglomeration in stored powders and suspensions. The breakage of such agglomerates, i.e., the *dispersion*, aims at the elimination of very coarse agglomerates that have adverse effects on the product quality (e.g. for paints and coatings), at the homogeneous distribution of primary particles, or at least at a general reduction of particle size in order to achieve certain quality requirements (e.g. colour, viscosity or stability).

This section addresses the mechanisms of agglomerate breakup and the relevant dispersion techniques. It reviews experimental reports on the dispersion of pyrogenic powders with focus on empirical scaling laws of dispersion. Finally, the section discusses the applicability of various characterisation techniques to monitor the dispersion of colloidal suspensions.

5.1.1 Introduction to the Dispersion of Colloidal Suspensions

The dispersion of agglomerates requires that they experience a sufficiently high mechanical stress τ_{disp} that eventually exceeds the agglomerate strength σ_{agg}. The ratio of the two quantities, the fragmentation number Fa, is thus a measure on the effectiveness of the dispersion process (Rwei et al. 1990; Bałdyga et al. 2008):

$$\text{dispersion criterion: } Fa = \tau_{disp} / \sigma_{agg} > Fa_{cr}. \tag{5.1}$$

A successful deagglomeration additionally requires sufficiently long periods of stress t_{stress} or a sufficiently large number of stress events N_{stress}. This is because the breakup of agglomerates is usually preceded by agglomerate deformation (e.g. Higashitani et al. 2001) and/or fatigue processes that weaken the original agglomerate strengths (Kusters et al. 1993).

The strength of an agglomerate σ_{agg} is determined by the adhesive forces F_{ad} among its primary (constituent) particles, its internal structure (e.g. porosity, fractal dimension), and the size of the agglomerate and/or primary particles. In general, σ_{agg} can refer to different types of stress. However, principal aspects (e.g., the impact of agglomerate size and structure) are commonly discussed for the tensile strength σ_T (Bika et al. 2001). Rumpf (1970) examined the tensile strength of homogeneous, isotropic, and randomly-packed agglomerates. In this case, there is no impact of the agglomerate size, but rather of its internal structure, i.e., on the agglomerate porosity ε_{agg} and on the average number of contacts N_C to neighbouring particles (coordination number), as well as on the primary particle size x_p:

$$\sigma_{\mathrm{T}} = \frac{1 - \varepsilon_{\mathrm{agg}}}{\pi} N_{\mathrm{C}} \frac{F_{\mathrm{ad}}}{x_{\mathrm{p}}^2}. \tag{5.2}$$

This equation holds true for arbitrarily shaped primary particles, which may be compact individuals (as considered by Rumpf) or aggregates of even finer particles (Bałdyga et al. 2008). The equation can be further refined by considering the interdependence of porosity ε_{agg} and coordination number N_C. Rumpf proposed a simple relationship ($N_C \times \varepsilon_{agg} \approx 3.1 \approx \pi$) which is based on experimentally determined coordination numbers in fixed beds of spheres (Smith et al. 1929) and thus obtained:

$$\sigma_{\mathrm{T}} = \frac{1 - \varepsilon_{\mathrm{agg}}}{\varepsilon_{\mathrm{agg}}} \frac{F_{\mathrm{ad}}}{x_{\mathrm{p}}^2}. \tag{5.3}$$

Rumpf's model is frequently employed when computing the strength of agglomerates (e.g. Tang et al. 2001; Bałdyga et al. 2008), but it has also received criticism. According to Kendall (1988), its major inadequacy is the assumption of a simultaneous failure of all particle bonds when experiencing external stress, whereas in reality the agglomerate breakup is the ultimate consequence of a sequential particle separation (cracking). Moreover, fracture is always initiated at existing flaws, which are structural inhomogeneities, e.g. small, compact aggregates within large agglomerates. Based on the adhesion of elastic spheres, Kendall (1988) derived an alternative expression for the tensile strength:

$$\sigma_{\mathrm{T}} = 15.6\, \Gamma_{\mathrm{c}}^{5/6} \Gamma^{1/6} \cdot (x_{\mathrm{p}} \cdot l_{\mathrm{flaw}})^{-1/2} (1 - \varepsilon_{\mathrm{agg}})^4, \tag{5.4}$$

where Γ_c and Γ denote the fracture and interfacial energy (e.g. 14 J/m^2 and 0.6 J/m^2 for TiO_2) and l_{flaw} the length of the flaws (in the absence of flaws: $l_{flaw} = x_p$).

The 4th power dependency on $(1 - \varepsilon_{agg})$, a further deviation from Rumpf's equation, results from the observation that the elasticity of regular packings of monosized, elastic spheres obeys this relationship (Kendall et al. 1987). However, this specific result is not universally valid (Bika et al. 2001). Kendall (1988) himself provided a more general interpretation of Eq. (5.4) According to him, there is a twofold impact of the agglomerate porosity ε_{agg} on strength σ_{agg}: Firtsly, because the void volume considerably reduces the elastic constants (Mackenzie 1950). Secondly, because the coordination number N_C and the packing density $(1 - \varepsilon_{agg})$ are correlated. For this correlation, Kendall assumed a simple quadratic dependency ($N_C \propto (1 - \varepsilon_{agg})^2$), which clearly contradicts the observation that for large fractal aggregates ($\varepsilon_{agg} \rightarrow 1$), the coordination number N_C converges to approximately 2 (cf. Sect. 4.2.2.4). Hence, when separating both contributions of ε_{agg} and assuming proportionality between the adhesion force and the interfacial energies (Bika et al. 2001), one can rewrite Kendall's solution as:

$$\sigma_{\mathrm{T}} \propto (1-\varepsilon_{\mathrm{agg}})^n N_{\mathrm{C}} \frac{F_{\mathrm{ad}}}{x_{\mathrm{p}}^2} \cdot \sqrt{\frac{x_{\mathrm{p}}}{l_{\mathrm{flaw}}}}, \tag{5.5}$$

where $n = 1, 2$. Hence, the central deviation from Rumpf's solution is the consideration of flaws; in the absence of flaws both solutions coincide.

Note that for fractal agglomerates, the porosity $\varepsilon_{\mathrm{agg}}$ is a function of the agglomerate size x_{agg} ($1-\varepsilon_{\mathrm{agg}} \propto x_{\mathrm{agg}}^{d_{\mathrm{f}}-3}$), which means that the tensile strength σ_{T} decreases as the agglomerates grow. Moreover, the porosity of fractal agglomerates cannot be regarded as uniform, but shows strong local variations (cf. Sect. 4.2.2). Even for homogeneous agglomerates there is no perfect uniformity of porosity $\varepsilon_{\mathrm{agg}}$ and coordination number N_{C}. In particular, there is an obvious difference between the agglomerate surface and its interior regions; the former being less densely packed than the latter. Hence, the (tensile) strength of an agglomerate cannot be considered constant throughout the whole agglomerate volume. This has consequences for the way an agglomerate is dispersed by external stress; three fundamental types are distinguished (Park et al. 1993; Bałdyga et al. 2009):

- **erosion** of the agglomerate surface, which results in a steady size reduction of the agglomerate size mode and the appearance of a fine size mode, which is related to the eroded primary particles or aggregates; erosion is dominant for small stresses ($1 < Fa < 100$, Bałdyga et al. 2008);
- **fragmentation** of the agglomerate, which yields several fragments, the size of which being in the same order of magnitude; fragmentation occurs at high stress;
- **shattering** of the agglomerate, which means rupture into a large number of fragments considerably smaller than the original agglomerate; shattering is expected at extremely high stresses ($Fa > 10^4$, Bałdyga et al. 2009).

Additionally, the type of deagglomeration may depend on the type of stress that is exerted to the agglomerate. Deagglomeration in colloidal suspensions mainly results from the interaction with the continuous phase, i.e. from viscous forces and pressure fluctuations. The stress on the agglomerates is, thus, closely related to the flow field:

- laminar *shear flow*: pressure and shear forces on rotating agglomerates;
- laminar *elongational* (also: *converging* or *strain*) *flow*: excessive stretching of the agglomerates;
- *turbulent flow*: dominant stress depends on the agglomerate size as compared to the smallest eddies; pressure fluctuation or shear forces;
- *cavitation field*: intense, local agitation of the flow field by collapsing bubbles; stresses similar as in turbulent flow.

A more detailed discussion of deagglomeration in the different flow fields is given below. In any case it is very common to relate the result of a deagglomeration process to the power density (volume specific dissipation rate) P_V of the flow field. This quantity is regarded to determine the "steady state" of dispersity, i.e. the size distribution for very long dispersion times ($t_{\mathrm{disp}} \to \infty$). When aggregates of colloidal particles are dispersed, the dispersion time t_{disp} is usually much shorter than necessary for achieving the "steady state" and, therefore, becomes an additional factor that affects the resulting size distribution. Both factors, P_V and t_{disp}, are frequently summarised in the energy density E_V (or E_m), which strongly correlates with the mean or maximum particle size of continuous or short batchwise processes of emulsification, deagglomeration or grinding (e.g. Karbstein and Schubert 1995; Pohl et al. 2004; Kwade and Schwedes 2002).

Last but not least, the dispersion of colloidal suspensions, in particular the breakup of colloidal aggregates, increases the total interfacial area, thereby induces changes in the bulk properties of the liquid phase (e.g. salt content, pH) and may, thus, change the interfacial properties. In the case of highly intense dispersion processes, like stirred media milling, such changes may be even due to mechano-chemical activation of the solid surface (Sen et al. 1999; Stenger et al. 2004). Consequently, it is necessary to monitor and to regulate the bulk and interfacial properties of the colloidal suspension during the dispersion process. Otherwise re-agglomeration of the dispersed particles is fairly likely (e.g. Vasylkiv and Sakka 2001; Mende et al. 2004).

5.1.2 Mechanisms of Deagglomeration in Viscous Media

Deagglomeration can be caused by different types of mechanical stress. However, their relevance depends on the agglomerate properties (size, inertia) as well as on the medium viscosity. For instance, deagglomeration by capture bet-ween two surfaces or impact on a wall is only of minor relevance for the dispersion of colloidal suspensions. In this case, the deagglomeration typically results from the interaction with flow fields, which induces normal and shear stresses on the agglomerates and causes their deformation. The exact mechanism of deagglomeration then depends on the flow regime, the type of the flow field and the agglomerate size.

5.1.2.1 Laminar Flow

A flow field is laminar, when it is governed by the viscous properties. That applies to very slow fluid velocities or very high viscosities, which are typical for polymer composites or highly concentrated suspensions (e.g. paints). There are three ideal types of laminar flow: uniform flow (e.g. experienced by settling particles), shear flow (e. g in rheometers or pipes), and elongational flow (e.g. in nozzles and diffusers). Only the latter two are relevant for deagglomeration.

Particles that move in a laminar flow field with velocity gradient $\dot{\gamma}$ experience shear and normal stresses which vary along/across the surface and induce particle rotation and deformation. The rotation of spheres is stable with an angular velocity of $\omega = 1/2 \times \dot{\gamma}$ (Jeffery 1922; Trevelyan and Mason 1951), whereas aspherical particles or agglomerates rotate in a quasi-periodic or even chaotic manner (Blaser

2000, 2002). In any case, rotation means fluctuating hydrodynamic forces on the surface elements of the particle. The maximum stress for spherical particles can be calculated as (Raasch 1962):

$$\tau_{\mathrm{disp,max}} = \frac{5}{2}\tau_{\mathrm{lam}} = \frac{5}{2}\eta\dot{\gamma}, \tag{5.6}$$

where τ_{lam} is the shear stress in the undisturbed flow field. Note that the shear stress on the particle is independent from the particle size. The parameter τ_{lam} can be used to calculate the power density P_V, which allows a comparison with other types of dispersion:

$$P_V \propto \dot{\gamma}\tau_{\mathrm{lam}} \propto \eta\dot{\gamma}^2. \tag{5.7}$$

Additionally to P_V, the effect of dispersion, i.e., the degree of deagglomeration, depends on the number of revolutions N_{stress} (i.e. numbers of stress alternations). That is because weakening and loosing of particle bonds in the agglomerate and the subsequent agglomerate deformation do not occur instantaneously, but require some time. The number of rotations is derived from $N_{\mathrm{stress}} = f \cdot t_{\mathrm{stress}} \propto \dot{\gamma} \cdot t_{\mathrm{stress}}$ (Krekel 1966; Reichert 1973).

In elongational flow, the velocity gradient is parallel to the direction of flow. This leads to a stretching of particles by tensile stress and may eventually lead to rupture. In contrast to laminar shear, there is no permanent rotational motion; the particle thus experiences a quasi-static load.

The two types of laminar flow conditions were investigated by Taylor (1934) with regard to emulsification. For that purpose, he studied the deformation of single droplets in ideal shear and elongational flow. He was able to show that the latter is much more effective for emulsification when the viscosity ratio is high ($\eta_{\mathrm{droplet}}/\eta > 4$). Taylor's experimental set-up was used by Blaser (2000) for studying the dispersion of large flocs (µm-range) in laminar flow fields. He observed significant deformations (mainly bending for shear flow and stretching for elongational flow) and showed that the experimental data can be roughly predicted by approximating the flocs with ellipsoid particles. In a later paper, Blaser (2002) presented detailed theoretical results for the behaviour of ellipsoidal particles in laminar flow fields. Higashitani et al. (2001) conducted a numerical study on the dispersion of fractal and regularly packed agglomerates by laminar flow fields: Similar to Blaser's experiments, they observed a strong deformation (apparently stretching) of the agglomerates before fragmentation took place. The resulting fragments keep the deformation as long as they stay in the flow field and appear more compact than the original agglomerate. The numerical simulation (for agglomerates of monosized particles with uniform adhesive forces) showed that the average number of particles in the fragments anticorrelates with the hydrodynamic stress. That means that increasing stress coincides with a transition from *fragmentation* to *shattering*. *Erosion*, the third type of deagglomeration, was not observed in this study. Finally, the authors concluded that "under usual flow conditions" the elongational regime "is more effective to break up aggregates than simple shear flow".

5.1.2.2 Turbulent Flow

Turbulence occurs in any sufficiently rapid flow when the fluid inertia exceeds its molecular friction. It is the typical flow regime when suspensions of relatively low viscosity are dispersed—e.g. by stirring or in nozzles. Turbulent flow is characterised by multiscale eddy structures that erratically move through the flow field and cause local fluctuations in velocity and pressure. The velocity fluctuations can be quantified by the effective velocity difference $\overline{\Delta u'^2}$ over a distance Δr (Kolmogorov 1958). For the inertial subrange of microturbulence, it amounts to:

$$\overline{\Delta u'^2} \propto (P_V/\rho)^{2/3} \cdot \Delta r^{2/3} (\Delta r > 20 \times l_D), \tag{5.8}$$

where P_V is the volume specific dissipation rate (power density) and l_D the Kolmogorov scale of microturbulence ($l_D = (\eta^3/\rho^2 P_V)^{1/4}$). In the viscous subrange, i.e., within the smallest, laminar eddies, the velocity difference scales with distance and is affected by the fluid viscosity η:

$$\overline{\Delta u'^2} \propto P_V/\eta \cdot \Delta r^2 (\Delta r \leq 5 \times l_D). \tag{5.9}$$

The deagglomerating effect of turbulence relies on these velocity gradients, which lead to pressure fluctuation, local deformation, and oscillatory motion for particles in the inertial subrange and which also represent laminar shear for particles in the viscous subrange. The stress related to the pressure fluctuation (or turbulent shear) in the inertial subrange can be considered as a kind of Reynolds stress; it is, thus, calculated as:

$$\tau_{ti} = \rho \overline{\Delta u'^2} \propto \rho^{1/3} P_V^{2/3} x_{\mathrm{agg}}^{2/3} (\Delta r > 20 \times l_D). \tag{5.10}$$

The decisive parameter for the deagglomeration within the viscous subrange is the shear rate within the smallest micro-eddies, which can be approximated by means of Eq. (5.6). Consequently, one obtains:

$$\tau_{tv} \propto \eta \dot{\gamma}_{\mu e} \propto P_V^{1/2} \eta^{1/2} (\Delta r \leq 5 \times l_D). \tag{5.11}$$

The proportionality coefficients in Eqs. (5.10) and (5.11) are approximately 2 and 0.4,[1] respectively (Liepe 2003). That usually means that the stress on large agglomerates that experience dynamic pressure fluctuations in the inertial subrange is much larger than the laminar shear stress on small agglomerates that fall into the viscous subrange.

[1]Liepe gives a value 0.15, but assumes that $\tau_{tv} = \eta \dot{\gamma}_{\mu e}$, which deviates from Eq. (5.6) for laminar shear.

Consider an aqueous suspension with a specific dissipation rate (power density) of 1 W/kg or 1 kW/m^3, which coincides with a Kolmogorov scale l_D of 32 μm. The turbulent stress acting on a particle with 10 × l_D would be 9 Pa acc. to Eq. (5.10) and 0.4 Pa acc. to Eq. (5.11).

Deagglomeration in turbulent flows, therefore, starts with a rapid rupture of large flocs or agglomerates within the inertial subrange and ends up with a comparatively slow size reduction within the viscous subrange.

Turbulent flow can be realised in different experimental set-ups at varying intensity. The experiments in Table 5.1, for instance, refer to power densities of less than 100 mW/m^3 up to 10^4 kW/m^3. Consequently, there is a wide range of achievable size distributions (from a few microns up to a few mm). It is, therefore, more interesting to understand the scaling laws of dispersions, which require not only a model for the size dependency of stress, but also for that of the agglomerate strength. Finally, one can derive a scaling law for the maximum agglomerate size that can "survive" the dispersion process after a sufficiently long treatment:

$$x_{\max} \propto P_V^{-\alpha}. \tag{5.12}$$

Commonly, it is assumed that this scaling law applies to any "average" particle size (e.g. mean or median) as well. The papers cited in Table 5.1 contain theoretical models for the scaling laws as well as experimental data for comparison. Even though the models considered both subranges of microturbulence, only the viscous subrange was covered by experiments. Apart from Nagel and Kürten (1976), who focussed on the dispersion of bubbles, all other authors studied the dispersion of flocculated suspensions. It is interesting to see that all authors could essentially confirm their theoretical models, although the differences between them are evident. Nevertheless, for the dispersion of large micrometre-sized flocs in pure turbulent flow, the exponent of the scaling law (5.12) was found to be in the range of 0.27–0.37.

Table 5.1 Theoretical and experimental exponents α for power law dependency $x_{\max} \propto P_V^{-\alpha}$; references: Nagel and Kürten (1976), Tambo and Hozumi (1979), Henzler and Biedermann (1996), Kobayashi (2004)

Reference	Theoretical prediction		Experimental conditions		Experimental results	
	α_{inertial}	α_{viscous}	Turbulence	P_V (kW/m^3)	Size param.	α_{viscous}
Nagel	0.4		Jet nozzle	10–10^4	x_{ST}	0.40 (bubbles)
Tambo	0.5–0.4	0.38–0.33	Stirrer	10^{-5}–10^{-3}	$x_{\max,0}$	0.37..0.30 (flocs)
Henzler	1–0.31 (d_f: 3–1.5)	0.5–0.29 (d_f: 3–1.5)	Stirrer	0.001–1	$x_{1,0}$	0.33 (flocs)
Kobayashi	0.25	0.25	Stirrer	10^{-4}–1	$x_{99,3}$	0.27 (flocs)

5.1.2.3 Cavitation

Cavitation is the phenomenon of bubble formation and collapse in liquid flows of high velocity or in intensive ultrasonic fields. It results from the evaporation of the liquid phase or from the release of dissolved gases at zones of low pressure (Webster 1963). It occurs when the cavitation number falls below a critical value:

$$N_{\mathrm{cav}} = \frac{p - p_{\mathrm{sat}}}{\frac{1}{2}\rho u^2} \leq N_{\mathrm{cav,cr}}, \tag{5.13}$$

where p_{sat} denotes the saturation pressure of the liquid phase, ρ the liquid density and u the characteristic flow velocity. In real liquids, the "ideal" critical value ($N_{\mathrm{cav,cr}} = 0$) can be shifted to positive or negative values. This is because the cavitation is affected by dissolved gases, by the presence of nucleation centres, and by the degree of turbulence (Oertel 2004, pp. 493–497).

Once having been created, the cavitation bubbles start to grow until they reach flow regions of high pressure ($p > p_{\mathrm{sat}}$) or—in the case of ultrasonication—until the bubbles have reached a critical size. In contrast to this growth, the collapse of bubbles is a highly rapid, almost instantaneous process which creates shock waves and micro-jets with diameters of 1 μm to 10 μm and velocities up to several hundreds m/s (Bałdyga et al. 2009). Particle agglomerates that are exposed to these very intensive velocity and pressure fluctuations experience high mechanical forces and are prone to fragmentation (Thoma et al. 1991; Kusters et al. 1993). Additionally, cavitation may cause particle-particle or particle-wall impacts (Doktycz and Suslick 1990; Park et al. 1993) and thereby also induce deagglomeration. These are, however, considered as second-order effects (Kusters et al. 1994).

The hydrodynamic stress on a particle generated by bubble collapse can be estimated by (Bowden and Brunton 1961; Crum 1988):

$$\tau_{\mathrm{cav}} \propto \rho\, c_{\mathrm{m}}\, u_{\mathrm{cb}}, \tag{5.14}$$

with a proportionality coefficient of around 1. The quantity c_{m} is the velocity of the compressional wave (i.e. the sound velocity of the liquid medium) and u_{cb} is the velocity of the collapsing interface (i.e. the micro-jet velocity). The latter obeys the following relationship:

$$u_{\mathrm{cb}} \propto \sqrt{(p + \hat{p}_{\mathrm{US}} - p_{\mathrm{sat}})/\rho}, \tag{5.15}$$

where $\hat{p}_{\mathrm{US}}$ is the pressure amplitude of the ultrasonic field (Bałdyga et al. 2008).

The stresses calculated by Eq. (5.14) exist only in the close proximity of a collapsing bubble. The dispersing effect of cavitation, therefore, depends on the bubble concentration—or more exactly, on the number N_{cb} of collapsing bubbles per unit volume and unit time. This quantity, correspondingly the rate of deagglomeration, increases with rising power input through the flow field or by ultrasonic waves. However, different models were proposed with regard to the functional relationships (Kusters et al. 1993; Bałdyga et al. 2008, 2009).

5.1.3 Dispersion Techniques

In real dispersion processes, there is usually not just one prevailing type of flow field. Instead, the different types may exist in parallel (e.g. as laminar shear and elongation) or the type of flow is changing during the passage through the dispersion machine. This section reviews typical dispersion techniques that are used in industry and academia for colloidal suspensions. Experimental set-ups that are employed for dispersing flocs of micrometre particles (e.g. agitated vessels) or solid materials in polymer matrices (i.e. extruders) are not considered.

5.1.3.1 Disc Systems Under Laminar Operation

Disc systems are agitated vessels where the fluid motion is induced by quickly rotating discs with usually toothed rims. They are frequently employed for the preparation of stable suspensions with high solid contents, in particular for paints and coatings, and are recommended for highly viscous emulsions (Urban et al. 2006). Disc systems are used to suspend and homogenise particles in the liquid phase and facilitate their (subsequent) dispersion. The deagglomerating effect of such systems is clearly promoted by high suspensions viscosities which guarantee the typical, essentially laminar flow field as shown in Fig. 5.1.

The tangential velocities of the disc rim amount to 20–30 m/s under typical operating conditions. Decisive for dispersion is the maximum shear rate γ_{max}, which can be approximated by the disc's angular velocity and is usually in the order of 1000 s^{-1}. The maximum stress acting on the particles and the power density can be calculated according to Eqs. (5.6) and (5.7) respectively.

Consider a suspension with a dynamic viscosity of 1 Pa·s and assume a maximum shear rate of 1000 s^{-1} in a disc system disperser. This would coincide with a maximum shear stress of 2.5 kPa and maximum power densities of approx. 10^3 kW/m^3.

The stress on the particles is proportional to the suspension viscosity, which increases disproportionately strongly with the solid content. Best dispersion effects are, therefore, achieved at very high particle concentrations close to the maximum packing, which is qualitatively different to other types of dispersion machines. Conversely, disc systems show a rather poor performance at low suspension viscosities, which often coincide with low particle concentrations.

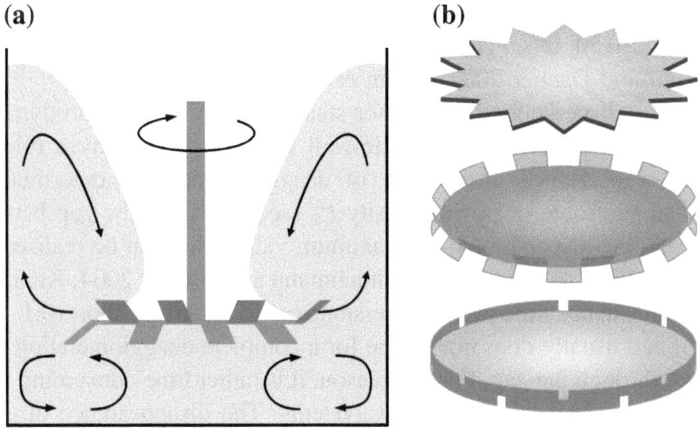

Fig. 5.1 **a** schematic flow field of disc systems at optimum operating conditions (acc. to Urban et al. 2006); **b** typical disc shapes

5.1.3.2 Rotor-Stator-Systems—Turbulent Flow

Rotor-stator systems consist of a rotating tool (rotor) and a coaxially fixed wall (stator). The suspension is brought into the thin annular gap between rotor and stator, where the particles experience intensive hydrodynamic stress. Two types of rotor-stator systems can be distinguished (Fig. 5.2).

In *colloid mills,* the suspension flows axially through the conically shaped slit between the two surfaces, which can be either smooth or toothed. The (adjustable) gap width typically lies in the range of a few hundred micrometres and the rotor operates with a speed of 1000–5000 min^{-1}, which corresponds to tangential velocities up to 40 m/s. Colloid mills are usually operated under turbulent conditions, but laminar flow is also possible.

Gear-rim dispersers are characterised by slotted rims of rotor and stator. The suspension is axially sucked in the centre of the rotor and, subsequently, centrifugally accelerated through the rims of the rotor and stator. The dispersing zone has a typical width between 0.5 μm and a few millimetres. Commercial machines

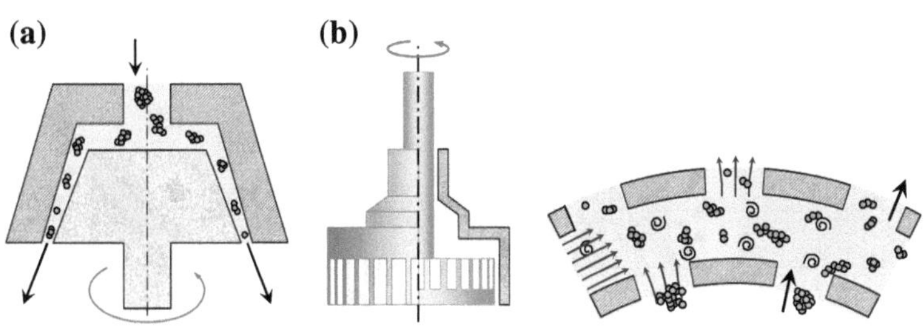

Fig. 5.2 Rotor-stator systems; **a** colloid mill; **b** gear-rim disperser

provide rotor speeds up to 25,000 min^{-1} (for small machines) and tangential velocities up to 50 m/s. The flow in the dispersing zone is highly turbulent (Schuchmann and Danner 2004; Urban et al. 2006).

The primary dispersing effect in rotor-stator systems is the hydrodynamic stress in turbulent flow, even though particle-wall collisions may play a role for large particles as well. Hence, the degree of deagglomeration is determined by the specific dissipation rate or power density P_V (see above) in the gap between rotor and stator, i.e. the dispersion zone. Maximum values that can be realised lie in the range of 10^5 kW/m^3 to 10^6 kW/m^3 (Schuchmann and Danner 2004; Kuntzsch 2004, p. 101). The residence time in the dispersion zone is relatively short (0.1–1 s, Urban et al. 2006) and usually does not suffice for a complete deagglomeration during the first passage through the gap. For this reason, it is rather time-consuming to disperse colloidal suspensions with rotor-stator systems. The disappearance of coarse size fractions >1 μm is achieved only by many cycles for continuous operation or by long dispersion times for batchwise operations (Pohl 2005, pp. 81–83). However, rotor-stator systems have received widespread acceptance, as they are easy to clean and the maintenance costs are low.

5.1.3.3 High Pressure Dispersion

High pressure systems use the kinetic energy of strongly accelerated flow field for the dispersion of colloidal suspensions and emulsions. There are different geometries of the dispersing unit, which can be classified as radial diffusers, counter-jet dispersers, and axial nozzle systems (Fig. 5.3a–c). These systems are operated at very high pressures (several hundred up to some thousand bar), which results in extremely high flow velocities in the smallest cross sections. High pressure dispersion systems are widely used for emulsification below 1 μm (Schuchmann and Danner 2004; Urban et al. 2006), whereas they only have academic relevance for the deagglomeration of solid particles (Wengeler et al. 2004; Teleki et al. 2008). This is due to the substantial abrasion of the dispersion units, which increases the cost and reduces the technical life-time. Sauter and Schuchmann (2008) have, therefore, proposed a modified high pressure post feeding (HPPF) system, where the particle suspension is injected in the turbulent zone behind a nozzle that was fed with pure continuous phase (Fig. 5.3d).

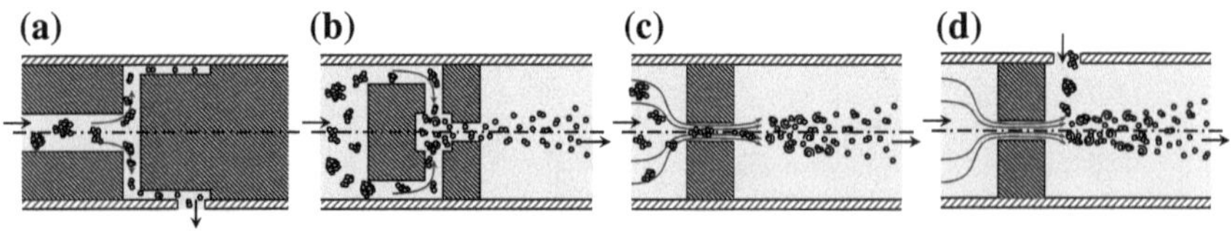

Fig. 5.3 High pressure (HP) dispersion systems. **a** Radial diffuser. **b** Counter-jet disperser. **c** Axial nozzle. **d** HPPF

The dispersing effect of HP systems is related to the intense turbulence in and behind the accelerating zone, as well as to cavitation (mainly for radial diffusers and nozzles) and laminar elongation at the entrance to the dispersion unit (especially for nozzles and counter-jet dispersers). The computation of the turbulent flow field in the dispersion unit commonly requires numerical tools, in particular when the effect of cavitation is to be adequately considered. Nonetheless, it is possible to provide simple estimates for the hydrodynamic stress on the particles because the power density P_V can be approximated by that of a pipe flow:

$$P_V = \frac{1}{2}\lambda\rho u^3 / d_\mathrm{h}, \tag{5.16}$$

where λ the loss coefficient, u the average fluid velocity, and d_h the hydraulic diameter.

Wengeler et al. (2006) described a typical HP dispersion system that is based on an axial nozzle with an internal diameter of 125 μm and a length of 10 mm. The system is operated at 1000 bar. Assuming that the pressure drop results entirely from the turbulent flow in this thin channel, one can calculate the mean flow velocity u (339 m/s), the Reynolds number (42357), the power density P_V (3.4×10^9 kW/m^3), and finally the Kolmogorov scale of microturbulence l_D (131 nm). Since the authors investigated the dispersion of pyrogenic powders with mean aggregate size below 1 μm, the deagglomeration is mainly due to laminar shear in the smallest eddies. Thus, eq. (5.11) applies, which gives an estimate of 20 kPa for the shear stress on the aggregates.

The high pressure drop in the dispersion units leads to cavitation at the nozzle outlets, which may considerably affect the flow pattern in that region. Simulations by Bałdyga et al. (2009) predict even an atomisation of the suspension under extreme conditions (2400 bar, 995 m/s). The stresses caused by the cavitation are much larger than the hydrodynamic ones. Yet, the hydrodynamic stresses already act at the nozzle entrance and, thus, contribute to the total deagglomeration process before it is eventually dominated by cavitation. The authors further discussed in detail the dispersion of agglomerated colloidal particles in HP systems for varying operating conditions.

5.1.3.4 Ultrasonication

An alternative to dispersion in flow fields is the application of highly intensive ultrasonic waves. This is realised in so-called ultrasonic baths or with ultrasonic horns (sonotrodes). The former belong to standard laboratory equipment and are widely used for the preparation and homogenisation of suspensions. However, the stress intensities in such baths are comparatively low, setting some practical limits for the dispersion of strong aggregates of colloidal particles. Even worse, there is a considerable variation of the ultrasonic energy density within such baths, which may impede a reproducible sample preparation. In contrast, sonotrodes guarantee reproducibility if the samples are well defined (i.e. with regard to solid content and suspension volume). Additionally, they facilitate the deagglomeration in the colloidal size range. Sonotrodes are usually optimised for aqueous solutions and suspensions and typically have diameters between a few millimetres and several centimetres. The generated sound fields work with frequencies in the range of 20–100 kHz and have a nominal power consumption of 50 W to approximately 1 kW. However, the effective power output depends on the weight and the acoustic impedance of the suspension as well as on the sonotrode geometry.

When operated in a batchwise mode, the induced turbulent field should be intensive enough to ensure the mixing of the complete sample volume, since only a small zone beneath the sonotrode tip actively contributes to the dispersion process. Alternatively, there are flow cells for continuous operation, in which the suspension flows axially onto the sonotrode tip and where the dispersion zone is usually defined by the cell geometry (Fig. 5.4). As a result of the high power consumption and, thus, energy dissipation, the samples should be cooled during ultrasonication.

The primary dispersing effect of ultrasonication is cavitation and the corresponding hydrodynamic stress by induced microturbulence (see above, Thoma et al. 1991; Kusters et al. 1993). Additional effects discussed in literature are particle-particle collisions (e.g. Park et al. 1993) and the hydrodynamic stress due to acoustophoretic motion of particles (cf. Sect. 2.3.6.2). The latter was regarded relevant for high sound frequencies (800 kHz) and coarse particles in the range >10 μm (Süß and Hanke 1969), but simulations for agglomerates of colloidal particles at typical frequency values (24 kHz) showed that this effect is orders of magnitude less important than cavitation (Bałdyga et al. 2008).

The maximum stress intensities acting on particles during ultrasonication are considered for energy and for force quantities. Some authors estimated the specific rate of dissipation (power density) based on caloric data, which yielded values in the range of 10^3–10^4 kW/m^3 (Pohl et al. 2004) to 10^5 kW/m^3 (Kuntzsch 2004, p. 100). The problem here is the definition of the dispersion zone, i.e. the volume where the ultrasonic field significantly contributes to deagglomeration. This zone

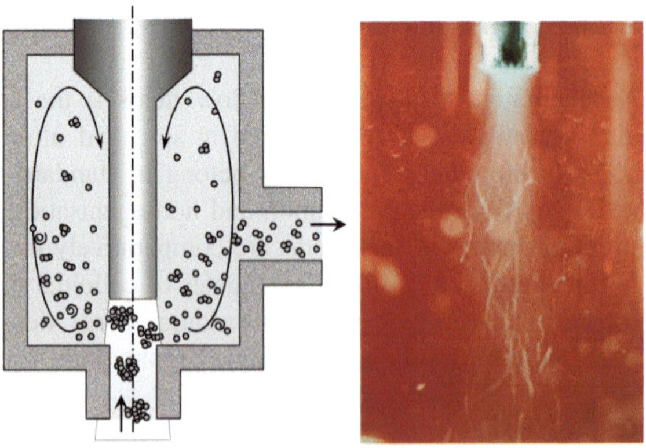

Fig. 5.4 Ultrasonic sonotrode with flow cell (*left*) and visualised cavitation zone (*right*; with ssion from Benno Wessely, copyright 1999)

cannot be exactly defined because the ultrasonic intensity decays exponentially with distance from the sonotrode. A different approach is to consider the hydrodynamic stress τ_{cav} in the cavitation field. Simulations by Bałdyga et al. (2008) for a continuously operated ultrasonic cell with a local dissipation rate P_V of 10^5 kW/m^3 yielded hydrodynamic stresses up to 100 MPa (by means of Eq. (5.14)).

Bałdyga et al. (2008) described a real ultrasonic device with nominal power consumption of 200 W and an effective power output of 46 W. The flow cell design leads to a cylindrical volume of 0.46 cm^3 beneath the sonotrode, where virtually all energy is dissipated. The cavitation induced hydrodynamic stresses in that zone varied between 5×10^7 Pa and 10^8 Pa. The flow rate is 10 cm^3/min, which means an energy density of 2.8×10^8 J/m^3.

An experimental method for the determination of the cavitation induced stress τ_{cav} was proposed by Thoma et al. (1991). It consists of the ultrasonication of glass bubbles (hollow glass spheres), which get partially ruptured according to the bubble strength distribution. This distribution can be quantified by mercury porosimetry, where bubbles are crushed when the pressure exceeds their strength. Comparing the bubble strength from before and after ultrasonic treatment reveals the cavitation pressure that prevails in the ultrasonic field. By means of this procedure, the authors determined cavitation stresses between 2 MPa to 80 MPa for a sonotrode (20 kHz, ∅ 12 mm) with power outputs from 2.5 W to 100 W (cf. Kusters et al. 1993). Hence, the experimental data on cavitation stress τ_{cav} are in the same order of magnitude as the calculated ones cited above.

In summary, ultrasonication facilitates the dispersion of colloidal suspensions at relatively high stress intensities. It relies on the presence of cavitation and on a deep penetration of the sound field into the suspension. The latter is guaranteed by low suspension viscosities (e.g. as for aqueous media with low solid content). A particular problem is the abrasion of the sonotrode tip, which depends on the solid phase, the particle size and shape, the sound intensity, as well as on the "abrasion history" of the tip. Abrasion particles from sonotrodes are coarse particles in the micrometre range and may spoil the quality of colloidal suspensions.

5.1.3.5 Stirred Media Milling

Stirred media mills are commonly used for the comminution of fine solid materials up to submicron or even nanometre regions (e.g. Mende et al. 2004). This is achieved by pumping the feed suspension through a highly agitated bed of a grinding medium (GM), which consists of coarse beads (e.g. from glass or ceramic materials) that are much larger (from 100 μm to a few mm, Schönert 2003) than the feed particles. Stirred media milling is used for fine grinding of crystalline solids, for the disintegration of organic materials (e.g. yeast cells), or for the dispersion of colloidal suspension (Kwade and Schwedes 2002).

The continuous agitation of the grinding media leads to frequent and fast collisions between the coarse beads. In the event of such a collision, the feed

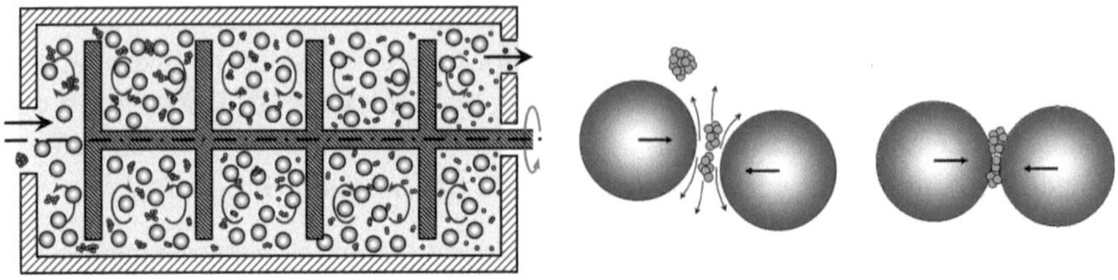

Fig. 5.5 Stirred media mill; set-up, and deagglomeration mechanisms (hydrodynamic stress, captured agglomerate

suspension is squeezed out from the contact region, which causes intense hydrodynamic stresses on the suspended feed particles. However, some of the feed particles remain in the contact zone and are eventually captured by the two beads, thereby experiencing high normal stresses (Fig. 5.5). Both mechanisms contribute to comminution; yet, the latter is mainly relevant for rigid, compact particles, while the former applies chiefly to deagglomeration (Kwade and Schwedes 2002).

The decisive parameters for the final size distribution are the average number of stress events N_{stress} of each feed particle, the stress intensity e_{st}, and the specific energy input E_m. The number of stress events refers either to the rapid flow between the surface of two approaching particles or to the particle capture between two beads in contact. It increases linearly with dispersion time t and rotational speed ω of the stirrer:

$$N_{\text{stress}} \propto \omega \cdot t / d_{\text{GM}}^{1 or 2}. \tag{5.17}$$

In this context, the term stress intensity means an energetic quantity; it essentially represents the kinetic energy of the GM beads, i.e.:

$$e_{\text{st}} = d_{\text{GM}}^3 \rho_{\text{GM}} u_{\text{st}}^2, \tag{5.18}$$

where u_{st} is the tip velocity of the stirrer. Typical values of the stress intensity lie in a range of 10^{-5}–10^{-2} Nm. The product of N_{stress} and e_{st} is proportional to the specific energy input E_m ($E_m = P/\dot{m}_{\text{solid}}$), which is related to the mass of the feed particles. Mende et al. (2004) reported values up to 10^5 kJ/kg for grinding of α-alumina to particle sizes below 100 nm. Given a certain energy input E_m, there is always an optimum stress intensity $e_{\text{st,opt}}$ at which the product particle size reaches a minimum. Similarly, the energy input E_m for achieving a certain size distribution can be minimised by applying optimum stress intensities $e_{\text{st,opt}}$. Hence, grinding media and stirrer speed allow for an optimum design of the dispersion process (Kwade and Schwedes 2002; Schönert 2003). Note that the final particle size distribution of the product can be also affected by re-agglomeration, in particular, when the relevant suspension parameters (e.g. pH) are not adequately controlled during the dispersion (Sommer et al. 2006).

5.1.4 *Comparison of Dispersion Procedures*

5.1.4.1 Crucial Parameters for the Dispersion Process

The defined dispersion of particle agglomerates and aggregates is an indispensable step when studying the macroscopic properties of colloidal suspensions or measuring their state of dispersion. In that regard, there is a vast number of publications that provide experimental data on the dispersion of colloidal suspensions. However, only a few of them systematically investigate the main factors of dispersion processes as well as to which extent a scaling and comparison of dispersion processes is possible. Based on these publications, one can identify a set of parameters that determine the effectiveness of dispersion:

- the dispersion time t_{disp}—usually normalised by the time period or frequency that the particles are really stressed (t_{stress} or f_{stress}, respectively), which yields a dimensionless dispersion time or a number of stress events;
- the power density P_V (specific energy dissipation) in the zone of dispersion, which is considered to determine the "steady state" of dispersion processes, i.e. for sufficiently large dispersion times ($t_{\text{disp}} \rightarrow \infty$);
- the specific energy input (E_V or E_m), which either refers to the total suspension volume or to the total particle mass, and is employed for relatively short dispersion times, i.e. the "steady state" is not being achieved;
- the maximum normal and tangential stresses τ_{disp} acting on the agglomerate surface.

Time, power input, and energy consumption are process parameters which are typically known or can, in principle, be directly measured during dispersion. In contrast, the microscopic stresses τ_{disp} on single agglomerates are usually quantified by means of appropriate models. Even though being interrelated, the four properties are rarely considered altogether. Instead, their usage for the characterisation of dispersion processes traditionally depends on the type of suspension treatment.

The power density P_V is the characteristic quantity of turbulent flow. It determines the size of the smallest eddies and the intensity of microturbulence. In addition, it is a measure of the "shear intensity" in laminar flows or the "intensity of cavitation" in ultrasonic fields (see above). The power input P in the dispersion zone can be derived from the pressure drop (e.g. in pipes and nozzles) or can be measured calorically (e.g. for rotor-stator systems and ultrasonication; Pohl 2005; Kuntzsch 2004). Additionally, P can be roughly approximated by the electric power consumption of the dispersion machine (e.g. for ultrasonication; Mandzy et al. 2005; Sauter et al. 2008), even though the real values may be lower by a factor of 2 to 5. A further source of uncertainty is the volume of the dispersion zone (V_{disp}), since the stress intensities are not uniformly distributed in dispersion apparatuses. In particular, this applies to agitated vessels, where the highest dissipation rates are obtained in the vicinity of the stirring instrument (Henzler and Biedermann 1996), and for ultrasonication, where only a small region below the sonotrode tip actively contributes to dispersion (Kuntzsch 2004, p. 100; Bałdyga et al. 2008).

The power density P_V was shown to be the crucial parameter for the "final" size distribution of large micrometre-flocs in turbulent flow (Nagel and Kürten 1976; Tambo and Hozumi 1979; Kobayashi et al. 1999). However, when dispersing agglomerates of colloidal particles with high agglomerate strength, such a final "steady state" is achieved only after long dispersion times. In that case, or when using continuously operating dispersion machines with short residence times in the dispersion zone, the resulting size distribution is affected by the dispersion time, or more generally by the specific energy consumption E_V:

$$\text{continuous operation: } E_V = P/\dot{V}, \tag{5.19}$$

and

$$\text{batchwise operation: } E_V = P \times t_{\text{disp}}/V_{\text{tot}}. \tag{5.20}$$

When the dispersion is realised by flow in pipes or nozzles (as for HP systems), the energy density E_V equals the pressure drop Δp.

The concept of energy density relies on the experimental finding that power density and dispersion time affect the resulting mean particle size in a similar manner; it applies to emulsification too (Karbstein and Schubert 1995; Schuchmann and Danner 2004). Besides this, it is in accordance with the classical laws of comminution, which correlate the change in particle size to the energy input per unit mass of solid (Richardson and Harker 2002, pp. 100–102).

Power and energy density provide a sound basis for the comparison of different dispersion mechanisms, techniques, and machines. However, only a minor part of the inserted energy is really used for the breakup of agglomerates. In order to gain insight into dispersion processes, to understand the type of deagglomeration (erosion, fragmentation, shattering) and to develop appropriate models, it is necessary to determine the maximum normal and tangential stresses τ_{disp} that the agglomerates experience. These stresses can be compared with the agglomerate strength (cf. Sect. 5.1.1). The relationships between τ_{disp} and P_V are principally known (e.g. Eqs. (5.6) and (5.7) for laminar flow, Eqs. (5.10) and (5.11) for turbulent flow). Bałdyga et al. (2008), for instance, used such relationships for comparing the hydrodynamic and cavitation stresses in cavitating flows. Rwei et al. (1990, 1991) studied the deagglomeration of carbon black pellets in laminar shear flow for varying shear stress and, thus, experimentally determined the cohesive strength of the pellets.

The two parameters, power density P_V and energy density E_V, characterise the dispersion process at their final and intermediate stages, respectively. A complete description of the dispersion kinetics is slightly more difficult and has to take into account the corresponding changes in size distribution and number concentration. In general, a first order decay is assumed for the coarse particle number concentration (Kusters et al. 1993; Sommer et al. 2006; Bałdyga et al. 2008). Accordingly, the initial decline of coarse particle concentrations should occur as exponential decay, as could be confirmed by experiment—e.g. by Rwei et al. (1991) for laminar shear and by Kusters et al. (1994) for ultrasonication. Similarly, the size reduction of agglomerates is frequently described as an exponential decay as well (e.g. Henzler and Biedermann 1996; Kobayashi et al. 1999):

$$\ln \frac{\bar{x} - \bar{x}_{\infty}}{\bar{x}_0 - \bar{x}_{\infty}} \propto -t, \tag{5.21}$$

where x_0 and x_{∞} denote the incipient and final mean particle size; with the latter being dependent on P_V. However, Eq. (5.21) is a rather rough approximation of experimental data, which often better fit to a stretched exponential (e.g. Henzler and Biedermann 1996; Urban et al. 2006 for turbulent and laminar flow, respectively). Other authors propose a power law dependency of type (Schuchmann and Danner 2004):

$$x_{50,3} = x_{\infty} + a \cdot t^n. \tag{5.22}$$

Simple kinetics as proposed by Eqs. (5.21) and (5.22) are especially critical if the type of deagglomeration is changing during the dispersion process. For instance, an agglomerate may be less compact than its fragments because it has a fractal structure and/or because it is deformed in the flow field. This may coincide with a transition from fragmentation to erosion, as could be observed by Pohl et al. (2004) for the deagglomeration of spray drying agglomerates. Kusters et al. (1993) reported on an initial time lag for the ultrasonic dispersion of dense agglomerates at low power densities. This indicates a fatigue process that first reduces the agglomerate strength below a critical value, beyond which the deagglomeration occurred.

In order to compare the deagglomeration kinetics for different stress intensities, one either uses dimensionless times, e.g. by multiplication with the laminar shear rate (Rwei et al. 1991), or the number of stress events, e.g. by multiplication with a characteristic frequency (Reichert 1973; Kwade and Schwedes 2002). For the comparison of different dispersion machines, the time scale is converted to the energy density E_V (Pohl et al. 2005).

5.1.4.2 Dispersion of Pyrogenic Powders

Pyrogenic powders are flame-made materials that are composed of fractal aggregates of nanosized particles from amorphous (e.g. SiO_2) or crystalline (e.g. TiO_2) materials. These aggregates are usually further assembled to coarse, micrometre-sized agglomerates (cf. Sect. 4.5). When the powders are used as filler (e.g. in paints) or abrasive additives (e.g. in polishing slurries), they have to be well dispersed in the continuous phase. Firstly, for destroying the large agglomerates, which could otherwise spoil the product's homogeneity or which could cause cracks on polished surfaces. Further, for uniformly distributing the aggregates in order to benefit from their high porosity and the fineness of primary particles (e.g.

for rheological properties). Last but not least, the final application may set an upper limit to the average particle size (e.g. for certain polishing qualities). Hence, there is a considerable practical interest on the dispersion behaviour of such powders. Most often, one or more of the following questions should be answered

- Does a lower size limit exist that corresponds to a given dispersion machine and its process parameters?
  This size may correspond to the size of originally formed primary particles or to the size of the first (primary) aggregates that were generated in the flame and are considered to have comparatively strong interparticle bonds.
- Which type of kinetics applies and is there a scaling law for the dispersion process?
  The dispersion time is proportional to the energy consumption of a dispersion machine and should, therefore, not exceed the needed value. If possible, the dispersion kinetics is described in terms of energy, which may allow up-scaling from laboratory to industrial dispersion processes.
- How efficient is the dispersion process?
  An important criterion for the selection of dispersion instruments/machines is the energy consumption required for a certain change in size distribution.
- Does the dispersion process coincide with a contamination of the suspension?
  Contamination may be caused by abrasion due to particle-wall collisions or interactions between cavities and the dispersion instrument. Size and concentration of the contaminant particles or the concentration of dissolved ionic species may impede the employment of certain dispersion techniques or at least require additional effort for "decontamination" (e.g. filtration).
- Which type of deagglomeration prevails—erosion, fragmentation, or shattering?
  Such knowledge can be used to select the dispersion techniques according to the product specification. Erosion, for instance, is less favourable when coarse particle fractions should be avoided.
- How large is the concentration of nanosized primary particles or aggregates after dispersion?
  Since the presence of nanoparticles, e.g., in food or pharmaceutical products, is considered as a potential risk by legal authorities, this aspect needs to be carefully studied for pyrogenic materials.

These questions are interrelated and essentially ask for the resulting size distribution of a given dispersion process or how it is affected by time or energy, respectively. Figure 5.6 shows a typical result for the deagglomeration of two pyrogenic materials (SiO_2 and Al_2O_3) by means of ultrasonication. The intensity weighted mean particle size (measured by DLS) is plotted versus the energy density (Eq. 5.20) for different suspension volumes (100–400 mL), particle concentrations (0.01–10 wt%), and power inputs (20–200 W, electric power consumption). There is obviously a strong correlation between x_{cum} and E_V; this implies that up-scaling, which is not affected by the ultrasonic power, of ultrasonic dispersion is possible. This might by due to the fact that the measured size values are much larger than the

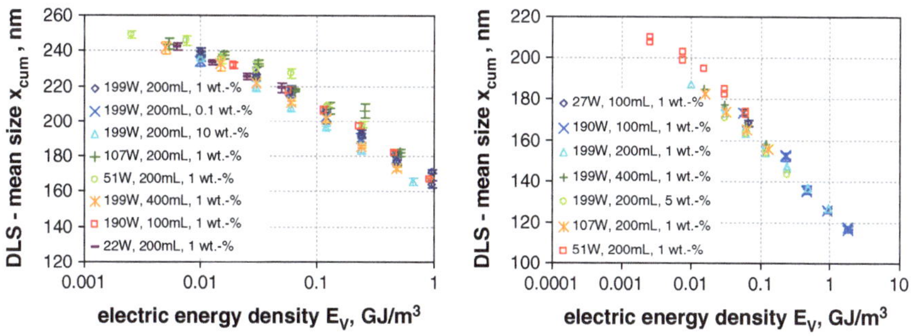

Fig. 5.6 Dependence of the mean particle size (x_{cum}) from the energy density (E_V) during ultrasonication of pyrogenic powders in water; variation of the suspension volume (100–400 mL), the particle concentration (0.1–10 wt%) and the electric power input (20–200 W); particle sizing with DLS; *left* SiO_2 (150 m²/g), *right* Al_2O_3 (100 m²/g)

primary particles of this type of material (approx. 20 nm in both examples). A "steady state" of dispersion was not observed for any experiment.

Strong correlations between the mean particle size and the energy density have also been reported for other dispersion techniques. They are frequently described by a power-law:

$$\bar{x} \propto E_V^{-\alpha}, \tag{5.23}$$

where $\bar{x}$ is some average value of particle size (e.g. a volume weighted median or an intensity weighted mean) and E_V denotes the energy density. Typical values of the exponent α are listed in Table 5.2. They are taken from several experimental studies on the dispersion of pyrogenic materials. In all of them, the final mean particle sizes varied between 0.1 µm and 1 µm, which are much finer than the coarse micrometre-sized agglomerates in powder and considerably larger than the primary

Table 5.2 Experimental exponents α for power law dependency (232); dispersion of pyrogenic powders with rotor-stator systems (RS), high pressure systems (HP), and ultrasonication (US); the applied power density E_V and achieved mean particle size x_{mean} are indicated, values for α were derived from data or cited from source; references: Mandzy et al. (2005), Pohl et al. (2004), Pohl et al. (2005), Sauter and Schuchmann (2008), Sauter et al. (2008)

Reference	Disp. techn.	Material	x_{mean} µm³	E_V GJ/m³	α	Source
Pohl et al. (2005)	RS	SiO_2, 200 m²/g	0.2–0.3	1–10	0.07	Derived
Pohl et a. (2005)	HP	SiO_2, 200 m²/g	0.1–0.2	0.1–0.3	0.12	Derived
Sauter and Schuchmann (2008)	HPPF	SiO_2, 200 m²/g	0.1–0.22	0.04–2	0.08–0.15	Derived
Pohl et al. (2004)	US	SiO_2, 200 m²/g	0.1–0.2	0.1–10	0.08	Cited
Pohl et al. (2004)	US	SiO_2, 90 m²/g	0.1–0.2	0.1–2	0.16	Cited
Pohl et al. (2005)	US	SiO_2, 200 m²/g	0.1–0.2	0.4–30	0.10–0.12	Cited
Mandzy et al. (2005)	US	TiO_2, 50 m²/g	0.15–0.27	0.02–2	0.09	Derived
Sauter et al. (2008)	US	SiO_2, 200 m²/g	0.1–0.22	0.01–3	0.1	Cited
Figure 5.6	US	Al_2O_3, 100 m²/g	0.1–0.2	0.003–3	0.1	Derived

particles. It can be seen that the values of the exponent α are relatively low (approx. 0.1) irrespective of the dispersion technique. That means that high energy densities are required to disperse the pyrogenic powders into the submicrometre size range or to reach even the nanometre region.

An exponent $\alpha = 0.1$ in Eq. (5.23) means that a size reduction by a factor of 2 requires an energy increase by a factor of approx. 1000. Note that for the dispersion of other materials, e.g., flocs of micrometre-sized PS latex (Kobayashi et al. 2004, laminar elongation), micrometre-sized dense agglomerates of crystalline SiO_2 (Wengeler et al. 2004, rotor-stator and HP systems) and for the grinding of crystalline materials in stirred media mills (e.g. Kwade and Schwedes 2002; Mende et al. 2004), much higher values of α are obtained.

An extrapolation of the experimental data (e.g. Pohl et al. 2004; Mandzy et al. 2005; Sauter et al. 2008) indicates that the specific energy input has to exceed 10 GJ/m^3, if mean particle sizes below 100 nm are requested. However, significant volume fractions of nanosized particles can already be generated at moderate energy densities (e.g. Pohl 2005, p. 103). Teleki et al. (2008) showed that the nature of such fine size fraction depends on the strength of the interparticle bonds which are affected by the conditions of the pyrolisis process. A non-commercial pyrogenic TiO_2 powder made in a spray-fed flame could be partially dispersed to the primary particles with HP systems, whereas in commercial SiO_2 and TiO_2 products the nanoparticle fractions were considerably larger than the primary particles.

In conclusion, the practical experience of numerous dispersion studies allows only an incomplete answer to the above formulated questions. It is obviously possible to disperse pyrogenic powders well below 1 μm with various conventional dispersion techniques (cf. Sect. 5.1.3). However, a further reduction of the size distribution requires a disproportionately large amount of energy. This can be described by a power-law between mean particle size and the energy density. There is no serious report on a "steady state" that corresponds to a certain power density as in the case of dispersing micrometre-flocs in turbulent flows (cf. Sect. 5.1.2). This may be related to the fact that fragments of a pyrogenic aggregate maintain the fractal structure of the original aggregate. This holds true as long as the aggregates are considerably larger than the primary particles. However, a dispersion to mainly primary particles or, for example, particles and aggregates below 100 nm, requires extremely high energy inputs and has only been partially observed.

As a result of the high energy consumption, it is useful to know the energy efficiency of the different dispersion machines. According to Pohl et al. (2005), best efficiency is obtained with high pressure systems, while ultrasonication requires up to 5 times more energy to reach the same size distribution. Rotor-stator systems, which are frequently used for the homogenisation of suspended particles and the deagglomeration of coarse agglomerates, perform worst with regard to energy consumption.

Last but not least, the evaluation of dispersion processes relies on the measured size distributions or characteristic particle sizes. Therefore, it cannot be separated from the employed characterisation technique, which may show different sensitivity towards fine and coarse particles. The difficulties in monitoring the dispersion progress in suspensions of pyrogenic powders will be addressed below.

5.1.5 Characterising the Progress of Dispersion

The dispersion of agglomerated or aggregated particle systems in a liquid solvent is usually quantified as reduction of particle size. More specifically, this means a shift of the whole size distribution to the lower end of the size axis, a decrease of characteristic size parameters (e.g. medians, means, or maximum sizes), and a decline of the coarse particle fractions. To monitor the progress of dispersion, thus, requires particle sizing techniques. However, the capability of such techniques to adequately reflect the progress of dispersion is rather varying. Even though this aspect has not gained much attention in the literature, there are a few studies that report on discrepancies between different sizing techniques in the context of dispersion.

Hinze (2001, pp. 63–66) discussed the deagglomeration of large micrometre-sized, non-fractal agglomerates of monodisperse, submicron silica particles by ultrasonication. While laser diffraction clearly showed a reduction of the mean particle size from the micrometre into the submicrometre region, ultrasonic spectroscopy was virtually insensitive to the presence of the agglomerates. Pohl (2005, pp. 96–97) compared the performance of DLS, electroacoustic mobility spectroscopy (Sect. 2.3.6.3) and angular light scattering in the context of fumed silica dispersion. According to him, angular light scattering was insensitive to changes in size distribution when the mean particle size fell below a critical value.

Such discrepancies have manifold origins. They are related to the principal differences with regard to the covered size range (e.g. below or above 1 μm), with regard to the probed particle property (e.g. hydrodynamic or acoustophoretic mobility) and with regard to the type of quantity in which the size fractions contribute to the measurement signal (e.g. by volume or by scattered light intensity). Last but not least, they are affected by the numerical algorithms used for signal analysis. These differences were already partly addressed in Sect. 4.5 for well-dispersed suspensions of pyrogenic powders. The focus of this section lies on agglomerated colloidal suspensions, which undergo a dramatic shift in size distribution from the micrometre range to the submicron or even nanoscale. The experimental practice is reviewed and the applicability of different characterisation techniques is discussed for selected examples

5.1.5.1 Measurands Frequently Used for Characterising the Progress of Dispersion

Studies on the mechanisms and efficiency of dispersion processes are usually conducted under well-defined conditions, which eases sampling and sample preparation. Frequently, the most appropriate characterisation techniques can be selected from a pool of techniques. Typically off-line techniques, such as dynamic light scattering (DLS), are employed and the complete particle size distribution is recorded. Conversely, monitoring dispersion processes in an industrial environment requires a rapid and robust response to the state of dispersion of dense particle systems. Often, online or inline probes are used that measure mean particle sizes or merely a macroscopic suspensions property. Altogether, a number of different measurands are utilised in dispersion monitoring:

- complete particle size distribution
- characteristic "average" values of the particle size distribution
- maximum particle size
- weight /concentration of coarse size fraction
- photometric quantities (turbidity)

Size distributions can be determined by various characterisation techniques. In the context of dispersion monitoring, laser diffraction (LD)—with or without extensions for static light scattering measurements—is frequently employed, in particular when the primary particles are well above the Rayleigh scattering region (i.e. $x_p \gg 100$ nm; e.g. Wengeler et al. 2004; Sommer et al. 2006) or when large micrometre-sized agglomerates are present (i.e. in the case of low dispersion energy input; Pohl et al. 2004; Maier et al. 2006). LD results are always presented as volume weighted distribution (Q_3), which are very close to its intrinsic type of quantity (cf. Sect. 2.3.4.3). A further, widespread method is DLS—especially for size measurements below 1 μm. Prerequisite for its application is the absence of significant fractions of coarse agglomerates. DLS is the preferred technique for monitoring the dispersion of pyrogenic powders (e.g. Pohl et al. 2005; Wengeler et al. 2006). The method intrinsically yields intensity weighted size distribution (Q_{int}, cf. Sect. 2.3.5.1). However, some authors (e.g. Teleki et al. 2008) represent DLS results as volume weighted distribution (Q_3), which had been derived by means of Mie's solution (1908) for sphere scattering. It was shown in Sect. 4.5.3 that this may lead to a major misinterpretation in the case of fractal aggregates (like for pyrogenic powders), for which the intensity weighted distribution are very close to the (solid) volume weighted distribution. Electroacoustic and acoustic sizing techniques (Sect. 2.3.6) offer an alternative to the optical methods, in particular for highly concentrated suspensions of inorganic materials (e.g. Stintz et al. 1998; Pohl et al. 2004). Additionally, some authors use sedimentation analysis (Süß and Hanke 1969; Thoma et al. 1991).

The change of the size distribution during dispersion allows insight into the type of deagglomeration. Moreover, the resolution into single size fractions facilitates the parameterisation and validation of population balance models on the deagglomeration (e.g. Sommer et al. 2006; Bałdyga et al. 2009). Last but not least, the amounts micrometre-agglomerates or nanoparticles can be quantified. However, the experimental uncertainty of complete size distributions is relatively large—in particular with respect to distribution width and rare size classes. Additionally, it often suffices to quantify the progress of dispersion by just one robust parameter–typically some "average" size of the distribution. This parameter (e.g. the volume weighted median $x_{50,3}$) is usually derived from the calculated size distribution. However, in the case of DLS, the preferred "average" is the intensity weighted harmonic mean (x_{cum}) because it can be directly obtained from the autocorrelation curve (or signal spectrum) and is, thus, unaffected by the numerical effects in the course of data analysis (cf. Sect. 2.3.5.1).

Some applications require not only a general reduction of particle size, but additionally a limitation of the maximum particle size or the concentration of coarse particles—e.g. pigment dispersion (Ruscitti et al. 2008) or preparation of CMP slurries (e.g. Kuntzsch et al. 2003). These parameters can be measured by special instruments (*grindometer*, ISO 1524) or by counting methods (e.g. optical particle counting, cf. Xu 2000, pp. 182–222).

Eventually, characterising a dispersion process does not necessarily imply particle sizing. Since the state of dispersion affects several macroscopic suspension properties, an indirect dispersion monitoring is possible and may even provide the more robust and cheaper instrumentation. In that regard, photometric characterisation techniques—mainly based on turbidity—have gained practical relevance for dispersion monitoring (e.g. Süß and Hanke 1969; Reichert 1973). They rely on the size dependency of light scattering: When a solid material is dispersed from, for example, a few hundred micrometres into the nanometre region, the suspension's turbidity first increases and later decays rapidly in the Rayleigh scattering region (or remains constant in the case of absorbing materials such as carbon black).

This brief review on the practice of dispersion monitoring—from scientific publications only—has mentioned several measurands and various characterisation techniques. It is very likely that they show different capabilities to reflect changes in size distribution and that these capabilities depend on the particle system involved. However, the scientific literature is rather sparse on studies that thoroughly examine the performance of characterisation techniques. There is no data basis for a general evaluation; instead, selected examples with a focus on pyrogenic powders are presented below.

5.1.5.2 Dispersion Monitoring for Suspensions of Pyrogenic Powders

The following discussion is based on an experimental study of Nogowski et al. (2010), in which suspensions of fumed silica (125 m^2/g, 10 wt%) were intensively dispersed by ultrasonication. The progress of dispersion was monitored by means of:

- dynamic light scattering (DLS; 2 instruments with scattering angles at 90° and 173°)
- laser diffraction (LD; forward scattering setup <37°)
- optical particle counting (OPC; extinction based, $x_{min} = 2$ μm)
- optical spectroscopy (wavelength: 200–1000 nm)
- rheometry (Searle type, static mode, shear rate: 1–100 s^{-1})

Apart from the rheometric characterisation, all techniques required dilution of the original suspension.

The DLS and LD instruments yielded monomodal particle size distributions of similar polydispersity. For this reason, the mean particle sizes suffice for quantifying the progress of dispersion. There are distinct and systematic differences between the three instruments. DLS showed a steady reduction of the mean particle size (x_{cum}) from 220 nm to 160 nm (173°) and 245 nm to 175 nm (90°), while LD measured mean particle sizes between 500 nm and 400 nm. Although delivering a size distribution, the OPC was only used to determine the total particle number concentration above a certain size limit, which provides an additional insight into the dispersion process. The optical spectroscopy, which yields a turbidity spectrum, and the rheometric measurements served as indirect dispersion monitors. In both cases, it was possible to cover most of data variation within one parameter, i.e. in the turbidity at 400 nm wavelength (τ_{400nm}) and in the apparent viscosity at a shear rate of 1 s^{-1} ($\eta_{1/s}$), respectively

The relative changes in x_{cum}, $x_{50,3}$, τ_{400nm} and $\eta_{1/s}$ are depicted in Fig. 5.7 (left). It reveals that all methods are sensitive to changes in the state of dispersion, yet they correlate differently with energy density. The mean size from DLS, the turbidity, and the viscosity decrease monotonically with growing energy input. This is in accordance with the experimental results cited in Sect. 5.1.4. In contrast, LD does not recognise a progress in dispersion for energy densities above 0.2 GJ/m^3. This results from the specific optical setup (forward scattering), which is optimised for analysing diffraction patterns of coarse particles but is hardly sensitive to colloidal particles. The figure reveals that the viscosity was most strongly affected by the dispersion process. This is related to the fractal nature of aggregates and agglomerates of fumed silica, for which a decrease in size corresponds to a decrease of the apparent volume fraction, and thus to a decrease of suspension viscosity. At high solid contents, this effect is rather pronounced.

The right diagram in Fig. 5.7 shows the evolution of the number concentration of coarse particles, as measured with OPC. An initial, rapid decrease of this parameter represents the disappearance of agglomerates in the course of dispersion. Intriguingly, there is a generation of coarse particle during the dispersion process. This could be attributed to the abrasion of the sonotrode tip (Schießl 2007).

The experimental data of Nogowski et al. (2010) illustrate that rather different characterisation techniques, not necessarily those used for particle sizing, can be employed to monitor dispersion processes. However, their sensitivity may be limited to certain size ranges (like for LD) and definitely varies with the particle system. For instance, the observed behaviour for turbidity and viscosity is specific for the examined fumed silica suspensions and may considerably deviate when the disintegration of dense, regularly packed agglomerates is monitored. Besides this, it was shown that a combination of methods is necessary if different distribution parameters need to be monitored simultaneously.

5.1.5.3 The Limits of Particle Sizing Techniques When Employed for Dispersion Monitoring

Any measured sized distribution is distinguished by the probed particle property and the type of quantity in which the size fractions contribute to the measurement signal. This sets a principal limit for the comparability of particle sizing results. Even so, there is a common expectation that a substantial change in the size distribution should be accordingly reflected in the results of any sizing method. However, Fig. 5.7 shows inconsistency between laser diffraction (LD) and dynamic light scattering (DLS) for large dispersion energies. This can be mainly attributed to the fact that the dispersion process shifted the size distribution below the reliable measurement range of the laser diffraction instrument. Such a transition is rather likely when monitoring the dispersion of colloidal suspensions; additionally, it also concerns other particle sizing methods (cf. measurement ranges in Table 2.2). Even more, when dispersing coarse agglomerates of colloidal suspensions, one frequently encounters very broad size distributions in which micrometre and submicron particles are present in significant amounts. The characterisation of such particle systems is rather challenging with a high risk of misinterpreting the results of sizing instruments.

Two typical misinterpretations are discussed in the following; both are related with the dispersion of fumed silica by gear-rim dispersers and by ultrasonication. When such powders are suspended in aqueous solutions under little agitation, a fairly broad distribution of micrometre sized agglomerates is obtained (>1 μm). The strength of these agglomerates depends on the primary particle size, on surface properties and on powder pre-treatment and is, therefore, product specific. Nonetheless, there is a general experience that relatively high energy inputs are required to shift the size distribution completely to the colloidal size range (cf. Sect. 5.1.4).The left graph of Fig. 5.8 shows the volume weighted size distributions (sum function Q_3) of such a product as measured with LD for a relatively low and a

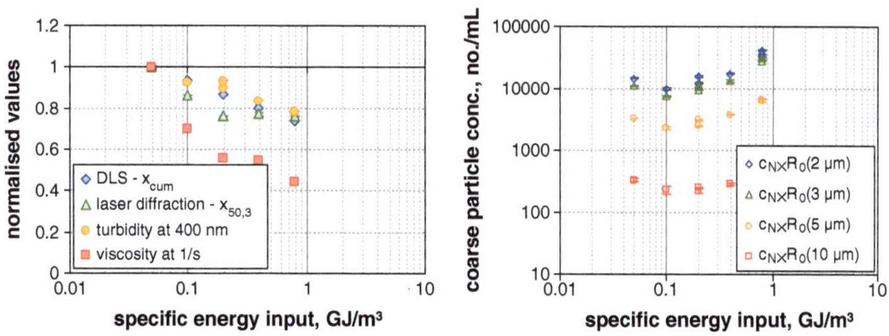

Fig. 5.7 Relative changes during the ultrasonication of fumed silica (125 m^2/g, 10 wt%); *left* average particle size from DLS and LD, turbidity at 400 nm wavelength and dynamic viscosity at shear rate of 1/s; *right* corresponding evolution of the coarse particle concentration as measured by OPC; cf. Nogowski et al. (2010)

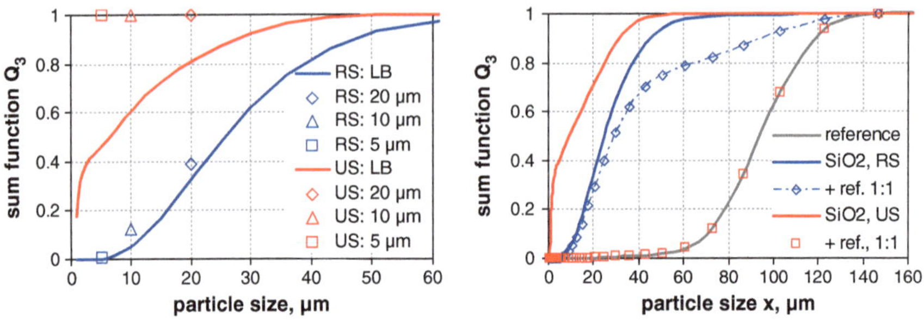

Fig. 5.8 Validation of laser diffraction results for differently dispersed fumed silica (*RS* rotor-stator system; *US* ultrasonication); *left* by comparison with micro-sieve analysis (underflow of 20, 10, and 5 µm sieves); *right* by LD measurements with and without a reference particle system (mixing ratio: 1:1 by weight)

relatively high energy input (i.e. for rotor-stator treatment and ultrasonication, respectively). The curves are compared with results from a micro-sieve analysis, which yields the mass retention for a defined cut size. It can be clearly seen that the results of microsieving and laser diffraction are very similar for weak dispersion. Both methods indicate that there are no significant volume (mass) fractions of particles below 5 µm, i.e. the size distribution lies within the measurement range of the LD instrument. In contrast, there is a substantial deviation between LD and microcsieving, after the suspension has been ultrasonically dispersed. While microsieving reveals that virtually the total particle volume was dispersed to sizes <5 µm, LD implies that more than 50 % per volume are larger than 5 µm. Such an overestimation of the coarse particle content is very typical for LD when the majority of the particles fall below the detectable measurement range. Indications for this kind of situation are large signals in the first size class and a poor reproducibility of LD results. A further interesting way to check the validity of LD results consists in adding a relatively coarse reference particle system to the suspension when conducting the laser diffraction measurements (Babick et al. 2005). Figure 5.8 (right) shows that this procedure yields bimodal size distributions for the low dispersion energies, where the single modes can be unambiguously traced back to the reference particles and the fumed silica. In contrast, the fumed silica “disappears” in the LD results of the mixture after ultrasonication (i.e. intense dispersion) has shifted the size distribution to the colloidal size range.

The risk of misinterpreting LD results when monitoring dispersion processes results from the method’s poor sensitivity to colloidal particles. Similar problems arise for DLS, yet this time from the poor sensitivity to coarse particle fractions. A good illustration of possible misinterpretations can be given with the example of weakly dispersed fumed silica suspensions. The particle sample discussed above in the context of LD showed no significant mass fractions below 5 µm when treated with gear-rim disperser (Fig. 5.8, left). However, when the same sample is analysed by means of DLS, a significant quantity of particles below 1 µm can be detected (Fig. 5.9, left). Similar size distributions are obtained for the underflow of the

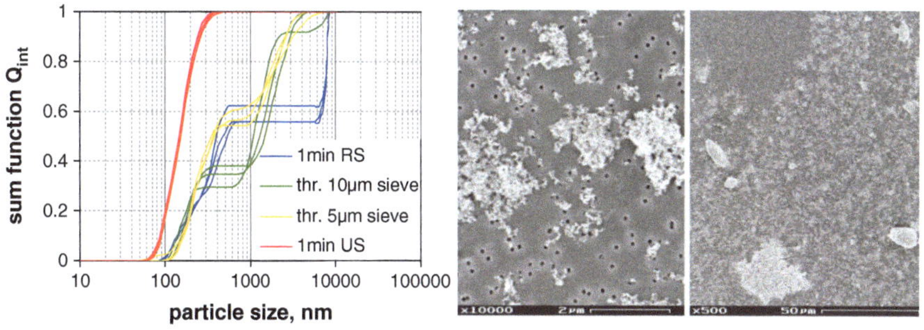

Fig. 5.9 DLS results for fumed silica suspensions corresponding to the LD results shown in Fig. 5.8 (*left*) and SEM images of the RS-treated suspensions, which reveal the existence of few submicrometre aggregates among large agglomerates ≫ 1 µm (*right*)

microsieves (10 µm and 5 µm), which indicates that the submicrometre fraction is real. Regardless, the repeatability of the DLS measurements is rather poor for all gear-rim dispersed samples, which impedes a detailed comparison (e.g. with regard to the maximum size). In contrast, a high repeatability was observed, after the suspension had experienced ultrasonication.

The examples clearly demonstrate a lack of reliability of DLS measurements for very broad size distributions, as they may occur when dispersing colloidal suspensions. Without validation by a second method (e.g. LD), it is difficult to decide on the accuracy of mean and maximum particle sizes from DLS measurements; poor repeatability may be an indicator for this. On the other hand, the general sensitivity of DLS for the diffusion of fine, colloidal particles remains unaffected by the possible presence of coarse agglomerate (in particular since the majority of micrometre particles and agglomerates is typically settled down before the DLS-measurement starts). Hence, DLS still facilitates an estimation of the size of the smallest particle fractions. In the example above, LD stated a minimum size of 4 µm for the weakly dispersed suspension (i.e. treated with the rotor-stator system), whereas DLS indicates size fractions down to 100 nm; electron microscopy could confirm the latter (Fig. 5.9, right).

5.2 Stability of Colloidal Suspensions

In general, *stability* means the quality of a substance or a system to remain at the same state. In practice, one needs to specify if "state" refers to the isotopes of elements, the chemical composition, the state of matter, or something else. When talking about the *stability of colloidal suspensions* or *suspension stability,* one usually addresses the size distribution of the particles, the homogeneity of the disperse system, and/or the structure of the suspension. Hence, the meaning of *suspension stability* is somewhat ambiguous. Moreover, a suspension can be permanently stable due to enthalpic or entropic reasons, or it may appear stable within relevant time periods (kinetic stability) because its physico-chemical state is changing very slowly. Last but not least, the classification of a suspension as stable or not is affected by the measurement technique employed to detect “instability”. That is because changes on a microscopic level are frequently detected indirectly via changes in the suspension’s macroscopic properties. This section reviews some general rules for estimating the microscopic or colloidal stability of suspensions and then focuses on measurement techniques for stability evaluation. Particular attention is given to the stability of binary suspensions.

5.2.1 Microscopic Stability

The stability of colloidal suspensions is important for their processing and the final product quality (shelf life). It is related to size and concentration of the particles and is essentially affected by the particle-particle interactions.

Very frequently, stability is referred to as the absence of particle aggregation (*microscopic* or *colloidal stability*), which means that the particles keep their individuality. Sometimes, stability refers to a constant (and homogeneous) distribution of the dispersed phase in the liquid (*macroscopic stability*). The two terms are not congruent. For instance, aggregation may produce a macroscopically stable gel, while non-aggregating particles may settle due to gravitation.

The following sections deal solely with microscopic stability, which is intrinsically a kinetic stability because aggregated systems are thermodynamically preferred. Yet, a colloidal suspension may appear stable for relevant time periods, which might comprise months (end products with long shelf life, e.g. CMP slurries) or merely hours (intermediate products or samples for particle sizing). In that regard, it is quite obvious that the rate of particle aggregation is an important parameter for the evaluation of the suspension stability. Additionally, the structural changes on the microscopic level that are induced by aggregation coincide with changes in the macroscopic suspension properties. Hence, such properties are employed for the stability evaluation as well. However, the sensitivities to incipient and advanced aggregation are very specific for a given measurement technique; they have to be properly regarded when interpreting stability measurements.

5.2.2 Criteria for Microscopic Stability

As defined above, microscopic stability means the absence of particle aggregation (coagulation). A sufficiently high degree of such stability is, therefore, achieved when repulsive forces between particles prevent them from collision. Of particular importance in this context is the repulsive double layer interaction that is related to the ion clouds surrounding charged particle surfaces.

Faraday (1857) demonstrated that adding small amounts of electrolyte to ruby gold sols induces turbidity, a colour shift from ruby to blue, and finally precipitation of gold particles, although he did not associate this effect with surface charge. At the turn of the 20th century, when the dispersed nature of colloidal suspensions had just been acknowledged, it became clear that colloidal particles can bear an electric charge (e.g. Zsigmondy 1898) similar to microscopic particles, the latter of which was already demonstrated by Quincke (1861). Hardy (1900) eventually related the surface charge of colloidal particles to their coagulation behaviour and, thus, to the suspension stability.

> This iso-electric point is found to be one of great importance. As it is neared, the stability of the hydrosol diminishes until, at the iso-electric point, it vanishes, and coagulation or precipitation occurs, the one or the other according to whether the concentration of the proteid is high or low, and whether the isoelectric point is reached slowly or quickly, and without or with mechanical agitation.
>
> (Hardy 1900)

The principal mathematical framework that quantifies the repulsive interaction between electric double layers as well as the attractive van-der-Waals interaction is the DLVO theory (Sect. 3.2.3). This theory calculates the free energy that is needed to approach two (particle) surfaces from infinity. The decisive parameter with regard to coagulation is the energy barrier V_{max}, i.e., the maximum energy required to bring the particles in contact (cf. Fig. 3.13, p. 103). It depends on the Hamaker function, the double layer properties, and the particle size. When the particles move solely due to their thermal energy (Brownian motion), a minimum value of V_{max} in the range of 10 k_BT to 20 k_BT is frequently reported to guarantee microscopic stability for long time periods (Healy et al. 1978; Bleier and Westmoreland 1991; Ramakrishnan et al. 1998; Mulvaney et al. 2000; McClements 2005).

Derjaguin and Landau (1993) derived a "criterion for stability" for strongly charged spheres ($\psi_0 e_0/k_BT \gg 1$) from the mathematical description of the interaction energy. This criterion is expressed as molar *critical coagulation concentration* (CCC) of an indifferent electrolyte, above which the interaction between the particles is purely attractive. For a symmetric electrolyte, one obtains:

$$ccc^{-1} \approx 1.3 \times 10^{-5} \cdot \frac{A_{131,0}^2 F^6 z_1^6}{(\varepsilon_m \varepsilon_0)^3 (RT)^5}, \tag{5.24}$$

where $A_{131,0}$ is the non-retarded Hamaker constant, z_1 the counter ion valency, ε_m the relative permittivity of the solvent, while ε_0, F and R denote the electric field constant, Faraday's constant and the universal gas constant, respectively. It appears that excessive charging of the surface has no effect on the suspension stability. Instead, this equation predicts a strong influence of the valency z of the electrolyte ions. In this respect, it is essentially confirmed by the early experimental findings of

Schulze (1882, 1883) which were reviewed and extended by Hardy (1900) and led to the *Schulze-Hardy rule*:

$$ccc_{z=1}^{-1} \div ccc_{z=2}^{-1} \div ccc_{z=3}^{-1} = K \div K^2 \div K^3. \quad (5.25)$$

This agreement with empirical data facilitated the quick acceptance of the DLVO theory within the colloidal community.

> If coagulative power be defined as the inverse of the concentration in grammemolecules per litre necessary to convert a given hydrosol into a hydrogel, then from Schulze´s measurements [Schulze 1882] the coagulative power of metals of different valency is: R′:R″:R‴ = 1:30:1650.
>
> (Hardy 1900)

The applicability of Eq. (5.24) breaks down for weakly charged particles ($\psi_0 e_0/k_B T \leq 1$) because the free energy of the EDL is then strongly related to the (apparent) surface potential ψ_0. For spherical particles with a thin double layer ($\kappa a \gg 1$) of Gouy-Chapman type, Derjaguin (1940) derived a criterion for the suspension stability that can be expressed as:

$$\frac{A_{131,0}\kappa}{4\pi\varepsilon_m\varepsilon_0\psi_0^2} \leq 2. \quad (5.26)$$

This equation predicts that the surface potential ψ_0 which is required to maintain the stability grows with the quadratic root of the Debye-Hückel parameter κ or with the fourth root of the ionic strength, i.e. salt concentration. For instance, in order to stabilise titania particles in an aqueous suspension of ionic strength of 0.001 M (i.e. the Debye length is 10 nm) a surface potential $\psi_0 \geq 17$ mV is required, whereas for an ionic strength of 0.1 M the surface potential ψ_0 has to exceed 53 mV.

The DLVO theory exclusively considers van-der-Waals forces and double layer interactions. Its predictive power is, therefore, lost when other types of interactions prevail or when the effect of (interfacial) additives is not properly accounted for. This might be the case when polymeric additives are used to induce or to prevent coagulation. Other non-DLVO interactions are the strong short-range repulsive interactions, frequently called *hydration interaction,* usually observed for high electrolyte concentrations (cf. Sect. 3.2.4). This type of interaction is commonly used to explain the surprisingly high stability of latex or titania suspensions (Healy et al. 1978; Molina-Bolívar et al. 1997; Yotsumoto and Yoon 1993a) or the extraordinary stability of very fine colloidal silica around the isoelectric point (Yotsumoto and Yoon 1993b; Paciejewska et al. 2007). Even though the origin of this stabilising interaction is still a matter of debate, there are numerous publications that aim to integrate it into a general interaction theory as well as into stability calculations (cf. discussion in Sect. 3.2.4.3).

5.2.3 Stability Ratio

The examination of the free energy of two interacting particles reveals the main factors that govern the suspension stability and provide criteria for classifying a colloidal system as stable or not. However, in order to understand how the particle-particle interactions quantitatively affect the suspension stability, it is necessary to consider their influence on the particle aggregation rate.

A first theory on the kinetics of aggregating systems was developed by Smoluchowski (1916), who considered collisions between diffusing particles and particles in shear flow which were thought to adhere when the particle centre distance r_{ij} falls below a critical minimum value $r_{ij,\min}$. Eventually, he derived the population balance equation, which describes the consumption of primary particles and the growth of aggregates (cf. Sect. 4.1.2). However, in the context of stability analysis, only the initial aggregation rate J between the primary particles is considered. For the absence of migration and shear, one obtains the following for monodisperse particles:

$$J = -\frac{\mathrm{d}c_{N,1}}{\mathrm{d}t} = K_{11}c_{N,1}^2 = 8\pi D_\mathrm{t}\, r_{ij,\min} c_{N,1}^2, \tag{5.27}$$

where $c_{N,1}$ and D_t denote the number concentration and the diffusion coefficient of the isolated primary particles, while $r_{ij,\min}$ is the minimum centre distance. The kernel K_{11} of the population balance has to be replaced by K_{ij} for non-uniform particles:

$$K_{ij} = 4\pi \cdot (D_{\mathrm{t},i} + D_{\mathrm{t},j}) \cdot r_{ij,\min} = \frac{8\mathrm{k_B}T}{3\eta} \cdot \frac{\bar{a}}{\bar{a}_\mathrm{harm}} \tag{5.28}$$

with the $\bar{a}$ and $\bar{a}_\mathrm{harm}$ being the arithmetic and harmonic means of the particle radii a_i and a_j.

Smoluchowski neglected any kind of interaction between the particles, which means that the diffusion of a particle is not affected by the presence of other particles and that each event of collision leads to a stable aggregate bond. Yet, as Fuchs (1934) argued for the stability of charged aerosols, any attractive or repulsive interaction affects the diffusive approaching of two particles and Smoluchowski's aggregation rate is changed by a factor W:

$$J = 8\pi D_\mathrm{t}\, c_{N,1}^2 r_{ij,\min}/W, \tag{5.29}$$

which depends on the free energy of the interacting particles:

$$W = r_{ij,\min} \int_{r_0}^{\infty} \exp\left(\frac{V_\mathrm{tot}}{\mathrm{k_B}T}\right) \cdot \frac{\mathrm{d}r}{r^2}. \tag{5.30}$$

The factor W describes the retardation of aggregation due particle interactions; it has values below 1 when attractive interactions prevail ($V_{tot} < 0$), whereas for repulsive interactions ($V_{tot} > 0$), W amounts to values above 1. For this reason, it is called *stability ratio*. A colloidal suspension can be considered as to be long-term stabilised when the stability ratio exceeds values of 10^6–10^9 (Hunter 1988, p. 239; Ramakrishnan et al. 1998).[2]

For the derivation of Eq. (5.30) Fuchs assumed that the diffusion coefficients of two approaching particles remain constant. However, due to viscous interaction the particles have a mutual impact on their mobilities: The closer the two particles approach, the larger the additional resistance is which is related to the squeezing of the fluid.

Honig et al. (1971) and Spielman (1970) independently worked on this aspect and included the viscous interactions in the calculation of the stability ratio W.

$$W_{ij} = r_{ij,\min} \int_{r_{ij,\min}}^{\infty} B(r) \exp(V_{\text{tot}}^{ij}/\text{k}_\text{B}T) \frac{\text{d}r}{r^2}, \tag{5.31}$$

where B is a hydrodynamic factor,

$$B = \frac{6\xi^2 + 13\xi + 2}{6\xi^2 + 4\xi}, \text{ with } \xi = \frac{r - r_{ij,\min}}{\bar{a}_{ij,\text{harm}}}. \tag{5.32}$$

In general, this hydrodynamic effect (lubrication effect) cannot be neglected. It is particularly important for thin double layers ($\kappa a \gg 1$) because the range of viscous coupling scales with the particle size a, whereas the double layer thickness scales with κ^{-1}. Similarly, the contribution of the hydrodynamic effect to the stability ratio is increased when the Hamaker constant A_{131} is decreased. In the absence of non-viscous interactions and external forces, the viscous coupling will even inhibit a particle collision.

> Spheres in a viscous continuum cannot come into contact with one another without the aid of an external attractive force which rapidly increases as the gap narrows.
>
> (Spielman 1970).

Fuchs' kinetic Eq. (5.29) is commonly related to the initial aggregation of interacting particles even though it actually refers to particle collisions. Yet, one usually assumes that any collision coincides with adhesion and, thus, aggregation, because for most particle systems the short range interactions are dominated by the attractive van-der-Waals forces. However, in the presence of short-ranging repulsive forces, e.g., due to hydration interactions or polymeric layers, adhesion is not an inevitable result of particle contact. In this case, a suspension can appear stable

[2]It should be noted that McGown & Parfitt (1967) gave a modified definition of the stability ratio, in which the rate of aggregation is not related to that of non-interacting particles (Smoluchowski, Eq. (5.27) but to that of uncharged particles subject to attractive van-der-Waals forces. In this case, the stability ratio differs from factor W in Eq. (5.30).

even at relatively low values of the stability ratio W. Nevertheless, short-ranging repulsion does not necessarily exclude aggregation as long as there is a secondary minimum of the interaction energy (cf. Fig. 3.13). If this is the case, the minimum collision distance $r_{ij,\min}$ has to be adequately chosen when calculating the stability ratio W. The stabilising effect of short-ranging repulsion can be quantified by an additional factor E in the r.h.s of Eq. (5.29)(cf. Eq. 4.1), which is described as *reactivity*, *sticking probability*, or *flocculation efficiency* (Smellie and La Mer 1958; Hogg 1999; Sandkühler et al. 2005; Oolsen et al. 2006; Paciejewska 2010).

Last but not least, it should be noted that Fuchs' equation ignores concentration effects, i.e. manybody interaction. This concerns the viscous interaction as well as long ranging double layer interaction. Additionally, the mean distance to the next interacting particle cannot be considered infinite for dense suspensions. Hütter (1999) showed by means of simulation that particle concentration adversely affects the stability ratio W in dense suspensions.

5.2.4 Experimental Evaluation of Suspension Stability

The previous sections showed that the stability of colloidal suspensions can be principally predicted from the particle and interfacial properties. In practice, however, such a calculation cannot replace the experimental determination. That is because the exact composition of a given suspension and the numerous factors affecting stability (e.g. zeta-potential) are rarely known altogether. Besides this, there is most often some uncertainty regarding the prevailing types of interaction and on the appropriate model for their calculation. Last but not least, experiments do not solely reflect the stability against aggregation, but rather allow, e.g., for the evaluation of the stability against Ostwald ripening, sedimentation, or phase inversion (for emulsions). Even more, they can reveal the suspension's stability behaviour for variation in ambient conditions (e.g. by means of freeze-thaw cycles). Similarly, the motivation to quantify stability is manifold. For instance, it may aim at predicting the shelf life of food and cosmetic products with colloidal particles, at monitoring changes in macroscopic properties that are caused by destabilisation, or at optimising the dosage of stabilising additives. In other cases, it is important to understand how colloidal particles behave in environmental media and human bodies (e.g. in the context of toxicological studies, Kuhlbusch et al. 2012; Nickel et al. 2013; Nogowski et al. 2013); or the stability experiments are employed for the parameterisation of models, e.g. for the hydration forces (Binks and Lumdson 1999; Molina-Bolivar et al. 1997; Yotsumoto and Yoon 1993a, b).

For some of these questions it is necessary to monitor the initial stages of particle aggregation, which are mainly measurable as slight changes of the particle size distributions. For others, the focus lies on changes in the macroscopic behaviour (e.g. rheology, sedimentation), which become manifest at later stages of aggregation. Such macroscopic changes are caused by the apparent increase of volume fraction of the dispersed phase (since liquid is "immobilised" within the aggregates) and may even result in phase transition, e.g. to gelation or to the formation of colloidal glass (Tanaka et al. 2004; Kätzel et al. 2007). In other cases, the stability should be evaluated indirectly via the changes of the interfacial properties (Greenwood 2003). Correspondingly, a large variety of measurement techniques is employed for characterising the stability of colloidal suspensions.

5.2.4.1 Stability from Monitoring the Size Distribution

Aggregation means coarsening of the size distribution. The rate of size changes can, therefore, be considered as a measure of the microscopic stability. In principle, one can use any particle sizing technique for the purpose of stability characterisation—provided that the sample preparation does not affect aggregation and that both preparation and measurement time are well below the characteristic time scale(s) of the aggregation process. For dilute suspensions, one usually employs optical sizing techniques.

For quantifying the initial aggregation rate J (Eq. 5.29) it is important to measure size distribution with high resolution. For colloidal particles well above the nanorange ($x > 100$ nm) this is possible with single particle light scattering (SPLS, an optical particle counting technique). This method was employed by Lichtenfeld et al. (2004) for the quantification of aggregation kinetics of colloidal PS latex and by Schelero et al. (2009) for studying the Ostwald ripening of submicron Pickering emulsions. The main advantage of SPLS is its capability to measure fractional number concentrations in narrow size classes, which facilitates the parameterisation of the aggregation kernel (cf. Eq. 4.1) and the determination of the initial aggregation rate J; it is, however, insensitive to very fine colloids.

An alternative for monitoring aggregation induced changes in the size distribution are static and dynamic light scattering (LS) techniques (Gregory 2009). These techniques can be applied to the whole colloidal size range. With multi-angle dynamic and static LS it is even possible to monitor the complete aggregation process from the initial dimer formation till the consolidation of a final gel (Manley et al. 2004, Manley et al. 2005). Such measurements reveal the changes in the size distribution as well as the structure of the aggregates. However, the resolution of the size distribution is rather poor and, unlike SPLS, the number concentration cannot be detected. That means that even though SLS and DLS can be employed to quantitatively describe the kinetics of aggregation, their results cannot be (directly) converted to the aggregation rate J or to kernels K_{ij} of the aggregation process. It is interesting to note that cross-correlation DLS techniques may offer additional sensitivity to aggregation since the coherent fraction of scattered light is implicitly measured as well (Geers et al. 2007).

In contrast to dilute suspensions, it is rather difficult to further dilute dense suspensions without affecting the interfacial properties, the particle interaction, and the qualitative aggregation behaviour. Even more, dilution of dense suspensions significantly changes the viscous interactions and, hence, may affect the size of aggregates that have already been formed. Consequently, dilution should be avoided when aggregation induced changes in size distribution are to be monitored in dense colloidal suspensions. That excludes most optical sizing techniques. Instead, acoustic sizing methods, like ultrasonic attenuation spectroscopy (cf. Sect. 2.3.6.2), can be used. Similar to SLS and DLS, the resolution of these techniques is not very high for colloidal particles and they cannot provide number concentrations as required for deducing the initial aggregation rate J. Even so, they accurately detect the coarsening of size distributions due to aggregation (Kippax et al. 1999) and other destabilisation mechanisms such as Ostwald ripening (Weiss et al. 1999).

5.2.4.2 Stability from Monitoring Scattering and Extinction of Radiation

The interaction of light with colloidal suspensions produces light scattering and light extinction (turbidity). Both phenomena are sensitive to particle size and concentration and can, therefore, detect changes in the suspension microstructure that are due to destabilisation. However, the mere values of the scattering intensity I_{sca} or the turbidity τ cannot be used to quantify the particle size distribution, although their temporal evolution may serve as useful indicators for the rate of aggregation. The meaningful interpretation of such experimental data necessitates some basic knowledge on light scattering (cf. Sects. 3.1.1 and 4.3).

For very fine, non-aggregated colloidal particles, the light scattering obeys Rayleigh's scattering law:

$$I_{sca} \propto c_{N,p} \cdot \frac{V_p^2}{\lambda^4} \cdot \left|m_r^2 - 1\right|^2 \propto c_{V,p} \cdot \frac{V_p}{\lambda^4} \cdot, \tag{5.33}$$

where λ denotes the wavelength, $c_{N,p}$ the number concentration, and $c_{V,p}$ the volume fraction of the primary particles (volume V_p). For aggregates of such nanosized particles, the Rayleigh-Debye-Gans (RDG) approximation can be used (cf. Sect. 4.3.1):

$$I_{sca} \propto c_{N,agg} \cdot \frac{N^2 V_p^2}{\lambda^4} \cdot S_{agg}(q) \propto c_{V,p} \cdot \frac{N V_p}{\lambda^4} \cdot S_{agg}(q). \tag{5.34}$$

From this, one can conclude that light scattering is very sensitive to the initial stages of aggregation, when the aggregates roughly obey the Rayleigh scattering (i.e. when $qx_g \lesssim 1$ and thus $S_{agg}(q) \approx 1$). At these stages the scattering intensity increases almost linearly with the aggregation number N. With proceeding aggregation and growing aggregate size, the scattering behaviour transforms into the fractal scattering regime, where the scattering intensity remains constant (cf. Eq. 4.45 and Fig. 4.9):

$$I_{\mathrm{sca}} \propto c_{N,\mathrm{agg}} \cdot \frac{N^2 V_{\mathrm{p}}^2}{\lambda^4} \cdot \frac{1}{\left(qR_{\mathrm{g}}\right)^{d_{\mathrm{f}}}} \propto c_{V,\mathrm{p}} \cdot \frac{NV_{\mathrm{p}}}{\lambda^{4-d_{\mathrm{f}}} R_{\mathrm{g}}^{d_{\mathrm{f}}}} \propto c_{V,\mathrm{p}} \cdot \frac{x_{\mathrm{p}}^{3-d_{\mathrm{f}}}}{\lambda^{4-d_{\mathrm{f}}}} \left(\mathrm{x}_{\mathrm{agg}}^{-1} < \mathrm{q} < \mathrm{x}_{\mathrm{p}}^{-1}\right). \tag{5.35}$$

Note that the transition from the Rayleigh scattering of the primary particles to the fractal RDG scattering of the aggregates is not only reflected in the absolute value of scattering intensity, but also in its wavelength dependency.

Scattering and turbidity measurements are exchangeable with regard to the sensitivity to particle aggregation—as long as the particles are optically non-absorbing ($\tau_{\mathrm{non\text{-}abs}} \propto I_{\mathrm{sca}}$). In contrast, there is a qualitative difference between the two quantities when the absorption of light cannot be ignored:

$$E_{\mathrm{abs}} \propto \tau_{\mathrm{abs}} \propto c_N \cdot \frac{V_{\mathrm{p}}}{\lambda} \cdot \mathrm{Im}\left(m_{\mathrm{r}}^2 - 1\right) \propto \frac{c_{V,\mathrm{p}}}{\lambda}. \tag{5.36}$$

In that case, the turbidity cannot reflect the initial steps of nanoparticle aggregation since the total particle volume remains unchanged.

Stability sensors based on optical transmission and backscattering are, e.g., used to monitor milk clotting in cheese-making (O'Callaghan et al. 2000; Mateo et al. 2010). In particular for studying aggregation processes, spectral turbidity measurements are employed. Gruy (2001) studied the formation of small aggregates in turbulent flow with this measurement technique, while Steinke et al. (2009) used it to monitor the dispersed phase in crossflow-filtration loop. Even though these papers explicitly focussed on non-stabilised systems, they inherently confirm the usefulness of the method to characterise the suspension stability. A further measurement technique that can be employed for this purpose is the turbidity or transmission fluctuation spectroscopy, which is sensitive to number fluctuations of micrometre-sized aggregates (Wessely et al. 1996; Gregory 2009). It thus detects the changes when the aggregation proceeds to its end.

In a similar manner to light, other types of radiation, e.g., electromagnetic waves in the X-ray domain, high-frequency electric fields, or acoustic waves, offer ways to monitor changes in composition and structure of suspensions. For instance, dielectric spectroscopy was used to investigate the moisture uptake and stability of cosmetic creams (Sutananta et al. 1996; Tamburic et al. 1996), and acoustic parameters (resonance frequency, attenuation, sound speed) were shown to correlate with sol-gel transition in suspensions of colloidal silica (Senouci et al. 2001), as well as with the phase transition of renneted milk (Bakkali et al. 2001).

5.2.4.3 Evaluation of Suspension Stability by Monitoring Sedimentation Behaviour

In experimental practice, the stability of a colloidal suspension is frequently evaluated by its (visible) sedimentation behaviour. This is mainly because the settling velocity of particles grows with size. Large, micrometre-sized aggregates ("flocs") settle quite quickly, which results in a separation of the dispersed from the liquid phase within a few minutes up to hours. In contrast, suspensions of very fine nanoparticles (<100 nm) reach a diffusion-sedimentation-equilibrium, in which the dispersed phase remains perpetually, yet not uniformly distributed throughout the whole sample volume (cf. Sect. 3.1.2).

The sedimentation behaviour of suspensions is not only affected by the size of the particles, but also by their concentration—or, more precisely, by the interparticle distance. Only for very dilute suspensions can this influence be ignored (*free settling*). In fact, the viscous interaction between the settling particles and the backflow of the displaced liquid retard the sedimentation process even for concentrations far below the maximum packing density and lead to a *hindered settling* of individual particles. In this regime the coarse particles still settle faster than the fine ones. Therefore, the interface between the original suspension phase and the supernatant is gradually broadening and the concentration-height profile reflects the particle size distribution (Mirza and Richardson 1979), similar to free settling (cf. Sect. 2.3.3.1).

However, if the interparticle distances are reduced in such a way that coarse particles cannot pass finer ones due to steric effects, then all particles are forced to settle with the same velocity and the sedimentation becomes equivalent to the flow through a fixed bed (*zone sedimentation*). As a result, the dispersed phase subsides with a sharp interface between the suspension phase and the supernatant. Zone sedimentation occurs for high particle concentrations or is a result of particle aggregation because aggregates "immobilise" liquid and, thus, increase the apparent volume of the dispersed phase (Fitch 1979).

Zone sedimentation of aggregating suspensions coincides with the formation of a particle network (gel) in the lower part of the sedimentation vessel. This gel is gradually consolidated under its own weight, which reduces its porosity and the sedimentation velocity. The gel is, therefore, called *compression* or *compaction zone*; it eventually comprises all particles of the suspension. The compression zone has a much higher porosity than the sediment of a well-stabilised suspension (Sobisch et al. 2006). For fast aggregation and very fine particles, the gelation can finish even before a visible sedimentation of the dispersed phase occurs.

From these principal sedimentation effects, one can employ the following relationships for the evaluation of suspension stability:

- aggregation leads to aggregates which settle faster than the primary particles,
- advanced aggregation in dense suspensions results in zone sedimentation with a sharp interface between the supernatant and the suspension phase, whereas this interface is diffuse for the hindered settling in stable suspensions; even more, zone sedimentation eventually leads to compression,
- the sediment porosity of a stable suspension is considerably lower than that of a flocculated one.

Stability evaluation via settling velocity or sediment porosity requires reference values of the stable suspensions and is usually conducted within parameter studies (e.g. pH variation). In contrast, the existence of zone sedimentation—a strong indication of destabilisation–can be directly elicited from the sedimentation experiment. It should, however, be noticed that zone sedimentation and compression can also occur in stable suspensions—provided that the particle concentration is sufficiently high. In any case, it is recommendable to counter-check the observations from sedimentation experiments with other experimental data or prior knowledge. For instance, an apparently stable suspension without visible separation of the dispersed phase may indicate gelation—in other words, aggregation of the primary particles, or a diffusion-sedimentation-equilibrium, which occurs for nanosized particles in the absence of aggregation.

In experimental practice, one usually examines the sedimentation behaviour of colloidal suspensions by means of centrifugation. This is done because colloidal particles settle very slowly under gravity.

For instance, a 1 μm silica particle settles in aqueous solutions with approximately 2 mm per hour; for a 0.1 μm silica particle the corresponding settling velocity is just 0.6 mm per day.

Centrifugation primarily accelerates the process of solid-liquid separation. However, with regard to suspension stability, there are even more differences between centrifugal and gravitational settling: First of all, centrifugation diminishes the relevance of diffusion, which is—unlike settling—not affected by the centrifugal speed. Furthermore, it reduces the time available for building aggregates. At the same time, however, it promotes aggregation because the different settling velocities give rise to particle collisions—in addition to those which arise by means of Brownian motion. Moreover, in centrifugal fields the settling objects experience relatively high shear forces which may affect the growth of aggregates. Last but not least, the excess pressure on the sediment is much higher than for gravitational settling and it depends on the centrifugal speed (Sobisch et al. 2006).

A powerful demonstration of the applicability of sedimentation analysis to the evaluation of suspension stability was provided by Paciejewska (2010), who investigated the stability behaviour of suspensions of pyrogenic oxides (cf. Sect. 4.5) with particular attention paid to binary systems. The sedimentation behaviour of these suspensions was measured with an analytical photocentrifuge that detects the optical particle concentration along the radial position in the centrifugal field (as transmission profiles). From these transmission profiles it was possible to derive an average initial settling velocity, an index for the sharpness of the interface between supernatant and suspension phases, as well as the height of the sediment and, thus, its porosity. Additionally, the initial sample transmission—a purely optical signal—was evaluated.

Figure 5.10 reproduces the pH dependency (pH = 2–10) of these parameters for three pyrogenic materials (SiO_2, TiO_2 and Al_2O_3), which where suspended in 0.01 M KBr solution. The corresponding IEPs of these suspensions were 4.0 (SiO_2), 7.2 (TiO_2), and 8.8 (Al_2O_3). Let us first focus on the results of TiO_2 and Al_2O_3: Around the IEPs one observes an increased initial settling velocity, which can be explained as increased size due to the coagulation of the pyrogenic aggregates. Moreover, the sharpness of the interface between supernatant and suspension

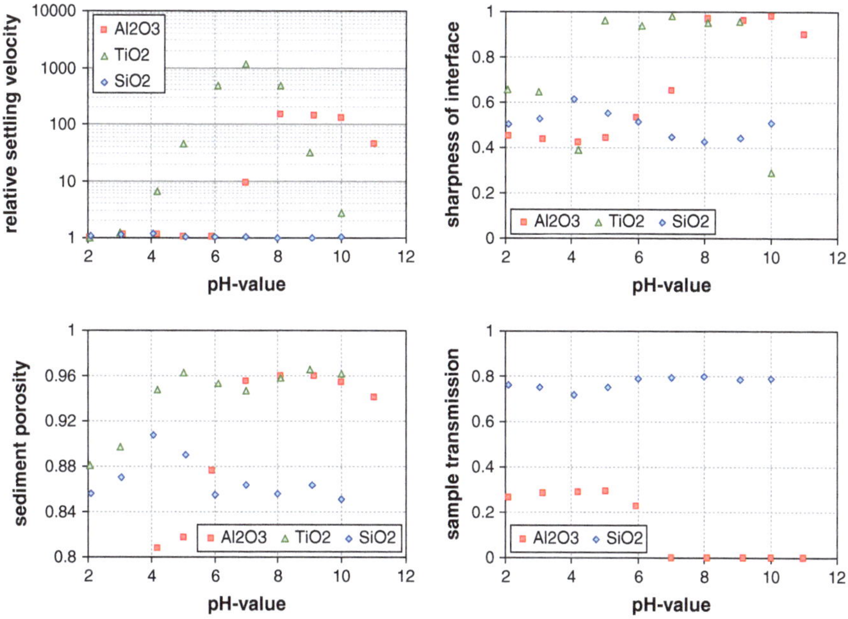

Fig. 5.10 Settling and optical properties of dense suspensions as a function of the pH for three pyrogenic powders (Al_2O_3, TiO_2 and SiO_2; 5 wt% in 0.01 M KBr solution), measured with an analytical photocentrifuge 24 h after sample preparation; reproduced from Paciejewska (2010, pp. 80–81); copyright 2010, with permission from K. Paciejewska

phase approaches 1, which would mean that the optical concentration discontinuously increases from zero (clear supernatant) to a maximum value (suspension phase). Such a value, therefore, indicates zone sedimentation. Additionally, one observes increased sediment poro-sities around the IEPs. Last but not least, the optical transmission of Al_2O_3 falls from approximately 30 % to zero when the pH is shifted to the region of the IEP. Hence, the three parameters from sedimentation and the optical parameter reflect the expected suspension destabilisation around the IEP. However, a second view on the data reveals that the significance of these four parameters for destabilisation is not completely identical. For instance, for TiO_2 one observes maximum values of the sediment porosity from pH 4 to 10, while the interface sharpness becomes maximal only in the range of pH 5 to 9. The initial settling velocity even reveals different degrees of coagulation in that pH-range. The SiO_2 data support the idea of a specific sensitivity of the four parameters to destabilisation: With regard to the initial sample transmission and the initial settling velocity, the SiO_2 samples are virtually stable over the whole pH range, whereas the interface sharpness and the sediment porosity values are significantly increased around the IEP and, thus, indicate the beginning of coagulation.

In summary, sedimentation analysis in gravitational or centrifugal fields provides a set of different parameters that are sensitive to incipient or advanced aggregation in colloidal suspensions. It thus offers a powerful tool to study the microscopic stability of colloidal suspensions.

5.2.4.4 Evaluation of Suspension Stability via Its Rheology

The rheology of a colloidal suspension reflects—similar to sedimentation—its microscopic properties; in particular, the particle concentration, the particle interactions, and the resulting suspension structure. It is, moreover, sensitive to aggregation phenomena (Quemada and Berli 2002). This sensitivity primarily results from the increase in the effective volume of the dispersed phase (the main factor determining the viscosity of suspensions) with ongoing aggregation because liquid is "immobilised" in the pores of the evolving aggregates or flocs. Since the aggregates usually grow in a fractal-like manner (cf. Sect. 4.1), the effective volume fraction of the dispersed phase as well as the suspension viscosity increase steadily. Hence, the temporal evolution of viscosity can be used to quantify the aggregation kinetics (Vogelsberger et al. 2000). When the aggregation is not affected by sedimentation or hydrodynamic agitation, the suspension eventually starts to gel and the viscosity diverges. Gelling suspensions are characterised by the yield stress τ_0 and by the viscoelastic properties, i.e. the storage modulus G' and loss modulus G'' for the linear visco-elastic regime.

Rheological experiments were used by Wittmar et al. (2012) for the stability analysis of functionalised nanoparticles in ionic liquid. The authors examined viscosity curves (i.e. viscosity vs. shear rate) as well as viscoelastic properties for suspensions of varying particle concentration but identical preparation history. The qualitative criterion for destabilisation was whether the suspensions did gel, i.e. whether the storage modulus G' was smaller the loss modulus G''. A different approach was proposed by Vogelsberger et al. (2000), who studied the nucleation and aggregation kinetics of silica, synthesised via TEOS (tetraethyl-orthosilicate) condensation and precipitation. For that purpose, the viscosity and the viscoelastic moduli were continuously monitored. The latter data were i. a. used to derive the rate of aggregation.

In many papers, the stability of dense colloidal suspensions is evaluated via the yield stress τ_0; the absence of a measurable yield stress is then interpreted as the absence of significant aggregation. Such an approach is mainly used in parameter studies, when the pH value, the salt concentration, or the dose of some additive is varied (e.g. Franks et al. 1999; Breyer and Sonnefeld 2004). Last but not least, there are various applications in which the onset of gelation has to be accurately detected (e.g. during cheese making). This can be monitored by commercial probes that use torsional or transversal low-frequency vibrations (O'Callaghan et al. 2000; Williams et al. 2001).

In conclusion, there are several rheology based techniques that can be employed for the evaluation of suspension stability. Most of them are only sensitive to advanced aggregation because they rely on viscoelasticity or the presence of yield stress. However, incipient aggregation can also be quantified by rheometry—provided that the dynamic viscosity is accurately measured.

5.2.4.5 Remarks on Using the Zeta-Potential for Explaining the Suspension Stability

The DLVO theory (Sects. 3.2.3 and 5.2.2) explains the stability behaviour of colloidal suspensions by the combined effect of van-der-Waals and double layer interactions. The decisive parameters for colloidal stability are the diffuse layer potential ψ_d and the thickness of the diffuse layer (Debye length $1/\kappa$). The diffuse layer potential ψ_d cannot be probed by experiment, but the zeta-potential serves as a good approximation (cf. Sects. 2.3.7 and 3.1.5.3). High zeta-potential values mean strong repulsion between dispersed particles and, thus, favour the stabilisation of colloidal suspensions. This effect is even more pronounced for thick double layers, which are formed for low electrolyte content. In contrast, low zeta-potential values, opposite signs of zeta-potentials, or high electrolyte concentrations promote frequent collisions of particles and eventually aggregation. The zeta-potential can, thus, be principally employed to predict the suspension stability. It is frequently determined as a function of the pH and/or the concentration of electrolytes, ionic surfactants, or polyelectrolytes (Greenwood 2003). However, zeta-potential values should be carefully used when evaluating the suspension stability. The wide-spread perception that there is a universally valid critical zeta-potential value $|\zeta_{cr}|$ which defines the transition from unstable to stable suspensions, was already rejected by Eilers and Korff (1940). This is because this transition is crucially affected by the electrolyte concentration. Moreover, the DLVO theory only accounts for two types of interactions. Further interactions, like steric and depletion interaction or hydration forces (cf. Sect. 3.2.4), may act in a stabilising manner irrespective of the zeta-potentials.

5.2.5 Stability Measurement—Aspects of Data Interpretation

The fact that the origin and mechanisms of particle aggregation are principally understood and can even be quantitatively predicted by means of aggregation kernels and stability ratio (cf. Sect. 5.2.3) does not imply that such a prediction is easy to accomplish or that the interfacial properties can be easily derived from stability measurements. Indeed, the stability theory ignores the origin of interparticle forces and is indifferent towards their temporal evolutions. If, however, stability measurements are evaluated with regard to the prevailing microprocesses, these aspects cannot be ignored. A profound stability analysis should, therefore, explore:

- the influence of particle concentration,
- the influence of time, and
- the significance of homoaggregation and heteroaggregation.

In stability measurements, the particle concentration may vary between a few ppm and several percents (by weight). The former values are relevant, for example, when investigating the behaviour of colloidal particles in environmental media or physiological solutions. The latter typically occur when colloids are used as main product components (e.g. pigments) or as fillers. Since the aggregation rate J is proportional to the squared particle concentration (Eq. 5.29) such concentration differences have dramatic effects on the observed suspensions stability.

Consider, for instance, the aggregation of colloidal particles at 1 ppmw and at 1 wt% for identical bulk and interfacial properties. The corresponding rates of aggregation J then differ by a factor of 10^8.

Additionally, the particle concentration decides the impact of particles on the solvent. High solid loadings affect the concentration of dissolved matter by adsorption, desorption, dissociation of surface groups, or dissolution of solid phase. That leads to changes in the solvent properties (e.g. pH value, ionic strength) as well as corresponding interfacial properties (surface charge). In contrast, the solvent can be considered unchanged for very low particle concentrations. This applies not only to the pH and the total electrolyte content; it further means that the concentration of surface active molecules (e.g. proteins, humic acids) in the solvent is not significantly changed by adsorption in the interface and that the concentration of dissolved solid material remains rather low for long time periods (cf. Sect. 3.1.7).

The time influence on the stability measurements is worth being investigated since the time spans that are relevant for stability vary considerably: The processing of colloidal suspensions is usually finished after less than a few hours, the transit through the gastrointestinal tract takes at maximum a few days, and the intended shelf-life of cosmetic products can amount to several months. Hence, the classification of colloidal suspensions with regard to stability strongly depends on the intended application or subjects of investigations. Moreover, colloidal suspensions do not only change by aggregation but also by various other microprocesses (e.g. adsorption/desorption of solutes, surface dissolution, surface precipitation, or Ostwald ripening). These microprocesses can either promote or adversely influence the aggregation; each of them possesses its own specific kinetic. The prediction of long-term stability from short-term observations is, therefore, rather ambitious.

The third aspect, homo- and/or heteroaggregation, concerns the homogeneity of the evolving aggregates. Only a few colloidal suspensions with industrial relevance contain just one particulate phase (e.g. CMP slurries); more typical are products that are composed of several dispersed phases (e.g. ceramic slurries, paints, or food products). Therefore, distinction should be made between the aggregation of different components (heteroaggregation) and that within one dispersed phase (homo-aggregation).

Similarly, one may be interested in the adhesion of colloidal particles to macroscopic surfaces (e.g. vessels, tubings, or environmental objects like soil grains).

The following paragraphs discuss the interdependency between particles and solvent properties and the effect of interfacial changes on selected examples. Again, they deal with pyrogenic powders, which are composed of colloidal aggregates with nanosized primary particles (i.e. the specific surface is relatively large). The stability of multicomponent colloidal suspensions will be discussed in the next Sect. (5.2.6).

5.2.5.1 Quantifying Stability as a Function of Solvent Properties

The stability of colloidal suspensions is frequently examined for a certain variety of suspension properties (e.g. solid content, liquid phase, concentration of ionic, or polymeric additives). That is, for instance, relevant for developing suspension formulas and preparation procedures for particle characterisation or for predicting the particle behaviour in environmental milieus. A typical problem of such parameter studies is that a variation in the concentration of the charge determining additive (e.g. a pH variation as in Fig. 5.10, p. 261) coincides with a significant variation of the total electrolyte content. This problem is most pronounced for dense suspensions ($\varphi_V \geq 1$ vol%) of very small particles (or particles with a high specific surface area).

To illustrate this effect, Fig. 5.11 depicts the total ionic strength in aqueous suspensions of fumed silica (5 wt%, 150 m^2/g), the pH of which varied between 2 and 10. The original pH of the suspensions after dispersing was approximately 4—depending slightly on the concentration of the background electrolyte (KBr). The figure clearly reveals, that a "salt-free" background could not be maintained when studying the influence of pH. A roughly uniform electrolyte background for all samples from pH 2 to pH 10 can be only guaranteed when the initial concentration of the background electrolyte was raised to 0.01 mol/L (Paciejewska 2010). The interdependency of the interfacial and solvent properties does not only exist when the surface charge originates from the dissociation of functional surface groups. Similar considerations are also necessary when ionic surfactants are used to adjust

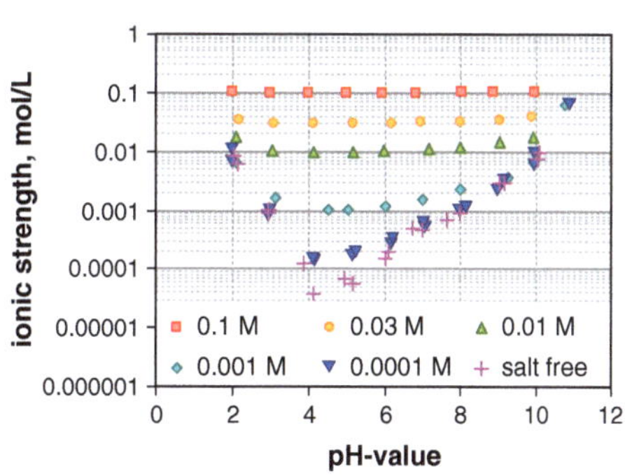

Fig. 5.11 Ionic strength in dense SiO_2 suspensions (150 m^2/g, 5 wt% in KBr solutions); reproduced from Paciejewska (2010, p. 51); copyright 2010, with permission from K. Paciejewska

the surface charge. This is because the adsorption isotherm for ionic surfactants in a salt-free medium differs from that in the case of excess electrolytes (Lyklema 2000, pp. 4.83-4.87). Last but not least, it should be noted that in the case of highly dilute suspensions ($\varphi_V \ll 1$ vol%) the impact of the solid phase on the solvent properties can be neglected and, thus, allow the independent variation of electrolyte content and pH over a much broader range. In that regard it is difficult to compare stability data of dense suspensions with the aggregation of isolated colloidal particles.

5.2.5.2 Effect of Sample Age on the Results of Stability Characterisation

Any qualitative or quantitative evaluation of suspension stability has to refer to a certain time period—either directly (e.g. stability for a certain number of days) or indirectly (e.g. as rate of aggregation). Moreover, the various microscopic and macroscopic suspension properties that are employed for stability characterisation exhibit different sensitivities for incipient and advanced aggregation processes or even phase transition (cf. Sect. 5.2.4). For that reason, it is difficult to compare different stability measurements or to transfer results from one experimental setup to another.

An illustration of this difficulty is given by Paciejewska (2010), who examined the changes of the sedimentation and the optical behaviour of fumed silica suspensions within the first week of aging (Paciejewska 2010, pp. 83–87). These suspensions (5 wt%) were prepared in aqueous solutions of KBr (0.01 M) with a uniform dispersion procedure. The pH value of the suspensions varied between 2 and 10. They were characterised with an analytical photocentrifuge, which reveals the settling behaviour, the sediment porosity and the initial sample transmission. Additionally, the zeta-potential was measured parallel to each sedimentation experiment. Figure 5.12 reproduces the zeta-potential curves, which were obtained for sample ages of 6, 24 h, and 7 days. The three curves are practically identical, which indicates that the electric double layer reached its equilibrium rapidly (i.e. within 6 h). Accordingly, the particle interaction and the tendency towards

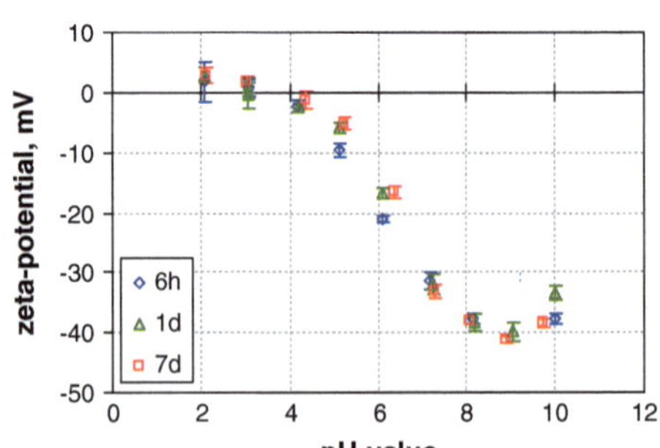

Fig. 5.12 Impact of sampling age on the zeta-potential of fumed SiO_2, 5 wt% in 0.01 M KBr-solution, error bars reflect the standard deviation of 5 measurements; reproduced from Paciejewska (2010, p. 84); copyright 2010, with permission from K. Paciejewska

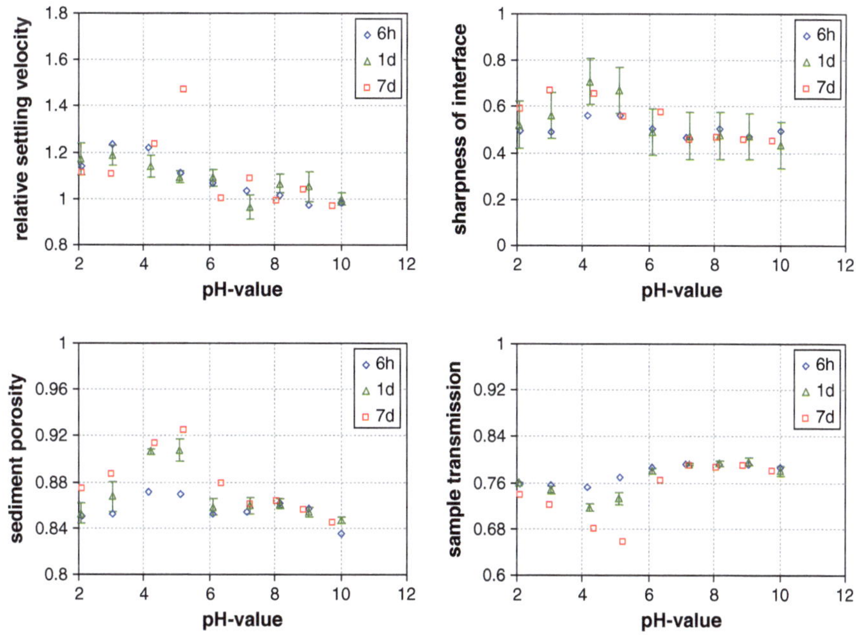

Fig. 5.13 Impact of sampling age on the settling and optical properties of dense fumed SiO_2 suspensions; 5 wt% in 0.01 M KBr solution; reproduced from Paciejewska (2010, pp. 84–85); copyright 2010, with permission from K. Paciejewska

aggregation should remain stable for the whole period of observation. The curves show low values of the zeta-potential for pH < 6 with an IEP in the pH-range from 3 to 4.

The results of the corresponding measurements with the analytical photocentrifuge are depicted in Fig. 5.13. Error bars are included for 24 h data, which reflect the standard deviation of 3 series of measurements with independently prepared stock suspensions.

After an aging time of 6 h, one observes a slight difference between the sample transmission in acid milieu (pH < 6) and the one for neutral and basic pH values, which may indicate an aggregation process that coincides with low zeta-potential values. Similarly, the initial sett-ling velocities of the suspensions are minimal for basic milieu and significantly larger for pH < 6, with maximum values for pH = 3–4, which corresponds to the IEP. Such a difference between the acid milieu and pH ≥ 6 is also visible, yet not really significant for the interface sharpness and the sediment porosity.

The two remaining curves—for 24 h and for 7 days—clearly reveal that virtually all notable changes occur for pH ≤ 6. With regard to the measurement uncertainty, they are less significant for the initial settling velocity and the interface sharpness than for sediment porosity and sample transmission. In total, the most dramatic changes are observed for pH levels around 5.

In her thesis, Paciejewska (2010) investigated the outstanding stability behaviour at pH 5 in detail. She was able to show that at this pH the sedimentation behaviour changes qualitatively between the first and the seventh day: An aggregation process takes place which coarsens the size distribution and is reflected by slight changes in settling velocities and sample transmission after 1 day. It proceeds within the first week to an extent in which large, highly porous and quickly settling agglomerates are formed. As a result, more than 50 % of the sediment volume (including pore volume) is formed when starting the centrifugation. That differs from all other measurements, which show a slow, steady sediment growth. It is interesting to note that this qualitative difference is not clearly reflected in the effective suspension parameters shown in Fig. 5.13.

The example shows that the various suspension properties specifically reflect the degree of destabilisation. The settling velocity, which directly reflects the increase of particle size during aggregation, is not necessarily the most appropriate parameter for the detection of aggregation because it is affected by other properties as well (e.g. by the effective agglomerate density). Parameters that are sensitive to particle interaction, like sediment porosity, may serve as a useful alternative or supplement.

5.2.6 *Stability of Binary Suspensions*

5.2.6.1 General Remarks on the Stability of Binary Suspensions

When a suspension consists of two particle components—a binary suspension—it can experience homoaggregation of each component and/or heteroaggregation between both components. Only in the absence of both types of aggregation can the suspension be considered microscopically stable. Several technical applications require such a state of total suspension stability, e.g. the preparation of mullite ceramic slurries made of silica and alumina particles (Fisher et al. 2001). Other technical applications, however, benefit from homoaggregation of only one particle component (e.g. the selective separation of multi-component suspensions, Snoswell et al. 2005b) or from heteroaggregation (e.g. sewage treatment, froth flotation of ore slurries, paper manufacturing). Heteroaggregation may be further used to produce core-shell particles (Viravathana and Marr 2000, Sun et al. 2004, Yates et al. 2005, Garcia-Perez et al. 2006, Cerbelaud et al. 2008). It is, therefore, worthwhile to look at the stability of binary suspensions in detail.

The microprocesses in binary suspensions are not only affected by the particle interactions, but also by the size ratio and the mixing ratio of the particulate components. A quantification of the suspension stability needs a separate consideration of the three principal pairs of aggregating materials, which is eventually expressed by three partial stability ratios—W_{11}, W_{22} and W_{12}:

$$W_{ij} = r_{ij,0} \int_{r_{ij,0}}^{\infty} \exp\left(\frac{V_{\mathrm{tot}}^{ij}}{\mathrm{k_B} T}\right) \cdot \frac{\mathrm{d}r}{r^2}. \tag{5.37}$$

The formation of dimers from particles of the components i and j obeys the following kinetic, when dominated by Brownian collisions:

$$\frac{\mathrm{d}c_{N,i+j}}{\mathrm{d}t} = 2(2-\delta_{ij})\pi \cdot c_{N,i}c_{N,j} \cdot r_{ij,0} \cdot (D_i + D_j)/W_{ij}. \tag{5.38}$$

If the particles of both components are approximately equal in size, an effective stability ratio for the whole suspension can be defined (Hogg et al. 1966):

$$\frac{1}{W_{\mathrm{eff}}} = \frac{\varphi_{N,1}^2}{W_{11}} + \frac{\varphi_{N,2}^2}{W_{22}} + \frac{2\varphi_{N,1}\varphi_{N,2}}{W_{12}}, \tag{5.39}$$

where $\varphi_{N,i}$ is the relative number frequency of particles of the component i in the whole particle population. The effective stability ratio is dominated by the least stable combination. It thus indicates the total stability behaviour of the binary suspension. In most instances, however, the partial stability ratios should be preferred because they allow for a differentiated discussion of homo- and heteroaggregation (Hogg et al. 1966).

5.2.6.2 The Stability of Aqueous Suspensions with Two Oxidic Components

In the case of aqueous suspensions of oxide materials, the stability is mainly determined by the van-der-Waals and double layer interactions (DLI) between the particles. The former is always attractive in the event of particle collision. Whether the DLI is attractive or repulsive depends on the signs of the surface charges, the absolute values of the zeta-potential, and the regulation capacity of the double layer. When all particles are identically charged, the DLI is repulsive. The most influential factors on the stability are the pH value, the total electrolyte concentration, and the valency of the ions. The pH range in which a binary suspension remains stable is typically different to the respective stability ranges of the two particle components.

Usually one assumes absolute microscopic stability of the binary system when the zeta-potentials of the two particle components are equal in sign and large in absolute value. He-teroaggregation is expected when both components are oppositely charged, while homoaggregation occurs when the attractive interaction forces between similar particles dominate due to low surface potentials (Raşa et al. 2004).

If both components are weakly charged and if their charges are of opposite sign, they experience homoaggregation as well as heteroaggregation. Simultaneous occurrence of homoaggregation of only one particle component as well as of heteroaggregation result from oppositely charged surfaces with one strongly charged and one weakly charged component. However, if the sign of the surface charge is equal for the two particle components, yet one is weakly charged and the other is strongly charged, then we predominantly observe the homoaggregation of the weakly charged component.

This simplified scheme for the description of particle aggregation is based on the assumption that the interfacial properties of the two components do not affect each other. It additionally ignores the consequences of charge regulation between interacting double layers (cf. Sect. 3.2.2). For instance, it is possible that double layer overlap induces a charge reversal on the surface of the weakly charged component and, thus, promotes heteroaggregation, where double layer repulsion is expected (Lyklema and Duval 2005).

A specific aspect of aggregation in binary suspensions is the morphology of heteroaggregates (cf. López-López et al. 2006). When dealing with suspensions that consist of only one dispersed phase, we know that a significant rate of aggregation coincides with low (microscopic) suspension stability. For binary suspensions the situation is more complex. For instance, heteroaggregation between a huge number of small particles and few coarse particles can result in agglomerates that resemble core-shell particles and that do not agglomerate with each other (Yates et al. 2005). A suspension of such aggregates and excess fine particles may then be considered microscopically stable as well.

Systematic experimental studies on the stability behaviour of binary systems are rare. Most frequently the focus lies on certain aspects, like the morphology of the evolving agglomerates (Snoswell et al. 2005a), or on technological solutions—e.g., for the selective separation of one particle component (Snoswell et al. 2005b)—which means that crucial suspension parameters, like pH or mixing ratio, remain constant during such investigation (cf. Islam et al. 1995 and references therein).

One of the few exceptions is a paper by Healy et al. (1973) which examined the influence of pH on the stability of binary suspensions of titania (TiO_2) and alumina (Al_2O_3). It showed that freshly prepared suspensions rapidly coagulated in the pH range between the IEPs of the two components, i.e. the pH range where the two oxides are oppositely charged. 24 h later, such a rapid aggregation could be observed only for a small pH-range around the IEP of alumina. This effect was explained as precipitation of $Al(OH)_3$ on the surface of the titania particles with the Al(III)-ions originating from the dissolution of the Al_2O_3 surface. In later publications, this hypothesis was able to be experimentally substantiated (Wiese and Healy 1975).

This kind of harmonisation of the interfacial properties in binary suspensions is a material-specific phenomenon which is related to the solubility of the dispersed phases and to the corresponding dissolution kinetic. However, the effect is not always observed. Ramakrishan et al. (1998) and Yates et al. (2005) also studied the heteroaggregation of alumina with other oxides (ZrO_2, SiO_2), without observing interfacial harmonisation and subsequent stabilisation. Similarly, Snoswell et al. (2005a) and Golikova and Chernoberezhskii (2005), who looked at the heteroaggregation of amorphous silica (relatively high solubility) with other solid particles, did not report on interfacial harmonisation or aging phenomena. In 2010, Paciejewska published a comprehensive study in which the stability of binary suspensions was examined both theoretically and experimentally, and where the conclusions of Healy et al. (1973) were explicitly addressed. The following part reproduces some key results of that study.

5.2.6.3 Stability for TiO_2/SiO_2 Binary Suspensions

Paciejewska (2010) investigated the stability behaviour of aqueous suspensions of pyrogenic powders. The study included binary suspensions from fumed silica (amorphous SiO_2) and pyrogenic titania (TiO_2—rutile/anatase mixture), which are composed of fractal aggregates of nanoparticles (cf. Sect. 4.5) and, accordingly, have high specific surface areas (in that case: 150 m^2/g and 50 m^2/g, respectively). The stability analysis was conducted for rather dense suspensions (standard solid content: 5 wt%). This, as well as the high specific surface area, led to a strong impact of the double layer formation (i.e. (de-)protonation of hydroxyl groups, ion adsorption, etc.) on the bulk properties of the continuous phase (i.e. pH value and ion concentration): In order to maintain a uniform ionic strength for the desired broad pH range (2–10), a relatively high concentration of background electrolyte was necessary (0.01 M KBr, cf. Figure 5.11). The stability of these suspensions was analysed by means of optical centrifugation analysis (cf. pp. 240–243).

Additional to the sedimentation behaviour, the zeta-potential was measured for each suspension. This was conducted by means of an electroacoustic measurement technique (Sect. 2.3.7.2). The technique yields an effective zeta-potential of the binary suspension, which is calculated from the electroacoustic raw signals by assuming effective particle properties (e.g. for the permittivity and density of the solid material). When the two particle components contribute independently to the electroacoustic signal and do not affect each other with regard to the interfacial properties, it is possible to calculate the effective zeta-potential from the zeta-potentials of the single components. The comparison between such calculated zeta-potential values with experimental ones allows a first evaluation of the interfacial phenomena in the binary suspension.

Figure 5.14 shows the experimental zeta-potentials of the binary TiO_2/SiO_2 suspensions for different mixing ratios (weight based) together with the calculated curves, which were derived from the experimental data of the pure components under the assumption of independent contribution to the electroacoustic signal. It can be clearly seen that all data points of the binary suspensions fall into the range that is spanned between the zeta-potential curves of the pure SiO_2 ($pH_{IEP} \approx 4$) and the pure TiO_2 ($pH_{IEP} \approx 7.2$) and that the closer the experimental data points of the mixtures are to the SiO_2-curve, the more SiO_2 the mixtures contain. Moreover, the experimental data of a given mixing ratio are closer to the SiO_2 curve than the calculated curves for independent contribution of SiO_2 and TiO_2 to the electroacoustic signal. For instance, the effective IEP of the mixture 1:1 is approximately pH 4.5, while a value of 6.2 is expected from calculation. The experimental curves of mixing ratio ≤ 1:3 (TiO_2:SiO_2) fall virtually on the pure SiO_2 curve. Hence, the

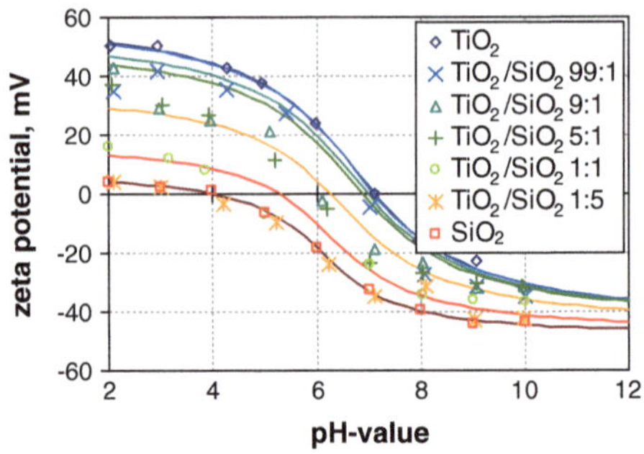

Fig. 5.14 Zeta-potential curves for binary suspensions from TiO_2 and SiO_2 of varying mixing ratio, 5 wt% in 0,01 M KBr-solution, sample age: 24 h; experimental data (*symbols*) and calculation for independent contribution (*solid lines*); reproduced from Paciejewska (2010, p. 99); copyright 2010, with permission from K. Paciejewska

zeta-potential measurements imply a harmonisation of the interfacial properties in such a way that SiO_2 molecules are transferred from SiO_2 surfaces onto the TiO_2 particles by dissolution, diffusion, and adsorption/precipitation.

If such an interfacial harmonisation exists, it has to affect the stability of the binary suspensions, which was evaluated by means of sedimentation. Two key results, the relative initial settling velocity and the total sediment porosity, are shown in Fig. 5.15. It can be seen that for neutral and slightly acidic solutions the sedimentation process is most rapid and yields highly porous sediments. Slowest sedimentation and minimum sediment porosities are observed for low pH values (pH ≤ ° 3). The impact of the suspension composition (mixing ratio TiO_2:SiO_2) is most distinct in the basic pH-range, where an increase of the SiO_2 fraction coincides with a clear decrease of the initial settling velocity. Additionally, it shifts the maximum destabilisation to lower pH-values. The addition of SiO_2 particles to a TiO_2 suspension, therefore, has a stabilising effect for basic pH-values. Even small amounts of SiO_2 can considerably reduce the settling velocity (cf. sample "99:1" at pH 9).

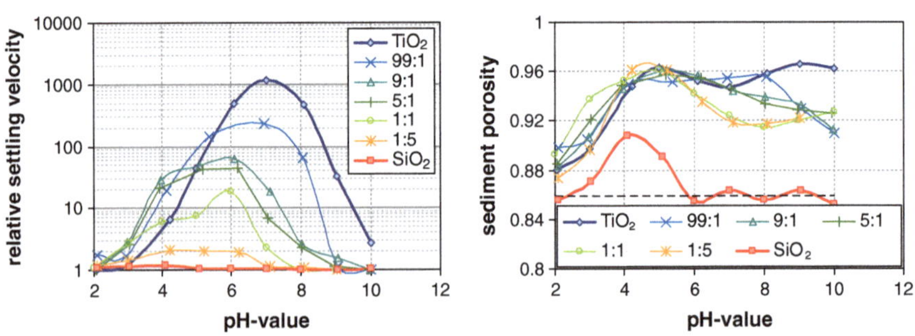

Fig. 5.15 Settling properties (normalised initial settling velocity and sediment porosity) for the suspensions shown in Fig. 5.14, measured with an analytical photocentrifuge at 250 g; reproduced from Paciejewska (2010, p. 102); copyright 2010, with permission from K. Paciejewska

Similarly, the sediment porosities are reduced, yet they do not reach the values of well-stabilised suspensions.

In order to understand the effect of SiO_2 on the suspension stability, Paciejewska (2010) examined TiO_2 suspensions which were prepared by suspending TiO_2 particles in the particle-free supernatant of aged SiO_2 suspensions. In addition to the electrokinetic and sedimentation measurements, she determined the concentration of dissolved silicic acid ($Si(OH)_4$) before and after adding the TiO_2 powder. She was able to show that a major part of the dissolved SiO_2 (3.3 mmol/L, 35 °C at pH ≈ 9.5) was adsorbed on the TiO_2 surface. For a suspension of 4.5 wt% TiO_2, which corresponds to a 5 wt% binary suspension with the mixing ratio TiO_2. SiO_2 = 9:1, this means that a maximum surface coverage of 1.4 μmol $Si(OH)_4$ per m^2 TiO_2 (with 50 m^2/g) or 0.855 molecules per nm^2 can be achieved. This is less than the hydroxyl concentrations of completely hydroxylised SiO_2 and TiO_2 surfaces (5 mol./nm^2 and 8 mol./nm^2, acc. to Jal et al. 2004 and Pelton et al. 2006). Hence, when dispersed in SiO_2 supernatant, the TiO_2 particles were only partially covered with adsorbed $Si(OH)_4$ (i.e. dissolved SiO_2).

Even so, the zeta-potential curve of the thus prepared TiO_2 suspensions significantly deviates from that of the pure TiO_2 suspension and resembles the one of the "9:1" binary suspensions—in particular with respect to the IEP (Fig. 5.16). This can be explained with the different dissociation of silanol (Si–OH) and titanol (Ti–OH) groups.

For instance, the surface charge densities of rutile TiO_2 and fumed SiO_2 at pH 9 are 7.5 μC/cm^2 and 40 μC/cm^2, respectively. These values correspond to 0.5 and 2.4 deprotonated hydroxyl groups per nm^2 (Lyklema 1995, pp. 3.159). Hence, at pH 9 the TiO_2 surface contributes less effectively to the surface charge, and thus to the zeta-potential, than the adsorbed silicic acid. In the acid milieu, the adsorbed silicic acid is only weakly de-/protonated (the IEP of the SiO_2 suspension is at pH 4) and, thus, slightly reduces the positive net charge of the TiO_2 surface.

The impact of the dissolved silica (i.e. silicic acid) on the zeta-potential of the TiO_2 particles has according consequences for the stability of the suspension prepared in supernatant (Fig. 5.17). In the acid pH range, the reduced value of the

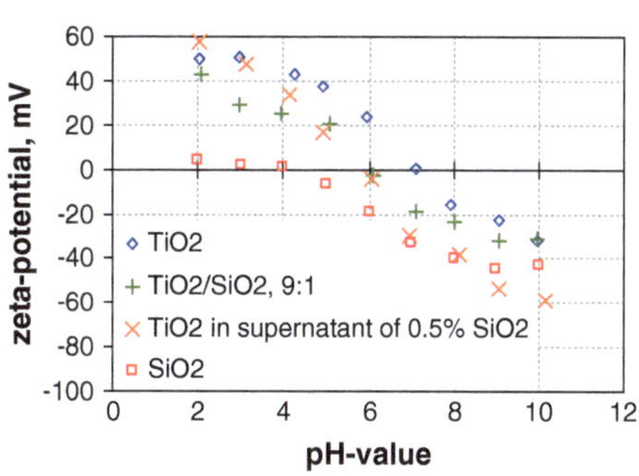

Fig. 5.16 Zeta-potential curves of different suspensions from TiO_2 and SiO_2, 5 wt% in 0.01 M KBr-solution, sample age: 24 h; reproduced from Paciejewska (2010, p. 124); copyright 2010, with permission from K. Paciejewska

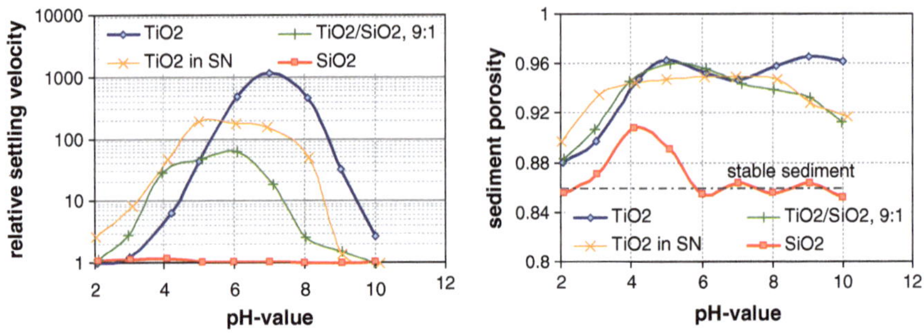

Fig. 5.17 Settling properties (normalised initial settling velocity and sediment porosity) for the suspensions shown in Fig. 5.16, measured with an analytical photocentrifuge at 250 g; reproduced from Paciejewska (2010, p. 125); copyright 2010, with permission from K. Paciejewska

(positive) zeta-potential promotes agglomeration, which results in an increased settling velocity and higher sediment porosity. Different to this, the dissolved silica increases the negative surface charge under basic conditions and, thus, decelerates agglomeration. Accordingly, the settling velocity and sediment porosity are reduced. Again, this behaviour is similar to the one of the binary "9:1" suspension.

Hence, the characterisation of the TiO_2 suspensions that were prepared in particle-free supernatant of SiO_2 suspensions, i.e., in saturated solutions of silicic acid, supports the idea of interfacial harmonisation in binary suspensions. However, the two situations differ in one important aspect: The addition of TiO_2 to a particle-free, saturated "silica solution" immediately leads to undersaturation and a restriction of the maximum surface concentration. In binary suspensions the SiO_2 particles act as a permanent source for dissolved silicic acid. Consequently, a complete surface coverage with adsorbed silicic acid or even precipitated silica can be achieved—provided that the SiO_2 content is sufficiently high and that the suspension had enough time to achieve an equilibrium state.

The dissolution of SiO_2 and the subsequent adsorption and precipitation of silicic acid on the TiO_2 surface are chemical reactions with a characteristic kinetic. Paciejewska was able to demonstrate that the stabilising effect of SiO_2 on the binary suspensions did not exist immediately after the suspension preparation, but was established 24 h later, which is in accordance with the above mentioned observations by Healy et al. (1973) on binary TiO_2-Al_2O_3 suspensions. That is, the effect of sample age on the measurable suspensions stability is twofold. It arises from the fact that coagulation affects the measurable suspension properties only after having reached a certain stage and additionally from the fact that the interface of suspended particles reaches their chemical equilibrium (e.g. with regard to dissociation of surface groups or the adsorption of ions and polymers) only after finite periods of time. The latter means that suspension stability is not only a function of the present suspension parameters (e.g. pH, or ionic strength), but can be considerably affected by previous processing steps (including dispersing procedure, temperature, or pH). In this respect, there is an influence of the "suspension history".

The stability of binary SiO_2-TiO_2 suspensions can be, for instance, clearly affected when the suspensions are first stored at basic pH values (≥9) for at least 24 h, before the final pH in the neutral or acid range is adjusted (Paciejewska 2010, pp. 124–126).

The main conclusion of Paciejewska's thesis is the necessity to consider the specific kinetics of interfacial phenomena when evaluating the stability of colloidal suspensions. This applies not only to binary, but to all kinds of colloidal suspensions. A major factor is the dissolution of the dispersed phase(s)—in particular if the solubility and the intrinsic dissolution rate are relatively large. Its relevance is especially pronounced for a large total surface area, which depends on the particle concentration and the specific surface of the particles and which determines the amount of substance that can be dissolved in a given period of time. For many nanoparticle ($x < 100$ nm) systems (e.g. additives for paints and coatings), it will not be permissible to ignore the influence of dissolution on the interfacial properties and even on suspension stability—independent from the Gibbs-Thomson effect, which becomes relevant at particle sizes below 10 nm (cf. Sect. 3.1.4).

References

Agglomerate Strength and Dispersion Mechanisms

J. Bałdyga, Ł. Makowski, W. Orciuch, C. Sauter, H.P. Schuchmann, Deagglomeration processes in high-shear devices. Chem. Eng. Res. Des. **86**(12), 1369–1381 (2008). doi:10.1016/j.cherd.2008.08.016

J. Bałdyga, Ł. Makowski, W. Orciuch, C. Sauter, H. P. Schuchmann, Agglomerate dispersion in cavitating flows. Chem. Eng. Res. Des. **87**(4), 474–484 (2009). doi:10.1016/j.cherd.2008.12.015

D.G. Bika, M. Gentzler, J.N. Michaels, Mechanical properties of agglomerates. Powder Technol. **117**(1–2), 98–112 (2001). doi:10.1016/S0032-5910(01)00318-7

S. Blaser, Flocs in shear and strain flows. J. Colloid Interface Sci. **225**(2), 273–284 (2000). doi: 10.1006/jcis.1999.6671

S. Blaser, Forces on the surface of small ellipsoidal particles immersed in a linear flow field. Chem. Eng. Sci. **57**(3), 515–526 (2002). doi:10.1016/S0009-2509(01)00389-X

F.P. Bowden, J.H. Brunton, The deformation of solids by liquid impact at supersonic speeds. Proc. R. Soc. London A **263**(1315), 433–450 (1961). doi:10.1098/rspa.1961.0172

L.A. Crum, Cavitation microjets as a contributory mechanism for renal calculi disintegration in ESWL. J. Urol. **140**(6), 1587–1590 (1988)

S.J. Doktycz, K.S. Suslick, Interparticle collisions driven by ultrasound. Science **247**(4946), 1067–1069 (1990). doi:10.1126/science.2309118

H.-J. Henzler, A. Biedermann, Modelluntersuchungen zur Partikelbeanspruchung in Reaktoren. Chem. Ing. Tech. **68**(12), 1546–1561 (1996). doi:10.1002/cite.330681205

K. Higashitani, K. Iimura, H. Sanda, Simulation of deformation and breakupof large aggregates in flows of viscous fluids. Chem. Eng. Sci. **56**(9), 2927–2938 (2001). doi:10.1016/S0009-2509(00)00477-2

G.B. Jeffery, The motion of ellipsoidal particles immersed in a viscous fluid. Proc. R. Soc. London A **102**(715), 161–179 (1922). doi:10.1098/rspa.1922.0078

H. Karbstein, H. Schubert, Developments in the continuous mechanical production of oil-in-water macro-emulsions. Chem. Eng. Process. **34**(3), 205–211 (1995). doi:10.1016/0255-2701(94)04005-2

K. Kendall, N.M. Alford, J.D. Birchall, Elasticity of particle assemblies as a measure of the surface energy of solids. Proc. R. Soc. London A **412**(1843), 269–283 (1987). doi:10.1098/rspa.1987.0089

K. Kendall, Agglomerate strength. Powder Metall. **31**(1), 28–31 (1988)

M. Kobayashi, Breakup and strength of polystyrene latex flocs subjected to a converging flow. Colloids Surf. A 235(1–3), 73–78 (2004). doi:10.1016/j.colsurfa.2004.01.008

A.N. Kolmogorov, Die lokale Struktur der Turbulenz in einer inkompressiblen zähen Flüssigkeit bei sehr großen Reynoldszahlen, in *Sammelband zur statistischen Theorie der Turbulenz*, eds. by A.N. Kolmogorov, H. Limberg, H. Goering (Akademie-Verlag, Berlin, 1958), pp. 71–76

J. Krekel, Zerkleinerung von Agglomeraten in Scherströmungen mit besonders hoher Schubspannung. Chem. Ing. Tech. **38**(3), 229–234 (1966). doi:10.1002/cite.330380307

K.A. Kusters, S.E. Pratsinis, S.G. Thoma, D.M. Smith, Ultrasonic fragmentation of agglomerate powders. Chem. Eng. Sci. **48**(24), 4119–4127 (1993). doi:10.1016/0009-2509(93)80258-R

K.A. Kusters, S.E. Pratsinis, S.G. Thoma, D.M. Smith, Energy-size reduction laws for ultrasonic fragmentation. Powder Technol. **80**(3), 253–263 (1994). doi:10.1016/0032-5910(94)02852-4

A. Kwade, J. Schwedes, Breaking characteristics of different materials and their effect on stress intensity and stress number in stirred media mills. Powder Technol. **122**(2–3), 109–121 (2002). doi: 10.1016/S0032-5910(01)00406-5

F. Liepe, Grundvorgänge und Mikroprozesse in turbulenten Strömungen, in *Handbuch der Mechanischen Verfahrenstechnik, Bd. 1*; Chapter 3.3, ed. by Schubert (Wiley-VCH, Weinheim, 2003), pp. 153–183

J.K. Mackenzie, The elastic constants of a solid containing spherical holes. Proc. Phys. Soc. B **63**(1), 2–11 (1950). doi:10.1088/0370-1301/63/1/302

S. Mende, F. Stenger, W. Peukert, J. Schwedes, Production of sub-micron particles by wet comminution in stirred media mills. J. Mat. Sci. **39**(16–17), 5223–5226 (2004). doi:10.1023/B:JMSC.0000039214.12131.58

O. Nagel, H. Kürten, Untersuchungen zum Dispergieren im turbulenten Scherfeld. Chem. Ing. Tech. **48**(6), 513–519 (1976). doi:10.1002/cite.330480604

H. Oertel (ed.), *Prandtl's essentials of fluid mechancs*, . In series: *Applied Mathematical Sciences*, vol. 158, 2nd ed. eds. by S.S. Antman, J.E. Mardsen, L. Sirovich (Springer, New York, 2004)

B. Park, D. Smith, S. Thoma, Determination of agglomerate strength distributions: part 4. Analysis of multimodal particle size distributions. Powder Technol. **76**(2), 125–133 (1993). doi:10.1016/S0032-5910(05)80019-1

M. Pohl, S. Hogekamp, N.Q. Hoffmann, H.P. Schuchmann, Dispergieren und Desagglomerieren von Nanopartikeln mit Ultraschall. Chem. Ing. Tech. **76**(4), 392–396 (2004). doi:10.1002/cite.200403371

J. Raasch, Die Bewegung und Beanspruchung kugelförmiger und zylindrischer Teilchen in zähen Scherströmungen, in *1. Europäische Symposium Zerkleinern, Frankfurt a.M., (Proceedings)*, eds. by H. Rumpf, D. Behrens (Verlag Chemie, Weinheim, 1962), pp. 138–150

H. Reichert, Desagglomeration organischer Farbpigmente in Scherströmungen hochzäher Flüssigkeiten. Chem. Ing. Tech. **45**(6), 391–395 (1973). doi:10.1002/cite.330450612

H. Rumpf, Zur Theorie der Zugfestigkeit von Agglomeraten bei Kraftübertragung an Kontaktpunkten. Chem. Ing. Techn. **42**(8), 538–540 (1970)

S.P. Rwei, I. Manas-Zloczower, D.L. Feke, Observation of carbon black agglomerate dispersion in simple shear flows. Polym. Eng. Sci. **30**(12), 701–706 (1990). doi:10.1002/pen.760301202

S. Sen, M.L. Ram, S. Roy, B.K. Sarkar, The structural transformation of anatase TiO_2 by high-energy vibrational ball milling. J. Mater. Res. **14**(3), 841–848 (1999). doi:10.1557/JMR.1999.0112

W.O. Smith, P.D. Foote, P.F. Busang, Packing of homogeneous spheres. Phys. Rev. **34**(9), 1271–1274 (1929). doi:10.1103/PhysRev.34.1271

F. Stenger, M. Götzinger, P. Jakob, W. Peukert, Mechano-chemical changes of nano sized alpha-Al_2O_3 during wet dispersion in stirred ball mills. Part. Part. Syst. Charact. **21**(1), 31–38 (2004). doi:10.1002/ppsc.200400902

N. Tambo, H. Hozumi, Physical characteristics of flocs – II. Strength of floc. Water Res. **13**(5), 421–427 (1979). doi:10.1016/0043-1354(79)90034-4
S. Tang, Y. Ma, C. Shiu, Modelling of mechanical strength of fractal aggregates. Colloids Surf. A **180**(1–2), 7–16 (2001). doi:10.1016/S0927-7757(03)00743-3
G.I. Taylor, The formation of emulsions in definable fields of flow. Proc. R. Soc. Lond. A **146** (858), 501–523 (1934). doi:10.1098/rspa.1934.0169
S.G. Thoma, M. Ciftcioglu, D.M. Smith, Determination of agglomerate strength distributions. Part 1. Calibration via ultrasonic forces. Powder Technol. **68**(1), 53–61 (1991). doi:10.1016/0032-5910(91)80063-O
B.J. Trevelyan, S.G. Mason, Particle motions in sheared suspensions. I. Rotations. J. Colloid Sci. **6** (4), 354–367 (1951). doi:10.1016/0095-8522(51)90005-0
O. Vasylkiv, Y. Sakka, Synthesis and colloidal processing of zirconia nanopowder. J. Am. Ceram. Soc. **84**(11), 2489–2494 (2001). doi:10.1111/j.1151-2916.2001.tb01041.x
E. Webster, Cavitation. Ultrasonics **1**(1), 39–48 (1963). doi:10.1016/0041-624X(63)90116-1

Dispersion Techniques and Dispersion of Pyrogenic Powders

J. Bałdyga, Ł. Makowski, W. Orciuch, C. Sauter, H.P. Schuchmann, Deagglomeration processes in high-shear devices. Chem. Eng. Res. Design **86**(12), 1369–1381 (2008). doi:10.1016/j.cherd.2008.08.016
J. Bałdyga, Ł. Makowski, W. Orciuch, C. Sauter, H. P. Schuchmann, Agglomerate dispersion in cavitating flows. Chem. Eng. Res. Des. **87**(4), 474–484 (2009). doi:10.1016/j.cherd.2008.12.015
H.-J. Henzler, A. Biedermann, Modelluntersuchungen zur Partikelbeanspruchung in Reaktoren. Chem. Ing. Tech. **68**(12), 1546–1561 (1996). doi:10.1002/cite.330681205
H. Karbstein, H. Schubert, Developments in the continuous mechanical production of oil-in-water macro-emulsions. Chem. Eng. Process. **34**(3), 205–211 (1995). doi:10.1016/0255-2701(94)04005-2
M. Kobayashi, Y. Adachi, S. Ooi, Breakup of fractal flocs in a turbulent flow. Langmuir **15**(13), 4351–4356 (1999). doi:10.1021/la980763o
M. Kobayashi, Breakup and strength of polystyrene latex flocs subjected to a converging flow. Colloids Surf. A 235(1–3), 73–78 (2004). doi:10.1016/j.colsurfa.2004.01.008
T. Kuntzsch, *Erfassung und Beeinflussung des Zustandes von Poliersuspensionen für das chemisch-mechanische Polieren (CMP) in der Halbleiterbauelementfertigung*. PhD thesis, Technische Universität Dresden, 2004
K.A. Kusters, S.E. Pratsinis, S.G. Thoma, D.M. Smith, Ultrasonic fragmentation of agglomerate powders. Chem. Eng. Sci. **48**(24), 4119–4127 (1993). doi:10.1016/0009-2509(93)80258-R
A. Kwade, J. Schwedes, Breaking characteristics of different materials and their effect on stress intensity and stress number in stirred media mills. Powder Technol. **122**(2–3), 109–121 (2002). doi:10.1016/S0032-5910(01)00406-5
N. Mandzy, E. Grulke, T. Druffel, Breakage of TiO_2 agglomerates in electrostatically stabilized aqueous dispersions. Powder Technol. **160**(2), 121–126 (2005). doi:10.1016/j.powtec.2005.08.020
S. Mende, F. Stenger, W. Peukert, J. Schwedes, Production of sub-micron particles by wet comminution in stirred media mills. J. Mat. Sci. **39**(16–17), 5223–5226 (2004). doi:10.1023/B:JMSC.0000039214.12131.58
O. Nagel, H. Kürten, Untersuchungen zum Dispergieren im turbulenten Scherfeld. Chem. Ing. Tech. **48**(6), 513–519 (1976). doi:10.1002/cite.330480604
B. Park, D. Smith, S. Thoma, Determination of agglomerate strength distributions: part 4. Analysis of multimodal particle size distributions. Powder Technol. **76**(2), 125–133 (1993). doi:10.1016/S0032-5910(05)80019-1

M. Pohl, S. Hogekamp, N.Q. Hoffmann, H.P. Schuchmann, Dispergieren und Desagglomerieren von Nanopartikeln mit Ultraschall. Chem. Ing. Tech. **76**(4), 392–396 (2004). doi:10.1002/cite.200403371

M. Pohl, *Benetzen und Dispergieren nativer und gezielt agglomerierter pyrogener oxidischer Partikel*. PhD thesis, Universität Fridericiana Karlsruhe (TH), 2005

M. Pohl, H. Schubert, H.P. Schuchmann, Herstellung stabiler Dispersionen aus pyrogener Kieselsäure. Chem. Ing. Tech. **77**(3), 258–262 (2005). doi:10.1002/cite200407020

H. Reichert, Desagglomeration organischer Farbpigmente in Scherströmungen hochzäher Flüssigkeiten. Chem. Ing. Tech. **45**(6), 391–395 (1973). doi:10.1002/cite.330450612

J.F. Richardson, J.H. Harker, *Coulson and Richardson's chemical engineering, vol. 2: particle technology and separation processes*, 5th ed. (Butterworth-Heinemann, Oxford, 2002). ISBN: 0 7506 4445 1

S.P. Rwei, I. Manas-Zloczower, D.L. Feke, Observation of carbon black agglomerate dispersion in simple shear flows. Polym. Eng. Sci. **30**(12), 701–706 (1990). doi:10.1002/pen.760301202

S.P. Rwei, I. Manas-Zloczower, D.L. Feke, Characterization of agglomerate dispersion by erosion in simple shear flows. Polym. Eng. Sci. **31**(8), 558–562 (1991). doi:10.1002/pen.760310804 see also: Correction. Polym. Eng. Sci. **31**(20), 1508 (1991)

C. Sauter, H.P. Schuchmann, Materialschonendes Hochdruckdispergieren mit dem high pressure post feeding (HPPF)-system. Chem. Ing. Techn. **80**(3), 365–372 (2008). doi:10.1002/cite.200700149

C. Sauter, M.A. Emin, H.P. Schuchmann, S. Tavman, Influence of hydrostatic pressure and sound amplitude on the ultrasound induced dispersion and de-agglomeration of nanoparticles. Ultrasonics Sonochem. **15**(4), 517–523 (2008). doi:10.1016/j.ultsonch.2007.08.010

K. Schönert, Zerkleinern, in *Handbuch der Mechanischen Verfahrenstechnik, Bd. 1*, ed. by H. Schubert (Wiley-VCH, Weinheim, 2003), pp. 299–382

H.P. Schuchmann, T. Danner, Emulgieren: Mehr als nur Zerkleinern. Chem. Ing. Tech. **76**(4), 364–375 (2004). doi:10.1002/cite.200406163

M. Sommer, F. Stenger, W. Peukert, N.J. Wagner, Agglomeration and breakage of nanoparticles in stirred media mills—a comparison of different methods and models. Chem. Eng. Sci. **61**(1), 135–148 (2006). doi:10.1016/j.ces.2004.12.057

W. Süß, H. Hanke, Über die Bedeutung von Kavitation und Grenzflächenreibung bei der Dispergierung suspendierter Feststoffe durch Ultraschall. Pharmazie **24**(5), 270–272 (1969)

N. Tambo, H. Hozumi, Physical characteristics of flocs—II. Strength of floc. Water Res. **13**(5), 421–427 (1979). doi:10.1016/0043-1354(79)90034-4

A. Teleki, R. Wengeler, L. Wengeler, H. Nirschl, S.E. Pratsinis, Distinguishing between aggregates and agglomerates of flame-made TiO2 by high-pressure dispersion. Powder Technol. **181**(3), 292–300 (2008). doi:10.1016/j.powtec.2007.05.016

S.G. Thoma, M. Ciftcioglu, D.M. Smith, Determination of agglomerate strength distributions. Part 1. Calibration via ultrasonic forces. Powder Technol. **68**(1), 53–61 (1991). doi:10.1016/0032-5910(91)80063-O

K. Urban, G. Wagner, D. Schaffner, D. Röglin, J. Ulrich, Rotor-stator and disc systems for emulsification processes. Chem. Eng. Technol. **29**(1), 24–31 (2006). doi:10.1002/ceat.200500304

R. Wengeler, F. Ruslim, H. Nirschl, T. Merkel, Dispergierung feindisperser Agglomerate mit Mikro-Dispergierelementen. Chem. Ing. Tech. **76**(5), 659–662 (2004). doi:10.1002/cite.200403391

R. Wengeler, A. Teleki, M. Vetter, S.E. Pratsinis, H. Nirschl, High-pressure liquid dispersion and fragmentation of flame-made silica agglomerates. Langmuir **22**(11), 4928–4935 (2006). doi:10.1021/la053283n

Characterisation of Dispersion Processes

F. Babick, M. Stintz, H. Barthel, M. Heinemann, Characterization of size and structure of fumed silica particles in suspensions, in *Organosilicon Chemistry VI—From Molecule to Material*, eds. by N. Auner, J. Weis (VCH-Wiley, Weinheim, 2005), pp. 882–887. ISBN-10: 3-527-31214-5

J. Bałdyga, Ł. Makowski, W. Orciuch, C. Sauter, H.P. Schuchmann, Agglomerate dispersion in cavitating flows. Chem. Eng. Res. Des. **87**(4), 474–484 (2009). doi:10.1016/j.cherd.2008.12.015

F. Hinze, *Beitrag zur Charakterisierung hochkonzentrierter Suspensionen mittels Ultraschallspektroskopie*. PhD thesis, Technische Universität Dresden, 2001

ISO 1524, *Paints, varnishes and printing inks—Determination of fineness of grind* (Beuth-Verlag Berlin, 2002)

T. Kuntzsch, U. Witnik, M. Hollatz, M. Stintz, S. Ripperger, Characterization of slurries used for chemical-mechanical polishing (CMP) in the semiconductor industry. Chem. Eng. Technol. **26** (12), 1235–1239 (2003). doi:10.1002/ceat.200303050

M. Maier, B. Hannebauer, H. Holldorff, P. Albers, Does lung surfactant promote disaggregation of nanostructured titanium dioxide? J. Occup. Environ. Med. **48**(12), 1314–1320 (2006). doi:10.1097/01.jom.0000215405.72714.b2

G. Mie, Beiträge zur Optik trüber Medien, speziell kolloidaler Goldlösungen, Ann. Phys. IV **25**(3), 377–445 (1908). doi:10.1002/andp.19083300302

A. Nogowski, F. Babick, M. Stintz, Several methods to quantify the dispersion progress of concentrated suspensions. On the CD-ROM: *WCPT6 2010, World Congress on Particle Technology, Nuremberg*, 26–29 Apr 2010. ISBN: 978-3-00-030570-2 (available from NürnbergMesse GmbH, Messezentrum, 90471 Nürnberg, Germany), paper 00284

M. Pohl, S. Hogekamp, N.Q. Hoffmann, H.P. Schuchmann, Dispergieren und Desagglomerieren von Nanopartikeln mit Ultraschall. Chem. Ing. Tech. **76**(4), 392–396 (2004). doi:10.1002/cite.200403371

M. Pohl, H. Schubert, H.P. Schuchmann, Herstellung stabiler Dispersionen aus pyrogener Kieselsäure. Chem. Ing. Tech. **77**(3), 258–262 (2005). doi:10.1002/cite200407020

M. Pohl, *Benetzen und Dispergieren nativer und gezielt agglomerierter pyrogener oxidischer Partikel*. PhD thesis, Universität Fridericiana Karlsruhe (TH), 2005

H. Reichert, Desagglomeration organischer Farbpigmente in Scherströmungen hochzäher Flüssigkeiten. Chem. Ing. Tech. **45**(6), 391–395 (1973). doi:10.1002/cite.330450612

O. Ruscitti, R. Franke, H. Hahn, F. Babick, T. Richter, M. Stintz, Application of particle measurement technology in process intensification. Chem. Eng. Technol. **31**(2), 270–277 (2008). doi:10.1002/ceat.200700465

K. Schießl, *Erarbeitung eines Konzeptes zur Einführung von einzelpartikelbewertenden Messverfahren für die Qualitätssicherung in der Herstellung von kolloidalen Dispersionen*. Internal Report. Technische Universität Dresden, Institut für Verfahrenstechnik und Umwelttechnik, 2007

M. Sommer, F. Stenger, W. Peukert, N.J. Wagner, Agglomeration and breakage of nanoparticles in stirred media mills—a comparison of different methods and models. Chem. Eng. Sci. **61**(1), 135–148 (2006). doi:10.1016/j.ces.2004.12.057

M. Stintz, F. Hinze, S. Ripperger, Particle size characterization of inorganic colloids by ultrasonic attenuation spectrometry, in *Handbook on ultrasonic and dielectric characterization techniques for suspended particles*, eds. by V.A. Hackley, J. Texter (The American Ceramic Society, 1998), pp. 219–228. ISBN: 1-57498-034-3

W. Süß, H. Hanke, Über die Dispergierung suspendierter Feststoffe durch Ultraschall. Pharmazie, **24**(4), 211–220 (1969)

A. Teleki, R. Wengeler, L. Wengeler, H. Nirschl, S.E. Pratsinis, Distinguishing between aggregates and agglomerates of flame-made TiO_2 by high-pressure dispersion. Powder Technol. **181**(3), 292–300 (2008). doi:10.1016/j.powtec.2007.05.016

S.G. Thoma, M. Ciftcioglu, D.M. Smith, Determination of agglomerate strength distributions. Part 2. Application to model agglomerates. Powder Technol. **68**(1), 63–69 (1991). doi:10.1016/0032-5910(91)80064-P

R. Wengeler, F. Ruslim, H. Nirschl, T. Merkel, Dispergierung feindisperser Agglomerate mit Mikro-Dispergierelementen. Chem. Ing. Tech. **76**(5), 659–662 (2004). doi:10.1002/cite.200403391

R. Wengeler, A. Teleki, M. Vetter, S.E. Pratsinis, H. Nirschl, High-pressure liquid dispersion and fragmentation of flame-made silica agglomerates. Langmuir **22**(11), 4928–4935 (2006). doi:10.1021/la053283n

R. Xu, *Particle characterization: Light scattering methods* (Kluwer Academic Publishers, Dortrecht, 2000). ISBN: 0-7923-6300-0

Basics of Suspension Stability

A. Bleier, C.G. Westmoreland, Effects of pH and particle size on the processing of and the development of microstructure in alumina–zirconia composites. J. Am. Ceram. Soc. **74**(12), 3100–3111 (1991). doi:10.1111/j.1151-2916.1991.tb04307.x

B. Derjaguin, On the repulsive forces between charged colloid particles and on the theory of slow coagulation and stability of lyophobe sols. Trans. Faraday Soc. **35**, 203–215 (1940). doi:10.1039/TF9403500203

B. Derjaguin, L. Landau, Theory of the stability of strongly charged lyophobic sols and of the adhesion of strongly charged-particles in solutions of electrolytes, Prog. Surf. Sci. **43**(1–4), 30–59 (1993). Reprinted from: Acta phys.-chim. **14**(6), 633–662 (1941). doi:10.1016/0079-6816(93)90013-L

M. Faraday, Experimental relations of gold (and other metals) to light, in *Experimental Researches in Chemistry and Physics*, ed. by J.W. Gentry (Taylor and Francis, London, 1857), pp. 391–443, The aerosol science contributions of Michael Faraday. J. Aerosol Sci. **26**(2), 341–349 (1995)

N. Fuchs, Über die Stabilität und Aufladung der Aerosole. Z. Phys. **89**(11–12), 736–743 (1934). doi:10.1007/BF01341386

W.B. Hardy, A preliminary investigation of the conditions which determine the stability of irreversible hydrosols. J. Phys. Chem. **4**(4), 235–253 (1900). doi:10.1021/j150022a001

T.W. Healy, A. Homola, R.O. James, R.J. Hunter, Coagulation of amphoteric latex colloids: reversibility and specific ion effects. Faraday Discuss. Chem. Soc. **65**, 156–163 (1978). doi:10.1039/dc9786500156

R. Hogg, The role of polymer adsorption kinetics in flocculation. Colloids Surf. A **146**(1–3), 253–263 (1999). doi:10.1016/S0927-7757(98)00723-7

E.P. Honig, G.J. Roebersen, P.H. Wiersema, Effect of hydrodynamic interaction on the coagulation rate of hydrophobic colloids. J. Colloid Interface Sci. **36**(1), 97–109 (1971). doi:10.1016/0021-9797(71)90245-1

R.J. Hunter, *Zeta potential in colloid science: principles and applications*. In series: *Colloid science*, vol. 2, 3rd ed. (Academic Press, London, 1988). ISBN: 0-12-361961-0

M. Hütter, Coagulation rates in concentrated colloidal suspensions studied by Brownian dynamics simulation. Phys. Chem. Chem. Phys. **1**(18), 4429–4436 (1999). doi:10.1039/a904615a

D.J. McClements, Theoretical analysis of factors affecting the formation and stability of multilayered colloidal dispersions. Langmuir **21**(21), 9777–9785 (2005). doi:10.1021/la0512603

D.N.L. McGown, G.D. Parfitt, Improved theoretical calculation of the stability ratio for colloidal systems. J. Phys. Chem. **71**(2), 449–450 (1967). doi:10.1021/j100861a041

J.A. Molina-Bolívar, F. Galisteo-González, R. Hidalgo-Álvarez, Colloidal stability of protein-polymer systems: a possible explanation by hydration forces. Phys. Rev. E **55**(4), 4522–4530 (1997). doi:10.1103/PhysRevE.55.4522

P. Mulvaney, L.M. Liz-Marzan, M. Giersig, T. Ung, Silica encapsulation of quantum dots and metal clusters. J. Mater. Chem. **10**(6), 1259–1270 (2000). doi:10.1039/b000136h
A. Oolsen, G. Franks, S. Biggs, G.J. Jameson, An improved collision efficiency model for particle aggregation. J. Chem. Phys. **125**(18), 184906 (2006). doi:10.1063/1.2387172
K. Paciejewska, F. Babick, M. Stintz, R. Lange, Conditioning of binary suspensions of colloidal metal oxides particles. On the CD-ROM: *PARTEC 2007, International Congress for Particle Technology, Nuremberg*, 27–29 Mar 2007 (available from NürnbergMesse GmbH, Messezentrum, 90471 Nürnberg, Germany), paper P12_03
K.M. Paciejewska, *Untersuchung des Stabilitätsverhaltens von binären kolloidalen Suspensionen*. PhD thesis, Technische Universität Dresden, 2010. urn: nbn:de:bsz:14-qucosa-65050
G. Quincke, Ueber die Fortführung materieller Theilchen durch strömende Elektricität. Ann. Phys. II **113**(8), 513–598 (1861)
V. Ramakrishnan, Pradip, S.G. Malghan, The stability of alumina-zirconia suspensions. Colloids Surf. A **133**(1–2), 135–142 (1998). doi:10.1016/S0927-7757(97)00135-0
P. Sandkühler, M. Lattuada, H. Wu, J. Sefcik, M. Morbidelli, Further insights into the universality of colloidal aggregation. Adv. Colloid Interface Sci. **113**(2–3), 65–83 (2005). doi:10.1016/j.cis.2004.12.001
H. Schulze, Schwefelarsen in wässriger Lösung. J. Prakt. Chem. **25**, 431–452 (1882)
H. Schulze, Antimontrisulfid in wässriger Lösung. J. Prakt. Chem. **27**, 320–332 (1883)
R.H. Smellie Jr, V.K. La Mer, Flocculation, subsidence and filtration of phosphate slimes. VI. A quantitative theory of filtration of flocculated suspensions. J. Colloid Sci. **23**(6), 589–599 (1958). doi:10.1016/0095-8522(58)90071-0
M.V. Smoluchowski, Drei Vorträge über Diffusion, Brownsche Molekularbewegung und Koagulation von Kolloidteilchen. Phys. Z. **17**, 557–571, 585–599 (1916)
L.A. Spielman, Viscous interactions in Brownian coagulation. J. Colloid Interface Sci. **33**(4), 562–571 (1970). doi:10.1016/0021-9797(70)90008-1
H. Yotsumoto, R.-H. Yoon, Application of extended DLVO-theory: 1. Stability of rutile suspensions. J. Colloid Interface Sci. **157**(2), 426–433 (1993a). doi:10.1006/jcis.1993.1205
H. Yotsumoto, R.-H. Yoon, Application of extended DLVO-theory: 2. Stability of silica suspensions. J. Colloid Interface Sci. **157**(2), 434–441 (1993b). doi:10.1006/jcis.1993.1206
R. Zsigmondy, Die chemische Natur des Cassiusschen Goldpurpurs. Liebigs Ann. Chem. **301**(2–3), 361–387 (1898)

Experimental Characterisation of Suspension Stability

F. Bakkali, A. Moudden, B. Faiz, A. Amghar, G. Maze, F. Montero de Espinosa, M. Akhnak, Ultrasonic measurement of milk coagulation time. Meas. Sci. Technol. **12**(12), 2154–2159 (2001). doi:10.1088/0957-0233/12/12/317
B.P. Binks, S.O. Lumsdon, Stability of oil-in-water emulsions stabilised by silica particles. Phys. Chem. Chem. Phys. **1**(12), 3007–3016 (1999). doi:10.1039/a902209k
T. Breyer, J. Sonnefeld, Rheological investigation in aqueous soluton of nanosized powders, in *Handling of higly dispersed powders*, eds. by E. Müller, C. Oestreich (Shaker Verlag, Aachen, 2004), pp. 124–131. ISBN: 3-8322-3244-3
H. Eilers, J. Korff, The significance of the phenomenon of the electrical charge on the stability of hydrophobic dispersions. Trans. Faraday Soc. **35**(3), 229–240 (1940). doi:10.1039/tf9403500229
B. Fitch, Sedimentation of flocculent suspensions: State of the art. AIChE J. **25**(6), 913–930 (1979). doi:10.1002/aic.690250602
G.V. Franks, S.B. Johnson, P.J. Scales, D.V. Boger, T.W. Healy, Ion-specific strength of attractive particle networks. Langmuir **15**(13), 4411–4420 (1999). doi:10.1021/la9815345
H. Geers, W. Witt, F. Babick, Stability analysis of emulsions and suspensions with photon cross-correlation spectroscopy. On the CD-ROM: *PARTEC 2007, International Congress for*

Particle Technology, Nuremberg, 27–29 Mar 2007 (available from NürnbergMesse GmbH, Messezentrum, 90471 Nürnberg, Germany), paper P14_08

R. Greenwood, Review of the measurement of zeta potentials in concentrated aqeous suspensions using electroacoustics. Adv. Colloid Interface Sci. **106**(1–3), 55–81 (2003). doi:10.1016/S0001-8686(03)00105-2

J. Gregory, Monitoring particle aggregation processes. Adv. Colloid Interface Sci. **147–148**, 109–123 (2009). doi:10.1016/j.cis.2008.09.003

F. Gruy, Formation of small silica aggregates by turbulent aggregation. J. Colloid Interface Sci. **237**(1), 28–39 (2001). doi:10.1006/jcis.2001.7432

U. Kätzel, T. Richter, M. Stintz, H. Barthel, T. Gottschalk-Gaudig, Phase transitions of pyrogenic silica suspensions: a comparison to model laponite. Phys. Rev. E **76**(3), 031402 (2007). doi:10.1103/PhysRevE.76.031402

P. Kippax, J.D. Sherwood, D.J. McClements, Ultrasonic spectroscopy study of globule aggregation in parenteral fat emulsions containing calcium chloride. Langmuir **15**(5), 1673–1678 (1999). doi:10.1021/la981351i

T. Kuhlbusch, C. Nickel, B. Hellack, S. Gartiser, F. Flach et al., In series: *Fate and behaviour of TiO2 nanomaterials in the environment, influenced by their shape, size and surface area*, ed. by D. Völker, Texte 25/2012 (Umweltbundesamt, Dessau-Roßlau, 2012). ISSN 1862-4804

H. Lichtenfeld, H. Stechemesser, H. Möhwald, Single particle light-scattering photometry—some fields of application. J. Colloid Interface Sci. **276**(1), 97–105 (2004). doi:10.1016/j.jcis.2004.03.038

J. Lyklema, *Fundamentals of Interface and Colloid Science III: Liquid-Fluid Interfaces* (Academic Press, San Diego, 2000). ISBN 0-12-460523-0

S. Manley, L. Cipelletti, V. Trappe, A.E. Bailey, R.J. Christianson et al., Limits to gelation in colloidal aggregation. Phys. Rev. Lett. **93**(10), 108302 (2004). doi:10.1103/PhysRevLett.93.108302

S. Manley, B. Davidovitch, N.R. Davies, L. Cipelletti, A.E. Bailey et al., Time-dependent strength of colloidal gels. Phys. Rev. Lett. **95**(4), 048302 (2005). doi:10.1103/PhysRevLett.95.048302

M.J. Mateo, D.J. O'Callaghan, C.D. Everard, M. Castillo, F.A. Payne, C.P. O'Donnell, Evaluation of on-line optical sensing techniques for monitoring curd moisture content and solids in whey during syneresis. Food Res. Int. **43**(1), 177–182 (2010). doi:10.1016/j.foodres.2009.09.023

S. Mirza, J.F. Richardson, Sedimentation of suspensions of particles of two or more sizes. Chem. Eng. Sci. **34**(4), 447–454 (1979). doi:10.1016/0009-2509(79)85088-5

C. Nickel, B. Hellack, A. Nogowski, F. Babick, M. Stintz, H. Maes, A. Schäffer, T.A. J. Kuhlbusch, In series: *Mobility, fate and behavior of* TiO_2 *nanomaterials in different environmental media*, eds. by D. Völker, P. Apel, Texte 76/2013 (Umweltbundesamt, Dessau-Roßlau, 2013). ISSN 1862-4804

A. Nogowski, F. Babick, C. Nickel, T. A. J. Kuhlbusch, M. Stintz, Size and stability of functionalised titanium dioxide as a function of pH and ionic background electrolyte. On the CD-ROM: *PARTEC 2013, International Congress for Particle Technology,* Nuremberg, 23–25 Apr 2013. ISBN: 978-3-00-040578-5 (available from NürnbergMesse GmbH, Messezentrum, 90471 Nürnberg, Germany), paper 317

D.J. O'Callaghan, C.P. O'Donnell, F.A. Payne, On-line sensing techniques for coagulum setting in renneted milks. J. Food Eng. **43**(3), 155–165 (2000). doi:10.1016/S0260-8774(99)00145-4

K.M. Paciejewska, *Untersuchung des Stabilitätsverhaltens von binären kolloidalen Suspensionen.* PhD thesis, Technische Universität Dresden, 2010. urn: nbn:de:bsz:14-qucosa-65050

D. Quemada, C. Berli, Energy of interaction in colloids and its implications in rheological modeling. Adv. Colloid Interface Sci. **98**(1), 51–85 (2002). doi:10.1016/S0001-8686(01)00093-8

N. Schelero, H. Lichtenfeld, H. Zastrow, H. Möhwald, M. Dubois, T. Zemb, Single particle light scattering method for studying aging properties of Pickering emulsions stabilized by catanionic crystals. Colloids Surf. A **337**(1–3), 146–153 (2009). doi:10.1016/j.colsurfa.2008.12.013

B. Senouci, S. Serfaty, P. Griesmar, M. Gindre, Acoustic resonance in tetramethoxysilane matrices: A new tool to characterize the gel formation. Rev. Sci. Instrum. **72**(4), 2134–2138 (2001). doi:10.1063/1.1355266

T. Sobisch, D. Lerche, T. Detloff, M. Beiser, A. Erk, Tracing the centrifugal separation of fine-particle slurries by Analytical centrifugation. Filtration **6**(4), 313–311 (2006)

L. Steinke, B. Wessely, S. Ripperger, Optische Extinktionsmessverfahren zur Inline-Kontrolle disperser Stoffsysteme. Chem. Ing. Tech. **81**(6), 735–747 (2009). doi:10.1002/cite.200800129

W. Sutananta, D.Q.M. Craig, J.M. Newton, The use of dielectric analysis as a means of characterising the effects of moisture uptake by pharmaceutical glyceride bases. Int. J. Pharm. **132**(1–2), 1–8 (1996). doi:10.1016/0378-5173(95)04226-1

S. Tamburic, D.Q.M. Craig, G. Vuleta, J. Milic, A comparison of electrical and rheological techniques for the characterisation of creams. Int. J. Pharm. **137**(2), 243–248 (1996). doi:10.1016/0378-5173(96)04528-0

H. Tanaka, J. Meunier, D. Bonn, Nonergodic states of charged colloidal suspensions: repulsive and attractive glasses and gels. Phys. Rev. E **69**(3), 031404 (2004). doi:10.1103/PhysRevE.69.031404

W. Vogelsberger, A. Seidel, R. Fuchs, Contribution to the determination of kinetic parameters of the sol-gel transformation by rheological measurements. J. Colloid Interface Sci. **230**(2), 268–271 (2000). doi:10.1006/jcis.2000.7095

J. Weiss, N. Herrmann, D.J. McClements, Ostwald ripening of hydrocarbon emulsion droplets in surfactant solutions. Langmuir **15**(20), 6652–6657 (1999). doi:10.1021/la981739d

B. Wessely, J. Altmann, S. Ripperger, The use of statistical properties of transmission signals for particle characterization. Chem. Eng. Technol. **19**(5), 438–442 (1996). doi:10.1002/ceat.270190509

P.R. Williams, R.L. Williams, R. Jones, A. Al-Hussany, S. Ravji, New techniques in sol-gel characterisation—mechanical measurements and fractal characteristics. J. Non-Cryst. Solids **293–295**, 731–745 (2001). doi:10.1016/S0022-3093(01)00852-3

A. Wittmar, D. Ruiz-Abad, M. Ulbricht, Dispersions of silica nanoparticles in ionic liquids investigated with advanced rheology. J. Nanopart. Res. **14**, 651 (2012). doi:10.1007/s11051-011-0651-1

H. Yotsumoto, R.-H. Yoon, Application of extended DLVO-theory: 1. Stability of rutile suspensions. J. Colloid Interface Sci. **157**(2), 426–433 (1993a). doi:10.1006/jcis.1993.1205

H. Yotsumoto, R.-H. Yoon, Application of extended DLVO-theory: 2. Stability of silica suspensions. J. Colloid Interface Sci. **157**(2), 434–441 (1993b). doi:10.1006/jcis.1993.1206

Stability of Binary Suspensions

M. Cerbelaud, A. Videcoq, P. Abélard, C. Pagnoux, F. Rossignol, R. Ferrando, Heteroaggregation between Al2O3 submicrometer particles and SiO_2 nanoparticles: Experiment and simulation. Langmuir **24**(7), 3001–3008 (2008). doi:10.1021/la702104u

M.L. Fisher, M. Colic, M.P. Rao, F.F. Lange, Effect of silica nanoparticle size on the stability of alumina/silica suspensions. J. Am. Ceram. Soc. **84**(4), 713–718 (2001). doi:10.1111/j.1151-2916.2001.tb00731.x

P. Garcia-Perez, C. Pagnoux, F. Rossignol, J.F. Baumard, Heterocoagulation between SiO_2 nanoparticles and Al2O3 submicronparticles; influence of the background electrolyte. Colloids Surf. A **281**(1–3), 58–66 (2006). doi:10.1016/j.colsurfa.2006.02.018

E.V. Golikova, Y.M. Chernoberezhskii, Investigation into the heterocoagulation of two-component disperse systems containing nanosized and submicron particles with different degrees of hydrophilicity. Glass Phys. Chem. **31**(3), 280–290 (2005). doi:10.1007/s10720-005-0057-1

T.W. Healy, G.R. Wiese, D.E. Yates, B.V. Kavanagh, Heterocoagulation in mixed oxide colloidal dispersions. J. Colloid Interface Sci. **43**(3), 647–649 (1973). doi:10.1016/0021-9797(73)90051-9

R. Hogg, T.W. Healy, D.W. Fuerstenau, Mutual coagulation of colloidal dispersions. Trans. Faraday Soc. **62**, 1638–1651 (1966). doi:10.1039/TF9666201638

A.M. Islam, B.Z. Chowdhry, M.J. Snowden, Heteroaggregation in colloidal dispersions. Adv. Colloid Interface Sci. **62**(2–3), 109–136 (1995). doi:10.1016/0001-8686(95)00276-V

P.K. Jal, S. Patel, B.K. Mishra, Chemical modification of silica surface by immobilization of functional groups for extractive concentration of metal ions. Talanta **62**(5), 1005–1028 (2004). doi:10.1016/j.talanta.2003.10.028

J.M. López-López, A. Schmitt, A. Moncho-Jordá, R. Hidalgo-Álvarez, Stability of binary colloids: kinetic and structural aspects of heteroaggregation processes. Soft Matter **2**(12), 1025–1042 (2006). doi:10.1039/b608349h

J. Lyklema, *Fundamentals of interface and colloid science II: Solid-liquid interfaces* (Academic Press, San Diego, 1995). ISBN 0-12-460524-9

J. Lyklema, J.F.L. Duval, Hetero-interaction between Gouy-Stern double layers: Charge and potential regulation. Adv. Colloid Interface Sci. **114–115**, 27–45 (2005). doi:10.1016/j.cis.2004.05.002

K.M. Paciejewska, *Untersuchung des Stabilitätsverhaltens von binären kolloidalen Suspensionen.* PhD thesis, Technische Universität Dresden, 2010. urn: nbn:de:bsz:14-qucosa-65050

R. Pelton, X. Geng, M. Brook, Photocatalytic paper from colloidal TiO_2—fact or fantasy. Adv. Colloid Interface Sci. **127**(1), 43–53 (2006). doi:10.1016/j.cis.2006.08.002

V. Ramakrishnan, Pradip, S.G. Malghan, The stability of alumina-zirconia suspensions. *Colloids Surf., A*, 133(1–2):135–142 (1998). doi:10.1016/S0927-7757(97)00135-0

M. Raşa, A.P. Philipse, J.D. Meeldijk, Heteroaggregation, repeptization and stability in mixtures of oppositely charged colloids. J. Colloid Interface Sci. **278**(1), 115–125 (2004). doi:10.1016/j.jcis.2004.05.020

D.R.E. Snoswell, T.J. Rogers, A.M. Howe, B. Vincent, Controlling porosity within colloidal heteroaggregates. Langmuir **21**(24), 11439–11445 (2005a). doi:10.1021/la051957o

D.R.E. Snoswell, J. Duan, D. Fornasiero, J. Ralston, The selective aggregation and separation of titania from a mixed suspension of silica and titania. Int. J. Miner. Process. **78**(1), 1–10 (2005b). doi:10.1016/j.minpro.2005.07.001

J. Sun, B.V. Velamakanni, W.W. Gerberich, L.F. Francis, Aqueous latex/ceramic nanoparticle dispersions: colloidal stability and coating properties. J. Colloid Interface Sci. **280**(2), 387–399 (2004). doi:10.1016/j.jcis.2004.08.014

P. Viravathana, D.W.M. Marr, Optical trapping of titania/silica core-shell colloidal particles. J. Colloid Interface Sci. **221**(2), 301–307 (2000). doi:10.1006/jcis.1999.6603

G.R. Wiese, T.W. Healy, Heterocoagulation in mixed TiO2–Al2O3 dispersions. J. Colloid Interface Sci. **52**(3), 458–467 (1975). doi:10.1016/0021-9797(75)90271-4

P.D. Yates, G.V. Franks, S. Biggs, G.J. Jameson, Heteroaggregation with nanoparticles: effect of particle size ratio on optimum particle dose. Colloids Surf. A **255**(1–3), 85–90 (2005). doi:10.1016/j.colsurfa.2004.12.035

Chapter 6
Summary and Conclusion

6.1 Main Ideas

This book has dealt with the physical properties of particle aggregates in colloidal suspensions. The presence of such aggregates is more the rule rather than an exception because very fine colloidal (i.e. nano-sized) particles have a strong tendency towards aggregation—be it in the liquid phase or during gas phase synthesis. The physical behaviour (e.g. light scattering, hydrodynamic mobility) of colloidal aggregates may seriously deviate from those of equally sized compact spheres. That is typically due to their relatively high porosity and their heterogeneous, fractal-like structure. In general, there are rather complex relationships between the morphology (size, shape, structure) and the physical properties of an aggregate. Yet, frequently one can apply approximate solutions to the average properties of aggregate populations, in particular when the primary particles are very fine. Since the macroscopic behaviour of colloidal suspensions (e.g. stability, rheology) does not only result from the properties of individual particles and aggregates, but also from the attractive and repulsive interaction between them, it is necessary to examine the interfacial properties as well. For colloidal aggregates, that means that one has to study the dependency of the interfacial properties on the aggregate morphology. This holds particularly true for the electric double layer within the aggregate, where overlap and charge regulation is inevitable.

The knowledge of the prevailing structure-property relations can considerably enhance the tailored preparation and the processing of colloidal suspensions. Additionally, characterisation techniques are needed which allow a direct or indirect evaluation of the state of dispersion. Again, this requires that the non-spherical, anisotropic, and fractal-like nature of colloidal aggregate is appropriately accounted for. It turns out that each measurement technique specifically reflects the morphology of particle systems and possesses their particular sensitivities to the different size ranges. In practice, this means that different measurement tasks (e.g. number concentration of nanoparticles, maximum particle size, and average

F. Babick, *Suspensions of Colloidal Particles and Aggregates*,
Particle Technology Series 20, DOI 10.1007/978-3-319-30663-6_6

particle size) are best solved by a combination of different measurement techniques. Such a combination can even reveal the structural properties of colloidal aggregates.

6.2 Summary of the Book

The book is divided into four main chapters. The first of them introduces important characterisation techniques which are used for particle sizing, describing aggregate structures, or quantifying interfacial properties in colloidal suspensions. Since there are numerous techniques, they are only briefly reviewed with regard to measuring principle and application experience. A few techniques that are widely used in practice, like dynamic light scattering (DLS) or optical centrifugation analysis (OCA), are discussed in more detail. A further chapter is devoted to the physico-chemical fundamentals of colloidal suspensions. Particular attention is drawn to the interfacial properties and the corresponding interactions between particles.

The centrepiece of this book is Chap. 4, which examines the structure-property relations of colloidal aggregates. This comprises a presentation of analytical tools for the quantification of aggregate morphology (including the fractal dimension), a section on the scattering behaviour of aggregates with a separate discussion on light scattering (which generally means intracluster multiple scattering), and an analysis of numerical tools for the computation of hydrodynamic aggregate properties. It is further shown how these tools can be employed for an elaborate characterisation of colloidal aggregates. The chapter closes with an investigation of the interaction between aggregates and reveals that the frequent approximation with the interaction of the closest pair of primary particles severely fails for thick double layers. Generally, simple solutions are rare for the computation of physical aggregate properties. If the aggregate properties are dominated by the interaction between its primary particles (e.g. by multiple scattering, hydrodynamic perturbations, or double layer overlap), numerical tools have to be employed that usually require huge computational costs. This imposes practical limits to routine application (e.g. in instrumentation software). Yet, the exact numerical solutions for numerical aggregate populations can provide functional relationships between morphological and physical aggregate properties, which facilitate the implementation of semi-empirical models.

Finally, the book deals with the dispersion of colloidal suspensions and considers their microscopic stability (Chap. 5). Apart from a brief review on dispersion mechanisms and dispersion techniques, the chapter examines empirical scaling laws for the dispersion of pyrogenic powders and discusses aspects of monitoring dispersion processes. The dispersion of colloidal suspensions is usually accompanied by measures to stabilise the colloidal particles. The suspension stability depends on the interaction forces between the particles and can even be predicted under certain conditions. Nonetheless, the suspension stability has to be evaluated on a basis of measurements. Aspects of the comparability and extrapolation of such measurements are addressed. In particular, it is shown with the example of binary suspensions that a momentary aggregation kinetic does not suffice to quantify suspension stability. The specific kinetics of interfacial phenomena (e.g. surface dissolution, ion adsorption, or surface precipitation) has to be examined as well.

6.3 Final Remarks

The investigation of colloidal suspensions is characterised by two peculiarities. First, it is generally not possible to separate the various physical effects (e.g. diffusion and sedimentation, or viscous interaction and double layer repulsion). Secondly, the presence of fractal aggregates—either due to ongoing aggregation or to particle synthesis—is rather likely. As a result of this, it is usually not possible to employ simple models that have been established for micrometre spheres. More sophisticated models based on a rigorous physical analysis may require more brainwork and will probably increase the costs of computation; it can, however, provide more accurate structure-property relations which enhance our understanding of products and processes as well as of the characterisation techniques. However, this approach is not easy, since the microscopic processes in colloidal suspensions are not yet completely understood and can be dramatically affected by chemical (i.e. material specific) reactions. Even more, when modelling colloidal properties, one typically encounters problems of different fields of science (e.g. laser physics, inorganic chemistry, particle sizing, fluid mechanics, and numerical mathematics). This is best coped with by interdisciplinary work. It is this cooperation between different domains of research which makes colloid science into one of the most fascinating scientific disciplines.

Appendix A
Glossary

A.1 Terms

The terminology of this book refers to subjects in colloid chemistry, material science, optics, particle technology, process engineering, and measurement technology. Naturally, the use of specialised terms within all of these different fields is not necessarily consistent. Although there are various efforts to harmonise terminology on an international and interdisciplinary level, one should be aware that a uniform "terminology of everything for everyone" with a complete and consistent set of unambiguous definitions is neither feasible nor practical. The following glossary explains the meaning of terms as used in this book, rather than providing exact definitions.

Agglomerate generic term for a contiguous cluster of particles which are held together by physical adhesion forces, chemical bonds, and/or form closure; the individual particles from which an agglomerate is formed are called →primary particles, particles that can be identified within an agglomerate are called →constituent particles, the agglomerate itself is sometimes called a secondary particle; note that agglomerates may possess a multiscale structure due to consecutive agglomeration processes (e.g. aggregation of primary particles in a flame, agglomeration of aggregates in the gas phase, compaction of agglomerates in powder bulk phase); agglomerates can be classified by the kind and strength of the particle bonds (e.g. →aggregates for strong adhesion forces and flocs for weak ones).

Agglomeration (MICROSCOPIC PROCESS) generic term for the build-up of particle →agglomerates from individual →primary particles; it refers primarily to the collision and adhesion of the primary particles, but may comprise re-structuring and consolidation of the agglomerates as well; the agglomeration is frequently replaced by more specific terms like →aggregation (for the agglomeration of colloidal particles), coagulation (for the agglomeration in liquid dispersions) or flocculation (for the agglomeration due to polymeric agents); note that these terms are not mutually exclusive.

F. Babick, *Suspensions of Colloidal Particles and Aggregates*,
Particle Technology Series 20, DOI 10.1007/978-3-319-30663-6

Agglomeration (UNIT OPERATION) generic term in process engineering for all unit operations which enlarge the particle size of a dispersed system by the creation of particle →agglomerates—e.g. spray drying, granulation in drums, and compaction of powders.

Aggregate →agglomerate with relatively strong interparticle adhesion, commonly applied to agglomerates of colloidal particles that are closely touched or even sintered; sometimes referred to agglomerates "that can not be dispersed further" or to agglomerates, in which particles are attached via faces rather than via contact points—the former "definition" reveals a poor understanding of deagglomeration processes, whereas the latter is particularly meaningful for pigment characterisation.

Aggregation formation of particle →aggregates from colloidal primary particles—e.g. during gas phase synthesis of solids or in the course of a crystallisation process; the resulting →aggregates are characterised by a relatively high mechanical strength.

Abbe limit minimum distance h_{min} between two geometric objects that can be resolved with classical optical instruments; depends on the wavelength λ, the refractive index of the medium m, and the aperture angle α ($h_{min} = \lambda/(2m \sin \alpha)$).

Chemical-mechanical polishing or *planarisation* (*CMP*) process of flattening and smoothing surfaces by rapidly moving a polishing pad over the surface on which a colloidal suspension is dispensed; smoothing is achieved by the combined effect of mechanical treatment and chemical reactions.

Coagulation formation of particle →aggregates from colloidal →primary particles in a liquid continuum—usually due to the lack of sufficiently strong double layer repulsion.

Colloids and *colloidal particles* refers nowadays to dispersed systems with submicrometre particles ($x \leq 1$ μm); characterised by high particle diffusion coefficients, slow sedimentation, or even sedimentation-diffusion equilibria, weak light scattering, significance of non-viscous particle interactions, suspension structuring, etc.

Originally applied by T. Graham to material systems that behaved differently than solutions (e.g. with regard to membrane penetration) and that were assumed to contain very fine particles invisible for light microscopes (i.e. *submicron* particles)—such material systems were glue (*colla*), jelly, or "metal hydrosols".

Constituent particles individual particles that can be identified within an →agglomerate; frequently identical to or identified with the →primary particles.

Dispersion (PROCESS) in general, the process of generating and/or homogeneously distributing a dispersed phase in a fluid continuum (e.g. by emulsification oil in water or by suspending powders); more specifically, the process of size reduction of the dispersed phase (e.g. as deagglomeration or homogenisation of emulsions); cf. →state of dispersion.

Dispersion (STATE OF MATTER) generic term for dispersed systems with a fluid continuum.

Dissipation conversion of energy into heat, which is not completely reversible; occurs in any real relaxation process and reduces the energy that can be used for mechanical work; the rate of dissipation is the product of driving force (e.g. temperature or velocity difference) and the resulting relaxation flux (e.g. heat or momentum flux).

Electric double layer (EDL) interfacial region between an electrically charged surface and the electrically neutral bulk phase of liquid solvent; the term refers to the layer of surface charges and the zone ("layer") adjacent to the surface containing a compensating counter-charge (i.e. an excess of counter-ions); the counter-charge is distributed within an immobile layer (Stern layer) immediately at the surface and within a diffuse layer where the ions experience Coulombic and Brownian forces; the detailed structure of the EDL is manifold and can be explained by models at different levels of sophistication (e.g. with regard to the charge distribution in the Stern layer, the transition from Stern to diffuse layer, or possible ion–ion-correlations in the diffuse layer); cf. Sect. 3.1.5.

Feret diameter distance between a pair of parallel tangents or planes on the two-dimensional image of a particle or its three-dimensional reconstruction, respectively; usually only the minimum and maximum Feret diameters are used for further evaluation.

Flocculation formation of weakly bound →agglomerates (*flocs*) from colloidal or micrometre →primary particles in a liquid continuum—frequently induced by polymeric additives (e.g. depletion or bridging flocculation).

Fractal aggregate, fractal agglomerate →aggregates or →agglomerates with a non-uniform distribution of the →constituent particles, which typically coincides with a very porous, branch-like morphology; fractal aggregates are characterised by a power-law decrease of the pair-correlation density function $g(x)$ (Eq. (4.8)) and a power-law relationship between mass and size (Eq. (4.9)), in which the exponent is less than the Euclidean dimension; fractal aggregates are not ideal fractal objects, but rather obey the fractal relationships only in a statistical sense (cf. Sect. 4.2.1).

Heterodyne detection method for the coherent detection of laser signals, which superposes the optical signal with a reference beam from a coherent light source (local oscillator); in contrast to →homodyne detection, the local oscillator is operated at a slightly different frequency than the optical signal; heterodyne measurements allow for velocity measurements via Doppler effect and are employed, e.g., in electrophoretic light scattering.

Homodyne detection method for the coherent detection of laser signals, which superposes the optical signal with a reference beam from a coherent light source (local oscillator); there is no frequency shift between the local oscillator and the optical signal; homodyne detection provides higher signal-noise-ratios than →heterodyne detection, but at the expense of a more sophisticated receiver design; in DLS instrumentation, homodyne detection is usually achieved by superposing the scattered light with the incident light; for ideal homodyne condition, the latter has to be much larger than the scattered light, otherwise a combination of homodyne and →self-beating (direct) detection prevails.

Hydrodynamic interaction or *viscous interaction* interaction between particles or particles and walls that are mediated by flow fields in a viscous fluid; decays with reciprocal distance ($\propto 1/r$), long-ranging interaction.

Inversion (or *deconvolution*) the transformation of a signal spectrum to the distribution of a quantity—frequently, the signal spectrum can be described as a linear integral equation of the requested distribution function; inversion is achieved by numerical algorithms that find a biased guess of the distribution function and minimise its dependency on the measurement error; the result is influenced in every instance by the employed inversion algorithm and its parameterisation; moreover, the inversion of measurement data cannot reveal all details of the distribution function (i.e. limited information content).

Legendre ellipse an ellipse that is equivalent to a two-dimensional figure with respect to area and area moment of inertia.

Lennard-Jones potential a mathematical model for the energy potential of two electrically neutral, interacting atoms or molecules; the Lennard-Jones potential accounts for short-ranging repulsive interaction (Born repulsion) and longer-ranging attractive interaction (dispersion forces, van-der-Waals attraction), which are described by a power-law approach with the exponents 12 and 6, respectively.

Mean size $\bar{x}_{k,r}$ can be understood as the kth root of the expectation value of the kth power of x (power mean) when weighted by a given →type of quantity (indicated by r). Mean values can be computed from the size distribution, which requires the knowledge of the density function $q_r(x)$ for the entire size range, or may be derived from a measurement signal of the whole particle system (e.g. from turbidity or from the Brownian relaxation of scattering signals). Mean values with $k = -1, 0, 1, 2$ are called harmonic mean, geometric mean, arithmetic mean, and quadratic mean (or rms value), respectively. Means are unambiguously defined for a given size distribution $q_r(x)$. However, the significance of means from two different types of quantity may formally agree—e.g. the (surface) area weighted arithmetic mean $\bar{x}_{1,2}$ is formally identical to the volume weighted harmonic mean $\bar{x}_{-1,3}$; both denote the Sauter diameter when x refers to the volume equivalent diameter.

Measurement method general way of employing a →measuring principle for the quantification of a measurand (e.g. particle size); its specific realisation is commonly called →measurement technique, even though the distinction between the two terms is not very strict; examples: (i) the settling velocity as a measure of size (=measurement principle) can be measured by gravitational or by centrifugal sedimentation (=measurement methods), (ii) the diffusion coefficient can be quantified by means of dynamic ultramicroscopy or by means of dynamic light scattering, (iii) sphere diameters from images can be obtained, e.g., by electron microscopy or by scanning force microscopy.

Measurement principle (physical) effect that is employed for the quantification of the measurand (e.g. particle size, zeta-potential); the dependency of the effect on the measurand has to be well-defined; examples: (i) settling velocity grows with particle size, (ii) diffusion coefficient is inversely proportional to size, (iii) 2D-images obtained by orthogonal projection allow the direct quantification of characteristic lengths and areas.

Measurement technique specific way of employing a →measuring principle for the quantification of a measurand (e.g. particle size); similar measurement techniques are sometimes grouped as one →measurement method; examples: (i) the settling velocity as a measure of size (=measurement principle) can be measured by gravitational sedimentation with an X-ray transmission detector (=measurement technique) or by centrifugal sedimentation in homogenously filled cuvettes with an optical transmission detector, (ii) the diffusion coefficient can be measured with dynamic light scattering by means of a self-beating autocorrelation detector or by means of →homodyne light detection and frequency analysis, (iii) sphere diameters can be derived from images of a scanning electron microscope, which were obtained without using the emitted secondary electrons.

Median size $x_{50,r}$ splits the particle population into two fractions of equal weight with regard to the given →type of quantity (indicated by r). One contains all particles with $x \leq x_{50,r}$, the other all particles with $x > x_{50,r}$. The median is, thus, derived from the cumulative function of the size distribution ($Q_r(x_{50,r}) = 0.5$).

Modal value $x_{\mathrm{mod},r}$: characterises the peak position in the density function $q_r(x)$ and can be interpreted as the most frequent particle size for the given →type of quantity (indicated by r); it is not affected by measurement uncertainty of rare particle fractions; many publications employ the modal value $x_{\mathrm{mod}^*,r}$ of the transformed density function $q_r^*(x)$; this is because the width of size classes in counting or fractionating methods follow a geometric series or because the resolution of spectroscopic methods is approximately constant with regard to $\log(x)$; regardless, the two parameters—$x_{\mathrm{mod},r}$ and $x_{\mathrm{mod}^*,r}$—are not identical; frequently, even the number of modes is different between the two corresponding density functions ($q_r(x)$ and $q_r^*(x)$).

Nanoparticle →particle with all external geometric extensions around or below 100 nm.

Ostwald ripening the growth of large particles due to the steady dissolution of small particles which have a higher solubility; Ostwald ripening leads to a steady coarsening of the particle size distribution; Ostwald ripening is employed in crystallisation technology for crystal purification or changing of the crystal modification; it is an unwanted effect when storing colloidal emulsions (set restriction to shelf-life).

Particle a piece of matter with defined boundaries to its environment (e.g. a dust particle in the air, a grain of sand at the beach, an oil droplet in water, an air bubble in jelly); contiguous clusters of particles (i.e. agglomerates) are also considered as (secondary) particles—provided that they do no contain all particles of the dispersed system; the physical boundaries between touching particles may vanish with time (e.g. by fusion), even if the outer shape still implies individuality—in that case it can be still meaningful to distinguish between the original particles (→constituent particle); note that in particle technology and colloid science, particles are regarded as the basic element of a dispersed system, i.e. they refer to objects on a level above atoms and molecules.

Particle property any property of a particle that is not (only) a bulk property of the dispersed phase; there are several, not necessarily excluding categories (geometric properties, dynamic properties, interactions with electromagnetic or sound fields); these properties are related to →particle size or, more generally, to the particle morphology, which also includes the particle shape (e.g. sphericity) and structural composition (e.g. fractal dimension).

Particle size ambiguous term which refers to a variety of measurands that correlate with the outer dimensions of a particle; particle size is either defined as a characteristic length in a two-dimensional projection of the particle (e.g. Feret diameter x_{Feret}) or as characteristic length of a regular body (e.g. sphere, circle, ellipse) that is equivalent to the particle with respect to a defined geometrical or physical particle property.

Primary particles particles from which an →agglomerate is formed; frequently identical to or identified with the →constituent particles of the agglomerate (or aggregate).

Pyrogenic powder particle systems produced in a hot gas phase reaction (e.g. flame synthesis); they are composed of relatively rigid, submicron →aggregates of nanosized →primary particles with a fractal-like morphology; these aggregates may further agglomerate to large, micrometre-sized clusters; pyrogenic materials are, e.g., carbon black, fumed silica, or fumed metal oxides (e.g. titania, alumina).

Quantile or percentile $x_{p,r}$ is related to the probability p (in percent) that a particle of the distribution is smaller or equal to $x_{p,r}$ ($Q_r(x_{p,r}) = p$ %).

Scattering (MICROSCOPIC EFFECT) diffuse re-direction of waves on objects (scatterers), which deviate from their environment with respect to the wave propagation properties (e.g. refractive index, acoustic impedance); the amplitude of the scattered wave (scattering field) can considerably vary with the angle observation—dependent on the ratio of object size to wavelength (expressed as ka); scattering can occur as a pure re-distribution of wave energy (elastic scattering) or may coincide with an energy transfer to heat or an other type of radiation (inelastic scattering).

Scattering (MACROSOSCOPIC PHENOMENON) response of a particle system to waves; it is related to the →scattering (MICROSCOPIC EFFECT) on single particles, yet is additionally affected by the concentration and distribution of particles in the continuous phase; in particle systems there is a certain degree of interdependence among the single scattering events (multiple scattering); the total scattering field of the system results from the interference of all individual scattering fields and is usually detected at relatively large distances from the particles (far field); scattering is related to the extinction (attenuation) of radiation.

Sedimentation (MICROSCOPIC PROCESS) motion of particles in viscous media due to gravitation or centrifugal fields (also: *settling*); the term refers frequently to the state in which the field forces are counterbalanced by the drag force; this state is almost instantaneously achieved in the case of colloidal suspensions; the sedimentation of an individual particle depends on its size, the density contrast, the rheological fluid properties, the field strength, and the viscous interactions with other particles.

Sedimentation (UNIT OPERATION) term in process engineering for unit operations which separate the dispersed phase from the continuous phase by means of gravity or centrifugation; the focus lies on the whole dispersed phase rather than on individual particles, which is why the separation is frequently supported by inducing particle agglomeration.

Self-beating detection direct detection of scattering signals in DLS measurements, i.e. without superposition with a reference beam; since the phase of the signal is not compared to a reference value, it is not possible to detect absolute velocities of the particles—instead, the relative motion between particles is probed; that means a self-beating detection is insensitive to a general drift of the particles (e.g. by convective flow).

State of dispersion morphology and spatial distribution (micro- and macroscopically) of the dispersed phase; most often referred to the size, shape, and structure of particles and particle aggregates as well as to the degree of an →agglomeration process.

Tyndall effect When light is transmitted through a suspension of fine colloidal particles, the light scattering on the particles suffices to make the light beam "visible" even though the particles are not. Similarly, very fine atmospheric particles allow us to "see" sunbeams in the forest. This phenomenon is called Tyndall effect. It was reported for the light transmission in colloidal gold sols by Faraday (1857, as cited in Gentry 1993) and later extensively studied by Tyndall for aerosols and colloidal sols. Tyndall found i.e. that the scattered light tends towards the blue end of light spectrum while transmitted light was more red or yellowish.

Type of quantity A cumulable quantity, like volume, number, or (ideally) scattering intensity of the particle system, which is used to quantify the frequency (weights) of particles having a certain size.

X-ray tube a vacuum tube with a cathode and anode of a defined material (e.g. copper); the cathode emits electrons which are accelerated by strong electric fields (voltages: 30–150 kV); when entering the anode the electrons are quickly decelerated, which produces a broad, continuous X-ray spectrum (*Bremsstrahlung*); additionally, some electrons interact with the electrons of the anode material, which creates a discrete X-ray spectrum of relatively high intensity, characteristic for the anode material.

Zeta-potential (or *electrokinetic potential*) electric potential at the shear (or slip) plane within an →electric double layer (EDL); the zeta-potential can be measured in case of a relative motion between the charged surface (and Stern layer) and the bulk liquid (including the diffuse layer)—e.g. by application of an electric field or due to sedimentation; the concept of a slip plane assumes a sharp transition between a stagnant liquid layer attached to the surface and a hydrodynamically mobile liquid phase which is located very close to the outer Helmholtz plane (OHP); the zeta-potential is, therefore, equal or lower in magnitude than the diffuse layer potential ψ_{d} at the OHP (cf. Sect. 3.1.5.1).

A.2 People

Petrus (Peter) Josephus Wilhelmus **Debye** (1884–1966) … was a Dutch physicist and physical chemist, who worked in the fields of quantum physics, X-ray analysis, microwave spectroscopy, and electrochemistry. Colloid science benefits from his contribution to X-ray and light scattering (in particular for aggregates—Eq. (4.39)—and concentrated suspensions—Eq. (2.24)), his work on electrolyte solutions (Debye-Hückel theory), as well as his remarks to electrophoresis and his research on polymers. He was awarded the Nobel Prize in Chemistry in 1936 "for his work on molecular structure through his investigations on dipole moments and the diffraction of X-rays and electrons in gases".

E. Courtens, Peter Debye—A life for science. Ferroelectrics **267**(1), 43–60 (2002). doi:10.1080/00150190211019

Boris Vladimirovich **Derjaguin** (1902–1994) … was a Russian chemists and physicist whose work is tightly connected with the development of the modern science of colloids and surfaces. Most prominent is his theory on colloidal stability, now known as DLVO theory, which he developed together with Landau and independently from Verwey and Overbeek. Additionally, he examined the adhesion and friction of solid bodies, developed the theory of capillary osmosis, started investigation into foam stability, worked on the direct measurements of molecular attraction, and contributed to the theory of electrokinetic phenomena.

N.V. Churaev, L.B. Boinovich, Boris Vladimirovich Dedication. Adv. Colloid Interface Sci. **104**(1–3), ix–xiii (2003). doi:10.1016/S0001-8686(03)00031-9

Albert **Einstein** (1879–1955) … was a German-born theoretical physicist who is mainly renowned for his special theory of relativity and its extension to the general theory of relativity. In addition to this, he worked on statistical mechanics and quantum theory and investigated the thermal properties of light. At the beginning of his scientific career he also set important landmarks for colloid science. This applies particularly to his explanation of Brownian motion, but is also valid for the calculation of suspension viscosity as well as his theory of critical opalescence. In 1921, he was given the Nobel Prize in Physics "for his services to theoretical physics, and especially for his discovery of the law of the photoelectric effect".

Michael **Faraday** (1791–1867) … was an English scientist who mainly contributed to the fields of electromagnetism and electrochemistry. Although he is acknowledged as an experimental physicist rather than a mathematician, he exerted a great influence on our modern concept of physics. In the Baker lecture of 1857 at the Royal Society, he talked about the preparation and stabilisation of colloidal gold sols and the optical effects that can be observed therein (e.g. colour changes during aggregation).

M. Faraday, Experimental relations of gold (and other metals) to light, in *Experimental Researches in Chemistry and Physics* (Taylor and Francis, London, 1857), pp. 391–443.

E. Hatscek, *Classical Works About Colloidal Solutions.* Schulze, Winssinger, Linder, Picton, Setmi, Sobrero, Ostwald Klassiker Series vol. 217, Leipzig, 1926.

J.W. Gentry, The aerosol science contributions of Michael Faraday. J. Aerosol Sci. **26**(2), 341–349 (1995). doi:10.1016/0021-8502(94)00110-K

Thomas **Graham** (1805–1869) … was a Scottish chemist who is best-remembered for his studies on the diffusion of gases and for the discovery of dialysis. The latter originated from his investigation of colloidal systems, which he defined as materials that do not contain visible particles (under the light microscope), yet behave differently to pure solutions. Graham created the term "colloid" (from latin *colla* = glue) and can be regarded as the founder of colloid science.

T. Graham, Liquid diffusion applied to analysis. Phil. Trans. R. Soc. Lond. **151**, 183–224 (1861). doi:10.1098/rstl.1861.0011

R.A. Smith, *The Life and Works of Thomas Graham* (John Smith & Sons, Glasgow, 1884).

Lev Davidovich **Landau** (1908–1968) … was a Soviet physicist who worked in several fields of theoretical physics, e.g. in quantum mechanics, superfluidity, and superconductivity. Additionally, he is renowned for his textbook series in physics which he created together with Lifshitz. His contribution to colloid science concerns the stability of colloids (DLVO theory). He was awarded the Nobel Prize in Physics in 1962 "for his pioneering theories of condensed matter, especially liquid helium".

Benoît **Mandelbrot** (1924–2010) … was a Polish-French mathematician who dealt with several math-related problems, e.g. in information theory, economics, and fluid dynamics. Mandelbrot won great renown through his publications on fractal objects, which inspired researchers of very different scientific fields. Soon after his seminal book on fractal structures in nature, science, and daily life, the idea of fractal dimension was adopted by colloid science to describe aggregation and aggregate morphology.

G.J. Klir, Benoit B. Mandelbrot (1924–2010): the founder of fractal geometry. Int. J. Gen. Syst. **40**(3), 231–232 (2011). doi:10.1080/03081079.2010.548133

R. Gomory, Benoît Mandelbrot (1924–2010). Nature **468**(7322), 378 (2010). doi: 10.1038/468378a

Wolfgang **Ostwald** (1883–1943) … was German-Baltic chemist and biologist; a son of the famous chemist Wilhelm Ostwald. Ostwald's research focussed on colloids, he worked, e.g., on the stability of colloids and developed a manometric sedimentometer for the quantification of coagulation in colloidal suspensions. Ostwald had a great deal of influence on the development of colloid science in

Germany. Last but not least, he introduced important terms like *dispersed phase* and *dispersion medium.*

Jean-Baptiste **Perrin** (1870–1942) … was a French physicist who worked on various subjects, for instance, on cathode rays, fluorescence, sound propagation, and the decay of radium. A major part of his scientific life was dedicated to colloids. His investigations on the Brownian motion of colloids by means of the previously developed ultramicroscope verified →Einstein's statistic explanation of this phenomenon and thereby confirmed the atomic nature of matter. Moreover, Perrin was able to closely limit the possible range of Avogadro's number and he discovered the sedimentation-diffusion equilibrium. "For his work on the discontinuous structure of matter and especially for his discovery of sedimentation equilibrium" he received 1926 the Nobel Prize in Physics.

C. Bigg, Evident atoms: visuality in Jean Perrin's Brownian motion research. Stud. Hist. Phil. Sci. **39**(3), 312–322 (2008). doi:10.1016/j.shpsa.2008.06.003

Marian **von Smoluchowski** (1872–1917) … was a Polish physicist whose research on discrete state matter is still highly valued in modern science. He is particularly acknowledged for his theory on Brownian motion, which he developed independently of →Einstein and which laid the foundation for the theory of stochastic processes. A similar rank is deserved by his discovery of density fluctuations in liquids and gases and their relevance for macroscopic scattering—most prominently explained by the phenomenon of critical opalescence. Both works proved very influential for the understanding of colloidal suspensions. Furthermore, he did pioneering work on the quantification of particle aggregation as well as in the field of electrokinetic phenomena.

A. Fuliński, On Marian Smoluchowski's life and contribution to physics. Acta Phys. Pol. B **29**(6), 1523–1537 (1998).

J. Lyklema, Electrokinetics after Smoluchowski. Colloids Surf. A **222**(1–3), 5–14 (2003). doi:10.1016/S0927-7757(03)00217-6

John William **Strutt**, 3rd Baron **Rayleigh** (1842–1919) … was an English physicist who left a rich legacy that covers problems in optics, acoustics, wave theory, electricity, thermodynamics, and statistical physics. In colloid science, his name is mainly encountered when describing the optical scattering of very fine particles, for which he developed a theoretical model that could explain all experimental data previously obtained by Tyndall on "precipitated clouds" (aerosols). In 1904, he received the Nobel Prize in Physics "for his investigations of the densities of the most important gases and for his discovery of argon in connection with these studies".

J.N. Howard, John William Strutt, third Baron Rayleigh. Appl. Opt. **3**(10), 1091–1101 (1964). doi:10.1364/AO.3.001091

P.N.T. Wells, Lord Rayleigh: John William Strutt, third Baron Rayleigh. IEEE Trans. Ultrason. Ferroelectr. Freq. Control **54**(3), 591–596 (2007). doi:10.1109/TUFFC.2007.281

The (Theodor) **Svedberg** (1884–1971) … was a Swedish chemist who early worked on colloidal materials and developed the ultracentrifuge, which became an important analytical technique for the characterisation of colloidal suspensions. He himself employed this technique to distinguish pure proteins. In 1926 he obtained the Nobel Prize in Chemistry "for his work on disperse systems".

T. Svedberg, Die Ergebnisse der Kolloid-Forschung. Ber. Dtsch. Chem. Ges. **47** (1), 12–38 (1914). doi:10.1002/cber.19140470104

Richard Adolf **Zsigmondy** (1865–1929) … was an Austrian-Hungarian chemist who developed the ultramicroscope in 1903 (together with Siedentopf), which allowed a visualisation of very fine colloidal particles (like in gold sols) for the first time and, thus, inspired scientist like →Perrin to do detailed studies of colloidal materials. A further achievement was the invention of the membrane filter (together with Bachmann). "For his demonstration of the heterogeneous nature of colloid solutions and for the methods he used, which are fundamental to modern colloid chemistry", he was rewarded the Nobel Prize in Chemistry in 1925.

Appendix B
Physical Basics

B.1 Hydrodynamics

B.1.1 Fluid Properties or Classification of Flow

> The chief peculiarities of fluids are these: 1st, they are heavy; 2nd, they expand in all directions unless they are confined; and 3rd, they allow themselves to be compressed more and more, according to the increased force applied.[1] (Bernoulli 1738)

This enumeration of Daniel Bernoulli lacks a fourth important property of fluids: *viscosity*. All of the properties may be more or less relevant for real flow phenomena. It is, therefore, convenient to classify flow regimes with regard to the prevailing fluid properties and to describe them in a correspondingly simplified manner:

- *ideal* flow, i.e neglect viscosity;
- *viscous* flow, i.e. account for fluid viscosity;
- *creeping* flow or *Stokes* flow, i.e. neglect fluid inertia and local acceleration;
- *laminar* flow, i.e. viscous properties prevail at all length scales;
- *turbulent* flow, i.e. viscosity prevails only at the microscale, whereas inertia dominates the macroscopic flow properties, which results in a complex, fluctuating flow pattern with eddies;
- *incompressible* flow—usually assumed for liquids;
- flow of fluids with or without *yield stress*;
- or more generally: *Newtonian* flow or *non-Newtonian* flow

---

[1]In Latin: *Fluidorum antem elasticorum præcipuæ affectiones in eo positæ sunt: 1°, ut sint gravia; 2°, ut se in omnes plagas explicent, nisi contineantur, et 3°, ut se continue magia magisqne comprimi patiantur crescentibus potentiis compressionis: ita comparatus est aër, ad quem potissimum presentes nostræ pertinent eogitationes.*

F. Babick, *Suspensions of Colloidal Particles and Aggregates*,
Particle Technology Series 20, DOI 10.1007/978-3-319-30663-6

The classification is accomplished by means of the rheological properties and/or by means of dimensionless numbers (e.g. Re-number). For colloidal suspensions, one usually assumes an incompressible, Newtonian liquid phase. The motion of colloidal particles in such liquid is dominated by the viscous properties (i.e. *Stokes flow*). In spite of this, the whole suspension may flow in a highly turbulent manner, or may show non-Newtonian rheological properties.

B.1.2 Fundamental Equations of Fluid Flow

The fundamental equations of fluid mechanics are the continuum mechanical conservation equations for mass, momentum, and energy. The mass conservation in the flow field $\mathbf{u}$ of a fluid with mass density ρ is expressed by the continuity equation:

$$\frac{\partial \rho}{\partial t} = -\mathrm{div}(\rho \mathbf{u}). \tag{B.1}$$

The conservation of momentum requires a set of three differential equations that account for the different space directions. For Cartesian coordinates, all three equations are comprised within the following notation:

$$\rho \frac{\mathrm{D}\mathbf{u}}{\mathrm{D}t} = \rho \left(\frac{\partial \mathbf{u}}{\partial t} + (\mathbf{u} \bullet \nabla)\, \mathbf{u} \right) = \rho \cdot \mathbf{f} + \nabla \bullet \mathbf{\Pi}, \tag{B.2}$$

where $\mathbf{f}$ denotes the effect of volume forces (e.g. gravity) and $\mathbf{\Pi}$ is the stress tensor, which reflects the effect of static pressure and viscous friction in the hydrodynamic field:

$$\mathbf{\Pi} = -p\mathbf{I} + \mathbf{\Pi}^{\mathrm{diss}}. \tag{B.3}$$

The *dissipative* (or *viscous* or *deviatoric*) part of the stress tensor ($\mathbf{\Pi}^{\mathrm{diss}}$) is a function of the rheological fluid properties and depends on the local shear rates. For Newtonian fluids, which are isotropic, purely viscous, and without rheological "memory", this dependency is linear (Cauchy-Poisson-law):

$$\mathbf{\Pi} = -p\mathbf{I} + 2\eta \mathbf{D} + \zeta(\nabla \bullet \mathbf{u}) \cdot \mathbf{I}, \tag{B.4}$$

where η and ζ denote the material properties—the dynamic and second (or volume) viscosity, respectively—and $\mathbf{D}$ is the rate of strain tensor:

$$\mathbf{D} = \frac{1}{2}\left(\nabla \circ \mathbf{u} + (\nabla \circ \mathbf{u})^T\right), \quad \text{i.e.}\, D_{ij} = \frac{\partial u_i}{\partial x_j} + \frac{\partial u_j}{\partial x_i}. \tag{B.5}$$

By combining Eqs. (B.2)–(B.5), one obtains the Navier-Stokes equations (named after Claude Louis Marie Henri Navier und George Gabriel Stokes) in their Cartesian notation:

$$\rho\left(\frac{\partial \mathbf{u}}{\partial t}+(\mathbf{u}\bullet\nabla)\,\mathbf{u}\right)=\rho\cdot\mathbf{f}-\nabla p+\eta\cdot\nabla^2\mathbf{u}+(\eta+\zeta)\nabla(\nabla\bullet\mathbf{u}). \tag{B.6}$$

These equations are non-linear; analytical solutions exist only for a few numbers of simple geometries and/or ideal conditions. For liquids, one usually assumes incompressibility, which eliminates the last term on the r.h.s of Eq. (B.6). For the flow past very small particles (i.e. for small particle Reynolds numbers), one can neglect the convective acceleration term ($(\mathbf{u}\bullet\nabla)\,\mathbf{u}$). Hence, the flow field around a steadily moving colloidal particle through a liquid medium is governed by the following equations (volume forces are neglected):

$$\mathrm{div}(\rho\mathbf{u})=0 \tag{B.7}$$

and

$$\eta\cdot\nabla^2\mathbf{u}=\nabla p. \tag{B.8}$$

In the context of particle motion it is quite common to describe the fluid field in curvilinear coordinates (e.g. spherical coordinates), which leads to a less elegant description of momentum conservation than given in Eq. (B.6) because the direction of the basis coordinates is not fixed (Landau and Lifshitz 1987, pp. 44–51).

The third fundamental quantity which obeys a conservation law is energy. The internal energy e, with enthalpy $h = e + p \cdot v$, of the fluid is affected by heat exchange due to temperature gradients, by volume work in compressible media, by the potential energy in an external field, and by the energy dissipation due to viscous friction. Its change in the flow field is described by a differential equation (Spurk and Aksel 2010, pp. 103–106)

$$\rho\frac{\mathrm{D}e}{\mathrm{D}t}=\rho\left(\frac{\partial e}{\partial t}+(\mathbf{u}\bullet\nabla)\,e\right)=\Phi-p\nabla\bullet\mathbf{u}+\nabla\bullet\lambda\nabla T, \tag{B.9}$$

where Φ is the rate of viscous dissipation:

$$\Phi=\prod_{ij}^{diss}D_{ij}. \tag{B.10}$$

B.1.3 Hydrodynamic Drag on Colloidal Particles in Viscous Fluids

The flow past a fixed particle or the motion of a particle through a quiescent fluid always coincides with a hydrodynamic force, which is caused by the pressure and viscous stress that acts on the particle surface:

$$F_{\mathrm{D}} = \int_S \mathbf{\Pi} \cdot \mathbf{n} \mathrm{d}S = \int_S \mathbf{\Pi}_{\mathbf{n}} \mathrm{d}S. \tag{B.11}$$

Moreover, the local variation of these quantities along the surface results in a hydrodynamic torque on the particle:

$$\mathbf{M} = \int_S \mathbf{n} \times \mathbf{\Pi}_{\mathbf{n}} \mathrm{d}S. \tag{B.12}$$

The hydrodynamic force and torque, thus, depend on the stress tensor $\mathbf{\Pi}$, or more specifically on its tangential component $\mathbf{\Pi}_{\mathrm{t}}$ since the flow on the surface has no normal component. The stress tensor can be derived from flow field $\mathbf{u}$ by means of Eq. (B.4), while the flow field is defined by the continuity Eq. (B.1) and the Navier-Stokes Eq. (B.6) as well as by the boundary conditions:

$$\text{on the surface}: \ \mathbf{u} - \mathbf{u}_{\mathrm{slip}} = \mathbf{v}_{\mathrm{p}}, \tag{B.13}$$

$$\text{in the undisturbed fluid}: \ \mathbf{u} \to \mathbf{u}^{\infty}, \ \left|\mathbf{r} - \mathbf{r}_{\mathrm{p}}\right| \to \infty. \tag{B.14}$$

The slip velocity $\mathbf{u}_{\mathrm{slip}}$ on the surface can be set to zero for Newtonian fluids, yet may be significant for Bingham fluids or for gas flow in the free molecular or transient regime. It can be approximated by the slip length l_{slip}, i.e. the penetration depth of an equivalent non-slip flow into the particle (Filippov 2000):

$$u_{\mathrm{slip}} = \frac{l_{\mathrm{slip}}}{\eta} \mathbf{\Pi}_{\mathrm{nt}}. \tag{B.15}$$

In the case of spherical particles, the flow field is described by spherical coordinates originating at the particle centre. The projection of the stress tensor onto the surface normal, i.e. onto the radial coordinate, is then:

$$\prod_{\mathrm{n}} = \prod_{r} = -\frac{\mathbf{r}}{r} p + \eta \left(\partial_r - \frac{1}{r} \right) \mathbf{u} + \frac{\eta}{r} \nabla (\mathbf{r} \bullet \mathbf{u}) \tag{B.16}$$

with the tangential component:

$$\frac{1}{\eta}\prod_{rt} = \left(\frac{1}{r}\partial_\vartheta u_r - \frac{1}{r}u_\vartheta + \partial_r u_\vartheta\right)\mathbf{e}_\vartheta + \left(\frac{1}{r\sin\vartheta}\partial_\varphi u_r - \frac{1}{r}u_\varphi + \partial_r u_\varphi\right)\mathbf{e}_\varphi. \quad \text{(B.17)}$$

B.1.4 Motion of Colloidal Particles in Viscous Fluids

The translational motion of particles principally obeys Newton's equation, which states that the change in particle momentum is driven by external forces:

$$m\frac{\mathrm{d}\mathbf{v}_\mathrm{p}}{\mathrm{d}t} = \sum \mathbf{F}_i, \quad \text{(B.18)}$$

which may comprise volume forces (e.g. gravity) as well as hydrostatic and hydrodynamic forces. Additionally, Brownian forces need to be considered in the context of colloidal particles. These are stochastic forces which result from the random collisions with the fluid molecules. They cannot be predicted for a given particle and a certain time, yet their time or population average effect is known from observation.

One approach to account for the stochastic force term was introduced by Langevin (1908). For the absence of volume forces, he transforms the x-component of Eq. (B.18)

$$m\frac{\mathrm{d}^2x}{\mathrm{d}t^2} = -6\pi\eta R_{\mathrm{h,t}} \cdot \frac{\mathrm{d}x}{\mathrm{d}t} + F_{\mathrm{stoch}} \quad \text{(B.19)}$$

by multiplication with the coordinate and employing the rules of differentiation:

$$\frac{m}{2}\frac{\mathrm{d}^2x^2}{\mathrm{d}t^2} - m\left(\frac{\mathrm{d}x}{\mathrm{d}t}\right)^2 = -3\pi\eta R_{\mathrm{h,t}} \cdot \frac{\mathrm{d}x^2}{\mathrm{d}t} + xF_{\mathrm{stoch}}. \quad \text{(B.20)}$$

The time average of this equation:

$$\frac{m}{2}\frac{\mathrm{d}^2}{\mathrm{d}t^2}\langle x^2\rangle - \langle m\dot{x}^2\rangle = -3\pi\eta R_{\mathrm{h,t}} \cdot \frac{\mathrm{d}}{\mathrm{d}t}\langle x^2\rangle \quad \text{(B.21)}$$

benefits from the fact that the average of the product of two independent variables (e.g. $\langle xF_{\mathrm{stoch}}\rangle$) vanishes. Moreover, the second term on the l.h.s. ($\langle m\dot{x}^2\rangle$) is just the thermal energy ($k_\mathrm{B}T$) which—after further transformation—leads to a linear differential equation for the mean square displacement along the x-axis:

$$\tau\frac{\mathrm{d}^2}{\mathrm{d}t^2}\langle x^2\rangle + \frac{\mathrm{d}}{\mathrm{d}t}\langle x^2\rangle = \frac{k_\mathrm{B}T}{3\pi\eta R_{\mathrm{h,t}}} \quad \text{with } \tau = \frac{m}{6\pi\eta R_{\mathrm{h,t}}} = \frac{2\rho a^2}{9\eta}. \quad \text{(B.22)}$$

The solution is:

$$\frac{\mathrm{d}}{\mathrm{d}t}\langle x^2\rangle = \frac{\mathrm{k_B}T}{3\pi\eta R_{\mathrm{h,t}}} + C \cdot \exp(-t/\tau). \tag{B.23}$$

For long observation times this converges to

$$\langle x^2\rangle = \frac{\mathrm{k_B}T}{3\pi\eta R_{\mathrm{h,t}}} \cdot t = 2D_{\mathrm{t}} \cdot t, \tag{B.24}$$

where D_{t} denotes the translational diffusion coefficient. The total mean square displacement $\langle|\Delta\mathbf{r}|^2\rangle$ for isotropic conditions is, therefore:

$$\left\langle|\Delta\mathbf{r}|^2\right\rangle = 6D_{\mathrm{t}} \cdot t. \tag{B.25}$$

Note, that the Brownian rotation can be treated analogously. The analysis then starts with the Euler's equation of angular momentum:

$$\mathbf{J}\frac{\mathrm{d}^2\varphi}{\mathrm{d}t^2} = -8\pi\eta R_{\mathrm{h,r}}^3 \cdot \frac{\mathrm{d}\varphi}{\mathrm{d}t} + \mathbf{M}_{\mathrm{stoch}}, \tag{B.26}$$

where $\boldsymbol{\varphi}$ is rotation angle, $\mathbf{J}$ is the tensor of inertia, and $\mathbf{M}_{\mathrm{stoch}}$ is the Brownian torque. The mean squared rotational displacement converges to a linear asymptote:

$$\left\langle|\Delta\varphi|^2\right\rangle = \frac{6\mathrm{k_B}T}{8\pi\eta R_{\mathrm{h,r}}^3} \cdot t = 6D_{\mathrm{r}} \cdot t, \tag{B.27}$$

in which D_{r} is the rotational diffusion coefficient.

References to Hydrodynamics

D. Bernoulli, *Hydrodynamica, sive de viribus et motibus fluidorum commentarii*, ed. by J.R. Dulseckeri, Argentorati (Strasburg), 1738, p. 200; as cited in: R.A. Smith, *The life and works of Thomas Graham* (John Smith & Sons, Glasgow, 1884).

A.V. Filippov, Drag and torque on clusters of N arbitrary spheres at low Reynolds number. J. Colloid Interface Sci. **229**(1), 184–195 (2000). doi:10.1006/jcis.2000.6981

L.D. Landau, E.M. Lifshitz, *Fluid Mechanics*, 2nd edn. (trans. by J.B. Sykes, W.H. Reid). In series: Course of Theoretical Physics, 2nd Engl. edn. (Pergamon, Oxford, 1987). ISBN 0-08-033933-6

L.D. Landau, E.M. Lifschitz, *Hydrodynamik*. In series: Lehrbuch der Theoretischen Physik, Bd. 6. (Akademie Verlag, Berlin, 1991). ISBN 978-3-8171-1331-6

P. Langevin, Sur la théorie du mouvement brownien. CR Hebd. Seance Acad. Sci. **146**, 530–533 (1908); reprinted in: Am. J. Phys. **65**(11), 1079–1081 (1997); with an introduction by D.S. Lemons (trans. by A. Gythiel).

J.H. Spurk, N. Aksel, *Strömungslehre. Einführung in die Theorie der Strömungen*, 8th edn. (Springer, Heidelberg, 2010). ISBN 978-3-642-13142-4

B.2 Optics

The optical properties of colloidal suspensions are primarily determined by the scattering and absorption of light at the individual particles. A first comprehensive explanation of this single particle scattering was provided by J.W. Strutt (Rayleigh, 1871a, b, c). Strutt discussed the colour and polarisation of light we receive from sky and, although he did not link light propagation to electromagnetism (or even Maxwell's equation), he had a clear conception of light as propagating "vibration" which also induces vibration of the scattering particles. His principal relationships and his quantitative analysis for light scattering does not only apply to the atmospheric molecules, but to all particles that are much finer than the wavelength of light. If this condition breaks, one has to consider the interference of waves that were scattered from different volume elements of the particle (Rayleigh-Debye-Gans scattering). In the general case one has additionally to account for the interaction between these waves. The incorporation of the latter in a mathematical framework defies a simple analytical solution. Lorenz (1898) and Mie (1908) proposed to develop the total scattering field of arbitrary spheres in terms of spherical harmonics. This principal idea is still employed even for more complex geometries.

Apart from particle size and shape, the scattering field is considerably influenced by the optical particle properties, i.e. by the dielectric function $\varepsilon(f)$ in optical domain or equivalently by the complex refractive index m. Indeed, Mie's paper (1908) on sphere scattering aimed at an explanation for the optical phenomena in colloidal gold sols (e.g. different colour of transmitted and reflected light), which can not be properly understood without adequate consideration of the wavelength dependency of the real and imaginary part of the refractive index.

The (microscopic) scattering at single particles results in macroscopic optical properties of the particle system (e.g. turbidity, intensity, and polarisation of the total scattering field). Rayleigh assumed independent scattering events, which is valid when the interparticle distance is much larger than the light wavelength. In such an event, the total scattering intensity is just the sum of the intensities of all scattering fields. For colloids, this applies only to very low concentrations. In dense colloidal suspensions the macroscopic scattering behaviour is influenced by density fluctuation and multiple scattering.

The next sections discuss only the scattering of highly diluted suspensions.

B.2.1 Basic Quantities

Since light is an electromagnetic wave, its scattering can be explained as radiation from oscillating dipoles, which are excited by the incident beam. The electric field $\mathbf{E}_{sca}$ of the scattered wave is, therefore, proportional to the dipole moment $\mathbf{p}$:

$$\mathbf{E}_{\text{sca}} \propto \mathbf{p} \tag{B.28}$$

with $\mathbf{p}$ being related to the incident field $\mathbf{E}_{\text{inc}}$ via the atomic polarisability α of the dipoles:

$$\mathbf{p} = \alpha \cdot \mathbf{E}_{\text{inc}}. \tag{B.29}$$

The equation applies to dipoles in free space, whereas in reality the particles might be embedded in an arbitrary medium with relative permittivity ε_{m}. The following sections consider only the additional scattering effect by the particles, which means only the excess dipole moment $\mathbf{p}_{\text{exc}}$ is of further relevance:

$$\mathbf{p}_{exc} = \varepsilon_{\text{m}}\alpha \cdot \mathbf{E}_{\text{inc}}. \tag{B.30}$$

Even though the atomic polarisability is a microscopic property of the atoms and molecules, it can be related to the macroscopic dielectric properties, i.e. to the relative permittivities of particles and medium, via the Clausius-Mossotti equation:

$$\alpha = \frac{3\varepsilon_0}{c_{N,\text{dipole}}} \cdot \frac{\varepsilon_{\text{p}} - \varepsilon_{\text{m}}}{\varepsilon_{\text{p}} + 2\varepsilon_{\text{m}}}. \tag{B.31}$$

Note that the relative permittivity depends on the frequency of the electromagnetic radiation and may have an imaginary part; therefore, it is frequently named a *complex dielectric function*. According to Maxwell's equation on electrodynamics, the velocity of wave propagation, thus the refractive index m, is determined by the dielectric and magnetic properties ($m^2 = \varepsilon\mu$). However, in the optical domain we can typically neglect the magnetic properties ($\mu \rightarrow 1$) and receive the following relationship[2]:

$$\varepsilon = \varepsilon' + \mathrm{i}\varepsilon'' = (m' + \mathrm{i}m'')^2 = m^2. \tag{B.32}$$

That means that the polarisability of an ideal dipole embedded in an unbound medium is proportional to the dipole volume and the optical contrast:

$$\alpha = 4\pi\varepsilon_0\alpha' = 3\varepsilon_0 V_{\text{dipole}} \cdot \frac{m_{\text{r}}^2 - 1}{m_{\text{r}}^2 + 2}, \tag{B.33}$$

where m_{r} denotes the relative refractive index. Note, that the polarisability volume α' is the identified with the polarisability within CGS notation.

[2]The sign in front of the imaginary part depends on the definition of the wave propagation. This text assumes for the field $E \propto \exp(\mathrm{i}\mathbf{kr} - \mathrm{i}\omega t)$, hence light absorption coincides with positive imaginary parts m''. In contrast, when the field propagation is described by $E \propto \exp(\mathrm{i}\omega t - \mathrm{i}\mathbf{kr})$, light absorption would correspond to negative m''.

Equations (B.29)–(B.33) regard the polarisability α as a scalar quantity. This holds true for isotropic materials and spherical dipoles. Anisotropy with respect to optical properties or particle shape inevitably means that the scattering of electromagnetic waves depends on the dipole's orientation. In this case the polarisability has to be expressed as tensor (cf. Bohren and Huffman 1983).

The scattering at particles is significantly affected by the polarisation of light. For the sake of convenience, only linear polarisation is considered within the next paragraphs; for elliptical polarisation, the reader is referred to the textbooks by van de Hulst (1981) and by Bohren and Huffman (1983). Linear polarisation means that the electric field vector of the incident light has a fixed orientation. In the general case, this vector does not lie in the scattering plane—which is defined by the light source, the particle, and the detector—but crosses it under the polarisation angle ϕ (anti-clockwise seen from the light source). The angle between the directions of propagation of the incident and scattered light is called scattering angle θ. These definitions facilitate a separate treatment of the electric field components, which are horizontally (parallel) or vertically (perpendicular) polarised with respect to the reference plane:

$$\text{incident light:} \begin{pmatrix} E_{||\text{inc}} \\ E_{\perp\text{inc}} \end{pmatrix} = E_0 \begin{pmatrix} \cos\phi \\ \sin\phi \end{pmatrix}, \tag{B.34}$$

$$\text{scattered light:} \begin{pmatrix} E_{||\text{sca}} \\ E_{\perp\text{sca}} \end{pmatrix} = -\frac{e^{ik(r-z)}}{ikr} \begin{pmatrix} S_2 & S_3 \\ S_4 & S_1 \end{pmatrix} \begin{pmatrix} E_{||\text{inc}} \\ E_{\perp\text{inc}} \end{pmatrix}, \tag{B.35}$$

where r is the radial coordinate starting from the particle centre and S_i are amplitude functions for the scattered field. The vector scattering amplitude is

$$\mathbf{V} = (S_2\cos\phi + S_3\sin\phi)\mathbf{e}_\theta - (S_4\cos\varphi + S_1\sin\phi)\mathbf{e}_\phi \tag{B.36}$$

with the direction

$$\mathbf{e}_\mathbf{V} = \sin\theta\cos\phi\,\mathbf{e}_r + \cos\theta\cos\phi\,\mathbf{e}_\theta - \sin\phi\,\mathbf{e}_\phi. \tag{B.37}$$

The effect or strength of the scattering field is usually quantified as scattering cross section C_{sca}, which is an area perpendicular to the incident light that receives as much light energy as is being scattered at the particle:

$$C_{\text{sca}} = \oint \frac{|\mathbf{V}|^2}{k^2}\,\mathrm{d}\Omega = \int_0^{2\pi}\int_0^{\pi} \frac{|\mathbf{V}|^2}{k^2}\sin\theta\,\mathrm{d}\theta\,\mathrm{d}\phi. \tag{B.38}$$

The scattering intensity in a certain direction (in the far field) can be expressed as partial scattering intensity $C_{\text{p,sca}}$, i.e. the total scattering cross section if the scattered field was uniform in all directions:

$$C_{\mathrm{p,sca}}(\theta,\varphi)=\frac{4\pi}{k^2}|\mathbf{V}|^2, \tag{B.39}$$

A special case is the scattering in the backwards direction with

$$C_{\mathrm{back}}=C_{\mathrm{p,sca}}(\pi)=\frac{4\pi}{k^2}|\mathbf{V}(\pi)|^2. \tag{B.40}$$

One result of scattering is the extinction of light, i.e. the loss of light intensity in the forward direction (θ = 0). The degree of extinction is determined by the amplitude of the forward scattered wave and is quantified as extinction cross section:

$$C_{\mathrm{ext}}=\frac{4\pi}{k^2}\mathrm{Re}(\mathbf{V}\cdot\mathbf{e_V})_{\theta=0}. \tag{B.41}$$

A second source for extinction is the absorption of light in the particle. The extent of absorption can be computed from the internal electromagnetic fields. It is, however, more convenient to derive it from the conservation of energy, i.e. the absorption cross section is:

$$C_{\mathrm{abs}}=C_{\mathrm{ext}}-C_{\mathrm{sca}}. \tag{B.42}$$

Last but not least, scattering affects the apparent velocity of light propagation in the suspension. That is because forward scattered light lags (or precedes) the incident light due to the different light propagation within the particle. The interference of the forward scattered wave with the incident wave, thus, affects the refractive index of the suspension m_{sus}. For dilute suspension, one obtains:

$$\frac{m_{\mathrm{sus}}}{m_{\mathrm{m}}}=1-\frac{2\pi}{k^3}\cdot c_N\cdot\mathrm{Im}(\mathbf{V}\cdot\mathbf{e_V})_{\theta=0} \tag{B.43}$$

B.2.2 Rayleigh Scattering

Rayleigh scattering applies to scattering events at very small particles as compared to the wavelength ($ka \ll 1$). Additionally, it is assumed that the scattering is absolutely coherent, which requires that the phase shift for wave propagation through a particle is negligibly small ($m_{\mathrm{r}}\,ka \ll 1$). If this holds true, one can directly relate the scattering amplitudes to the polarisability α of the particles:

$$\begin{pmatrix} S_2 & S_3 \\ S_4 & S_1 \end{pmatrix}=-\frac{\mathrm{i}k^3\alpha}{4\pi\varepsilon_0}\begin{pmatrix} \cos\theta & 0 \\ 0 & 1 \end{pmatrix}=-\mathrm{i}k^3\alpha'\begin{pmatrix} \cos\theta & 0 \\ 0 & 1 \end{pmatrix}. \tag{B.44}$$

The total scattering cross section of Rayleigh scatterers is:

$$C_{\text{sca}} = \frac{2}{3}k^4 \cdot |\alpha'|^2 = \frac{3}{2\pi} \cdot k^4 V^2 \left|\frac{m_{\text{r}}^2 - 1}{m_{\text{r}}^2 + 2}\right|^2 = \frac{8\pi}{3} \cdot k^4 a^6 \left|\frac{m_{\text{r}}^2 - 1}{m_{\text{r}}^2 + 2}\right|^2, \tag{B.45}$$

where V is the volume of the scatterer and a is its volume equivalent radius. The partial scattering cross section is computed from:

$$C_{\text{p,sca}} = k^4 |\alpha'|^2 \left(\sin^2\phi + \cos^2\phi \cdot \cos^2\theta\right), \tag{B.46}$$

which yields for vertical and horizontal polarisation

$$\text{I}_{\text{vv}}\colon C_{\text{p,sca}} = \frac{9}{4\pi} \cdot k^4 V^2 \left|\frac{m_{\text{r}}^2 - 1}{m_{\text{r}}^2 + 2}\right|^2 = 4\pi \cdot k^4 a^6 \left|\frac{m_{\text{r}}^2 - 1}{m_{\text{r}}^2 + 2}\right|^2 = \frac{3}{2} C_{\text{sca}}, \tag{B.47}$$

$$\text{I}_{\text{hh}}\colon C_{\text{p,sca}} = \frac{9}{4\pi} \cos^2\theta \cdot k^4 V^2 \left|\frac{m_{\text{r}}^2 - 1}{m_{\text{r}}^2 + 2}\right|^2 = 4\pi \cos^2\theta \cdot k^4 a^6 \left|\frac{m_{\text{r}}^2 - 1}{m_{\text{r}}^2 + 2}\right|^2 = \frac{3}{2} \cos^2\theta \cdot C_{\text{sca}}, \tag{B.48}$$

while for unpolarised light (or by averaging over the azimut)

$$\begin{aligned} C_{\text{p,sca}} &= \frac{9}{8\pi} \left(1 + \cos^2\theta\right) \cdot k^4 V^2 \left|\frac{m_{\text{r}}^2 - 1}{m_{\text{r}}^2 + 2}\right|^2 \\ &= 2\pi \left(1 + \cos^2\theta\right) \cdot k^4 a^6 \left|\frac{m_{\text{r}}^2 - 1}{m_{\text{r}}^2 + 2}\right|^2 = \frac{3}{4} \left(1 + \cos^2\theta\right) \cdot C_{\text{sca}} \end{aligned} \tag{B.49}$$

holds true. The absorption cross section amounts to

$$C_{\text{abs}} = 4\pi k \cdot \text{Im}(\alpha') = 4\pi \cdot ka^3 \text{Im}\left(\frac{m_{\text{r}}^2 - 1}{m_{\text{r}}^2 + 2}\right) = 3kV \cdot \text{Im}\left(\frac{m_{\text{r}}^2 - 1}{m_{\text{r}}^2 + 2}\right). \tag{B.50}$$

For the refractive index of the suspensions, one ends up with:

$$\frac{m_{\text{sus}}}{m_{\text{m}}} = 1 + 2\pi c_N \cdot \text{Re}(\alpha') = 1 + \frac{3}{2} c_V \cdot \text{Re}\left(\frac{m_{\text{r}}^2 - 1}{m_{\text{r}}^2 + 2}\right). \tag{B.51}$$

Note that for Rayleigh scattering, the extinction cross section C_{ext} can not be calculated by means of Eq. (B.41), which evaluates the phase lag of the scattered wave to the incident wave. Such an evaluation is meaningless for the a priori assumption of absolute coherence as employed in this approach. Instead, one computes the extinction cross section C_{ext} from C_{sca} and C_{abs} via the conservation equation of energy (B.42).

The above equations are based on the assumption of isotropic dipoles. They should not be used when the refractive index depends on orientation or when the particles are clearly anisometrically shaped (e.g. for cylinders and discs). In such a case the scattering becomes a function of orientation, even within the Rayleigh limit of scattering. It appears that the orientation averaged cross sections for scattering and absorption are larger than those of the volume equivalent sphere.

An ellipsoid will scatter more intensely than a sphere of the same volume. (Kerker 1969, p. 583).

Exact, analytical expressions for the scattering of very small ellipsoids were already derived by Lord Rayleigh (1897). Other particle shapes may require numerical solutions for Rayleigh scattering. Note that the calculation of the orientation averaged scattering behaviour benefits from symmetry relations (Bohren and Huffman 1983, pp. 154–157).

B.2.3. Rayleigh-Debye-Gans Scattering

The principal approach of Rayleigh scattering, which assumes coherent, yet independent scattering from all volume elements of the particles, can be straightforwardly extented to larger particles when these scattering events are weak and not interrelated. This is fulfilled when the optical contrast between particle and fluid phase remains rather low ($|m_r - 1| \ll 1$). A further condition is that the scattered waves do not significantly affect the incident field inside the particle, which applies for $2ka \cdot |m_r - 1| \ll 1$. The light scattering is, hence, reduced to an interference problem, which means that the scattering patterns are linear transformations of the spatial distribution of volume, i.e. of particle morphology.

Such a type of scattering is called Rayleigh-Debye-Gans (RDG) scattering. The amplitude functions of RDG can be written as:

$$\begin{pmatrix} S_2 & S_3 \\ S_4 & S_1 \end{pmatrix} = -\mathrm{i}k^3\alpha' \cdot \sqrt{P} \cdot \begin{pmatrix} \cos\theta & 0 \\ 0 & 1 \end{pmatrix} \approx -\frac{\mathrm{i}k^3(m_r - 1)}{2\pi} \cdot V\sqrt{P} \cdot \begin{pmatrix} \cos\theta & 0 \\ 0 & 1 \end{pmatrix}, \tag{B.52}$$

where P is the form factor:

$$P = V^{-2}\left|\int \mathrm{e}^{\mathrm{i}\delta}\mathrm{d}V\right|^2. \tag{B.53}$$

For spherical particles, this parameter is defined by:

$$P = \left(\frac{3(\sin x - x\cos x)}{x^3}\right)^2 = \left(\frac{3}{x}j_1(x)\right)^2 \quad \text{with}\, x = q \cdot a = \sqrt{\frac{5}{3}} \cdot q \cdot R_g. \tag{B.54}$$

Note that the form factor of non-spherical particles depends on the particle orientation (e.g. Kerker 1969, pp. 482–486).

The partial scattering cross section is

$$C_{\text{p,sca}} = \frac{1}{\pi} \cdot k^4 |m_{\text{r}} - 1|^2 V^2 P \cdot \left(\sin^2\phi + \cos^2\phi \cdot \cos^2\theta\right). \tag{B.55}$$

The total scattering cross section can be computed from C_{psca} by integrating over the solid angle. For small spheres with $P \approx 1 - \frac{1}{5}x^2$ one obtains:

$$C_{\text{sca}} = \frac{8}{3}\left(1 - \frac{2}{5}(ka)^2\right) \cdot k^4 V^2 \cdot |m_{\text{r}} - 1|^2. \tag{B.56}$$

The absorption cross section can be calculated with the Rayleigh solution (Eq. (B.50)) because the incident field is considered to remain unaffected:

$$C_{\text{abs}} = 2kV \cdot \text{Im}(m_{\text{r}}). \tag{B.57}$$

B.2.4 Mie Solution for Scattering at Arbitrary Spheres

The validity of Rayleigh and RDG scattering is restricted to very small and/or weakly scattering particles. For the general case, one must not ignore the interaction between the scattering waves from different volume elements inside the particle. Lorenz (1898) and Mie (1908) independently developed a scheme for computing the light scattering at arbitrary particles, which is based on a multipole expansion of the scattered field and the induced internal field. Based on these fields, it is possible to calculate the scattering amplitude functions S_1 and S_2 (the off-diagonal matrix elements S_3 and S_4 are zero due to the radial symmetry of spheres). For the far-field, the solutions are expressed as a series:

$$S_1(\theta) = \sum_{n=1}^{\infty} \frac{2n+1}{n(n+1)} (a_n \pi_n(\cos\theta) + b_n \tau_n(\cos\theta)), \tag{B.58}$$

$$S_2(\theta) = \sum_{n=1}^{\infty} \frac{2n+1}{n(n+1)} (b_n \pi_n(\cos\theta) + a_n \tau_n(\cos\theta)). \tag{B.59}$$

The angular functions π_n and τ_n are related to the associated Legendre polynomials of the first kind of degree n and order one (cf. Sect. C.1.7):

$$\pi_n = -P_n^1(\cos\theta)/\sin\theta, \tag{B.60}$$

$$\tau_n = -\mathrm{d}P_n^1(\cos\theta)/\mathrm{d}\theta. \tag{B.61}$$

They solely contain the impact of the scattering angle θ, while the expansion coefficients for the scattered field—the Mie coefficients a_n and b_n—reflect the impact of particle size and optical properties. For non-ferromagnetic particles (i.e. $\mu_P = \mu_F = 1$), which frequently applies to the optical domain, they are given as (notation acc. to Bohren and Huffman 1983):

$$a_n = \frac{\psi_n(\alpha)\psi'_n(\beta) - m_r\psi'_n(\alpha)\psi_n(\beta)}{\xi_n(\alpha)\psi'_n(\beta) - m_r\xi'_n(\alpha)\psi_n(\beta)}, \tag{B.62}$$

$$b_n = \frac{\psi'_n(\alpha)\psi_n(\beta) - m_r\psi_n(\alpha)\psi'_n(\beta)}{\xi'_n(\alpha)\psi_n(\beta) - m_r\xi_n(\alpha)\psi'_n(\beta)}, \tag{B.63}$$

where ψ_n and ξ_n are the Ricatti-Bessel functions of first and third kind, with the arguments being $\alpha = ka$ and $\beta = m_r ka$.

In contrast to Rayleigh or RDG scattering, one cannot neglect the impact of the induced internal field on the particle scattering if arbitrary particles are considered. The corresponding Mie coefficients can be calculated with:

$$c_n = \frac{m_r \cdot \mathrm{i}}{\xi'_n(\alpha)\psi_n(\beta) - m_r\xi_n(\alpha)\psi'_n(\beta)}, \tag{B.64}$$

$$d_n = \frac{-m_r \cdot \mathrm{i}}{\xi_n(\alpha)\psi'_n(\beta) - m_r\xi'_n(\alpha)\psi_n(\beta)}. \tag{B.65}$$

The amplitude functions S_1 and S_2 can be directly employed for calculating scattering intensity in a given direction (Eq. (B.39)). With further processing, they lead to equations for the cross sections for extinction and scattering and for the effective refractive index:

$$C_{\text{ext}} = \frac{4\pi}{k^2}\sum_{n=1}^{\infty}\frac{2n+1}{2}\mathrm{Re}(a_n + b_n), \tag{B.66}$$

$$C_{\text{sca}} = \frac{4\pi}{k^2}\sum_{n=1}^{\infty}\frac{2n+1}{2}\left(|a_n|^2 + |b_n|^2\right), \tag{B.67}$$

$$\frac{m_{\text{sus}}}{m_{\text{m}}} = 1 - \frac{3c_V}{4k^3a^3}\cdot\sum_{n=1}(2n+1)\mathrm{Im}(a_n + b_n). \tag{B.68}$$

The absorption cross section is derived from the internal field (Fu and Sun 2001, Arruda and Martinez 2010):

$$C_{\text{abs}} = \frac{4\pi}{k^2}\sum_{n=1}^{\infty}\frac{2n+1}{2}\mathrm{Im}\left(|c_n|^2\frac{\psi_n(\beta)\psi'_n(\beta^*)}{m_r} - |d_n|^2\frac{\psi_n(\beta^*)\psi'_n(\beta)}{m_r}\right). \tag{B.69}$$

Even though the series in Eqs. (B.58)–(B.68) are strictly infinite, it can be shown that sufficient numerical accuracy is achieved when the series are terminated above a critical number (Wiscombe 1980)

$$N_{\mathrm{W}} = ka + 4 \cdot \sqrt[3]{ka} + 2. \tag{B.70}$$

That means that for colloidal particles (x ≤ 1 μm), typically less than 20 terms need to be calculated, which can be easily accomplished with electronic help. However, for very large particles in the upper micrometre-range, the numerical problems concerned with computing the Mie coefficients need proper consideration (Stübinger 2010).

B.2.5. Generalised Multi-Particle Mie Theory for Clusters of Spheres

The generalised multi-particle Mie theory (GMM) is one way to calculate the scattering on aggregates (or more generally: clusters) of spheres. It was developed by Yu-lin Xu in a series of papers (e.g. Xu 1995, Xu 1997, Xu 1998a, Xu 1998b, Xu and Khlebtsov 2003) and is closely related to the Mie theory for single particle scattering. That means, essentially, that the electromagnetic fields are described by multipole expansions in spherical harmonics and that the scattering on a particle j within the cluster can be principally described by Mie's solution for spheres. For that purpose, one has to compute the incident field on j which results from the external field and the scattering fields from all particles $l \neq j$. In mathematical terms, one can, therefore, decompose the scattering field of particle j into contributions by the different multipole fields of the external wave as well as all other scattered waves. Since these fields are all interrelated but also superpose linearly, one eventually obtains a linear system of equations for the Mie scattering coefficients of all particles. A considerable complication results from the fact that the multipole expansion on particle j refers to a coordinate system centred in j, whereas the interacting scattering field from particle l is first described within the coordinate system l. It is, therefore, necessary to "translate" all equations so that they are expressed within a common reference system. The details of this theory are rather elaborate and can be retrieved from Xu's papers. This section provides only the basic equations of the theory.

An important result of GMM is the amplitude functions in the far-field limit (Xu and Khlebtsov 2003):

$$\left.\begin{matrix} S_1(\theta,\phi) \\ S_2(\theta,\phi) \\ S_3(\theta,\phi) \\ S_4(\theta,\phi) \end{matrix}\right\} = \sum_{l=1}^{N} \mathrm{e}^{-\mathrm{i}k\Delta^l} \sum_{n=1}^{N_W^l} \sum_{m=0}^{n} \sum_{p=1}^{2} f_m \cdot \left(a^l_{mnp} \cdot e^{\mathrm{i}\phi_m} + \left\{ \begin{matrix} -1 \\ -1 \\ 1 \\ 1 \end{matrix} \right\} \cdot (-1)^{m+p} a^l_{-mnp} e^{-\mathrm{i}\phi_m} \right) \cdot \left\{ \begin{matrix} \tilde{\tau}_{mn3-p} \\ \tilde{\tau}_{mnp} \\ \mathrm{i}\cdot\tilde{\tau}_{mnp} \\ -\mathrm{i}\cdot\tilde{\tau}_{mn3-p} \end{matrix} \right\} \tag{B.71}$$

with

$$f_m = (1+\delta_{0m})^{-1}, \tag{B.72}$$

$$\phi_m = (m-1)\cdot\phi + \beta, \tag{B.73}$$

$$\Delta^l = X^l \sin\theta\, \cos\phi + Y^l \sin\theta\, \sin\phi + Z^l \cos\theta, \tag{B.74}$$

$$a^l_{mn1} = a^l_{mn} \quad \text{and} \quad a^l_{mn2} = b^l_{mn}, \tag{B.75}$$

$$\tau_{mn1} = \tau_{mn}, \tag{B.76}$$

$$\tau_{mn1} = \tau_{mn} \quad \text{and} \quad \tau_{mn2} = \pi_{mn}, \tag{B.77}$$

$$C_{mn} = \sqrt{\frac{(2n+1)(n-m)!}{n(n+1)(n+m)!}}. \tag{B.78}$$

The angular dependency of the scattering field is contained within the functions π_{mn} and τ_{mn}, which are related to the associated Legendre polynomials of the first kind of degree n and order m (cf. Sect. C.1.7):

$$\pi_{mn} = (-1)^m P_n^m(\cos\theta)/\sin\theta, \tag{B.79}$$

$$\tau_{mn} = (-1)^m \mathrm{d}P_n^m(\cos\theta)/\mathrm{d}\theta. \tag{B.80}$$

The two parameters a^l_{mn} and b^l_{mn} denote the partial (Mie) scattering coefficients of the individual particles. They refer to the coordinate system l and are computed by solving the following system of equations:

$$a^j_{mn} = a^j_n \left[p^{jj}_{mn} - \sum_{l\neq j}^{N} \sum_{\nu=1}^{N_W^l} \sum_{\mu=-\nu}^{\nu} \left(a^l_{\mu\nu} A^{lj}_{mn\mu\nu} + b^l_{\mu\nu} B^{lj}_{mn\mu\nu} \right) \right], \tag{B.81}$$

$$b_{mn}^{j} = b_{n}^{j}\left[q_{mn}^{j,j} - \sum_{l\neq j}^{N}\sum_{\nu=1}^{N_W^l}\sum_{\mu=-\nu}^{\nu}\left(a_{\mu\nu}^{l}B_{mn\mu\nu}^{lj} + b_{\mu\nu}^{l}A_{mn\mu\nu}^{lj}\right)\right], \tag{B.82}$$

in which a_n^j and b_n^j are the Mie coefficients of isolated particles j (Eqs. (B.62) and (B.63)), $p_{mn}^{j,j}$ and $q_{mn}^{j,j}$ denote the expansion coefficients for the external field (within coordinates j), while $A_{mn\mu\nu}^{lj}$ and $B_{mn\mu\nu}^{lj}$ are the vector translation coefficients which transform the spherical coordinates from l to j.

Based on the far-field solution of the scattering amplitude functions, one can derive the following expressions for the extinction and scattering cross sections:

$$C_{\text{ext}} = \sum_{l=1}^{N} C_{\text{ext}}^{l} = \frac{4\pi}{k^2}\sum_{l=1}^{N}\sum_{n=1}^{N_W^l}\sum_{m=-n}^{n}\text{Re}\left(\left(p_{mn}^{l}\right)^{*}a_{mn}^{l} + \left(q_{mn}^{l}\right)^{*}b_{mn}^{l}\right), \tag{B.83}$$

$$C_{\text{sca}} = \sum_{l=1}^{N} C_{\text{sca}}^{l} = \frac{4\pi}{k^2}\sum_{l=1}^{N}\sum_{n=1}^{N_W^l}\sum_{m=-n}^{n}\text{Re}\left(\left(a_{mn}^{l}\right)^{*}a_{mn}^{(l)} + \left(b_{mn}^{l}\right)^{*}b_{mn}^{(l)}\right), \tag{B.84}$$

in which $(\;)^*$ means the complex conjugate, p_{mn}^{l} and p_{mn}^{l} denote the expansion coefficients for the total incident field on particle j (within coordinates j), while $a_{mn}^{(l)}$ and $b_{mn}^{(l)}$ are the total scattering coefficients of the cluster within the coordinates of l:

$$a_{mn}^{(l)} = \sum_{l=1}^{N}\sum_{\nu=1}^{N_W^l}\sum_{\mu=-\nu}^{\nu}\left(a_{\mu\nu}^{j}\tilde{A}_{mn\mu\nu}^{lj} + b_{\mu\nu}^{j}\tilde{B}_{mn\mu\nu}^{lj}\right), \tag{B.85}$$

$$b_{mn}^{(l)} = \sum_{l=1}^{N}\sum_{\nu=1}^{N_W^l}\sum_{\mu=-\nu}^{\nu}\left(a_{\mu\nu}^{j}\tilde{B}_{mn\mu\nu}^{lj} + b_{\mu\nu}^{j}\tilde{A}_{mn\mu\nu}^{lj}\right), \tag{B.86}$$

with a second set ($\tilde{A}_{mn\mu\nu}^{lj}$ and $\tilde{B}_{mn\mu\nu}^{lj}$) of vector translation coefficients.

The calculation of the absorption cross section requires the solution of the internal fields within the particles (i.e. c_n and d_n, cf. Eqs. (B.64) and (B.65))—in addition to the knowledge of all individual scattering fields:

$$C_{\text{abs}} = \sum_{l=1}^{N} C_{\text{abs}}^{l} = \frac{4\pi}{k^2}\sum_{l=1}^{N}\sum_{n=1}^{N_W^l}\sum_{m=-n}^{n}\left(\begin{array}{c}\left|c_n^l\right|^2\cdot\left|b_{mn}^l\right|^2\text{Im}\left(\dfrac{\psi_n(\beta^l)\psi_n'^{*}(\beta^l)}{m_r^l}\right)\ldots\\ -\left|d_n^l\right|^2\cdot\left|a_{mn}^l\right|^2\text{Im}\left(\dfrac{\psi_n^{*}(\beta^l)\psi_n'(\beta^l)}{m_r^l}\right)\end{array}\right). \tag{B.87}$$

B.2.6. Fraunhofer Diffraction

Fraunhofer diffraction applies to particles that are much larger than wavelength and for which all radiation that is incident on the particle surface is either absorbed or redirected by reflection and refraction. The only forward scattering results from the diffraction at the edges of the particle. In that sense, the particle can be replaced by an opaque, planar obstacle. According to Babinet's principle, this is complementary to the diffraction at an aperture of the same shape. The amplitude function for that case is given as:

$$S(\theta,\phi) = \frac{k^2}{4\pi}\oint e^{-ik\,\sin\theta(\xi\,\cos\phi+\eta\,\sin\phi)}\cdot(1+\cos\theta)\mathrm{d}\xi\,\mathrm{d}\eta, \tag{B.88}$$

where ξ, η denote the Cartesian coordinates of the aperture.

It appears that the total amount of diffracted light is just the same as blocked by the particle. Hence, the extinction cross section is only twice the geometrical cross section A_{proj} of the particle:

$$C_{\mathrm{sca}} = \int\limits_0^{2\pi}\int\limits_0^{\pi}\frac{|S(\theta,\phi)|^2}{k^2}\sin\theta\,\mathrm{d}\theta\,\mathrm{d}\phi = \oint \mathrm{d}\xi\,\mathrm{d}\eta = A_{\mathrm{proj}}, \tag{B.89}$$

$$C_{\mathrm{ext}} = \frac{4\pi}{k^2}\mathrm{Re}(S(0)) = \oint 2\mathrm{d}\xi\,\mathrm{d}\eta = 2A_{\mathrm{proj}}. \tag{B.90}$$

The diffraction pattern can be employed to characterise the morphology of particles. For spheres (or circular discs) the amplitude function $S(\theta)$ is described by the equation:

$$S(\theta) = \frac{k^2}{4\pi}\oint e^{-ik\xi\,\sin\theta}(1+\cos\theta)\mathrm{d}\xi\mathrm{d}\eta\cdot = \tfrac{1}{2}(ka)^2(1+\cos\theta)\frac{J_1(ka\,\sin\theta)}{ka\,\sin\theta}, \tag{B.91}$$

in which J_1 denote the Bessel function of the first kind of first order. The intensity of the diffracted light, thus, obeys:

$$I(\theta)/I_0 = \frac{(ka)^4}{(kr)^2}\cdot\left(\frac{J_1(ka\,\sin\theta)}{ka\,\sin\theta}\right)^2. \tag{B.92}$$

Fraunhofer diffraction is widely employed for the analysis of laser diffraction measurements, in particular for non-spherical particles in the micrometre-range.

B.2.7. Anomalous Diffraction

Anomalous diffraction denotes the forward scattering of large particles ($ka \gg 1$) that show a very low optical contrast to the continuous phase ($|m_r - 1| \to 0$) and for which the light absorption is virtually zero (i.e. real refractive index). In such a case, there is hardly any redirection of light by reflection and refraction; instead, there is a significant portion of light that is hardly affected in its direction of propagation when travelling through the particle. The forward scattering pattern is, thus, a combined effect of diffraction and transmission. Fraunhofer diffraction theory, which ignores the contribution of transmitted light, then needs ramification. Under the assumption of small scattering angles ($\cos\theta \approx 1$), the scattering amplitude is:

$$S(\theta,\phi) = \frac{k^2}{2\pi}\oint\left(1 - \mathrm{e}^{-\mathrm{i}kL(m_r-1)}\right)\mathrm{e}^{-\mathrm{i}k\,\sin\theta(\xi\,\cos\phi + \eta\,\sin\phi)}\mathrm{d}\xi\mathrm{d}\eta, \qquad \text{(B.93)}$$

where L is the path length a ray of light travels through the particle at the transverse coordinates ξ and η. The term $kL(m_r - 1)$ is, therefore, the phase lag this ray receives when passing the particle.

In general, the path length L depends on the shape and orientation of the particle. For the special case of spherical particles, L is a known function of the transverse coordinates. Equation (B.93) then transforms to (van de Hulst 1981, p. 184):

$$S(\theta) = (ka)^2\int_0^{\pi/2}\left(1 - \mathrm{e}^{-\mathrm{i}\rho\,\sin\tau}\right)J_0(z\,\cos\tau)\cos\tau\,\sin\tau\,\mathrm{d}\tau, \qquad \text{(B.94)}$$

where J_0 is the Bessel function of the first kind of zero order and the arguments are defined as:

$$\rho = 2ka\cdot(m_r - 1), \qquad \text{(B.95)}$$

$$z = ka\cdot\sin\theta \approx ka\theta. \qquad \text{(B.96)}$$

The scattering pattern of anomalous diffraction may considerably deviate from those observed in the Fraunhofer limit. This is of particular importance when analysing laser diffraction measurements (cf. Sect. 2.3.4.3).

Since anomalous diffraction is only relevant for forward scattering, it is not meaningful to discuss other cross sections than the extinction one. A first order approximation for real refractive indices was provided by van de Hulst (1981, p. 176):

$$C_{ext}/A_{proj} = 2 - \frac{4}{\rho}\cdot\sin\rho + \frac{4}{\rho^2}\cdot(1 - \cos\rho). \qquad \text{(B.97)}$$

This equation applies to extremely low optical contrast ($m_r - 1 \ll 1$). A wider range of applicability is achieved with the following modification (cf. van de Hulst 1981, p. 384):

$$C_{ext}/A_{proj} = 2 - \frac{16 \cdot m_r^2}{(m_r + 1)^2} \cdot \frac{\sin \rho}{\rho} + \frac{4}{\rho} \cdot (2 - m_r) \cdot (1 - \cos \rho) + 1.8 \cdot ka^{-2/3}. \tag{B.98}$$

Even though the anomalous diffraction theory provides no real benefit for the calculation of sphere scattering, which is best accomplished by Mie theory, it may help to better understand the functional relationships prevailing in this limit.

References to Light Scattering

T. J. Arruda, A. S. Martinez, Electromagnetic energy within magnetic spheres. J. Opt. Soc. Am. A 27(5), 992–1001 (2010). doi:10.1364/JOSAA.27.00099

C.F. Bohren, D.R. Huffman, *Absorption and Scattering of Light by Small Particles* (Wiley New York, 1983). ISBN 0-471-05772-X

Q. Fu, W.B. Sun, Mie theory for light scattering by a spherical particle in an absorbing medium. Appl. Opt. **40**(9), 1354–1361 (2001). doi:10.1364/AO.40.001354

M. Kerker, in *The Scattering of Light and Other Electromagnetic Radiation*. In series: E.M. Loebl (ed.) Physical Chemistry, vol. 16 (Academic Press, New York, 1969). ISBN 0-124-04550-2

L. Lorenz, in *Sur la lumière réfléchie et réfractée par une surface transparente*, ed by H. Valentiner. Oeuvres scientifiques de L. Lorenz (Tome Premier, Libraire Lehmann & Stage, Copenhague, 1898), pp. 403–529.

G. Mie, Beiträge zur Optik trüber Medien, speziell kolloidaler Goldlösungen. Ann. Phys. IV **25**(3), 377–445 (1908). doi:10.1002/andp.19083300302

L. Rayleigh, On the incidence of aerial and electric waves upon small obstacles in the form of ellipsoids or elliptic cylinders, and on the passage of electric waves through a circular aperture in a conducting screen. Phil. Mag. S. 5 **44**(266), 28–52 (1897). doi:10.1080/14786449708621026

J.W. Strutt (Lord Rayleigh), On the light from the sky, its polarization and colour. Phil. Mag. S. 4. **41**(271), 107–120 (1871a). doi:10.1080/14786447108640452

J.W. Strutt (Lord Rayleigh), On the light from the sky, its polarization and colour. Phil. Mag. S. 4. **41**(273), 274–279 (1871b). doi:10.1080/14786447108640479

J.W. Strutt (Lord Rayleigh), On the scattering of light by small particles. Phil. Mag. S. 4. **41**(275), 447–454 (1871c). doi:10.1080/14786447108640507

T. Stübinger, U. Köhler, W. Witt, Verification of Mie scattering algorithms by extreme precision calculations. On the CD-ROM: *WCPT6 2010, World Congress on Particle Technology, Nuremberg, 26.29.04* (2010). ISBN 978-3-00-030570-2 (available from NürnbergMesse GmbH, Messezentrum, 90471 Nürnberg, Germany), paper 00272.

H.C. van de Hulst, *Light Scattering by Small Particles* (Dover Publications, New York, 1981). ISBN 0-486-64228-3

W. Wiscombe, Improved Mie scattering algorithms. Appl. Opt. **19**(9), 1505–1509 (1980). doi:10.1364/AO.19.001505

Y.-L. Xu, Electromagnetic scattering by an aggregate of spheres. Appl. Opt. **34**(21), 4573–4588 (1995). doi:10.1364/AO.34.004573

Y.-L. Xu, Electromagnetic scattering by an aggregate of spheres: Far field. Appl. Opt., **36**(36), 9496–9508 (1997). doi:10.1364/AO.36.009496

Y.-L. Xu, Efficient evaluation of vector translation coefficients in multiparticle light-scattering theories. J. Comp. Phys. **139**(1), 137–165 (1998a). doi:10.1006/jcph.1997.5867
Y.-L. Xu, Electromagnetic scattering by an aggregate of spheres: asymmetry parameter. Phys. Lett. A **249**(1–2), 30–36 (1998b). doi:10.1016/S0375-9601(98)00708-7
R. Xu, *Particle Characterization: Light Scattering Methods* (Kluwer Academic Publishers, Dortrecht, 2000). ISBN 0-7923-6300-0
Y.-L. Xu, N. G. Khlebtsov, Orientation-averaged radiative properties of an arbitrary configuration of scatterers. J. Quant. Spectrosc. Radiat. Transfer 79–80, 1121–1137 (2003). doi:10.1016/S0022-4073(02)00345-X

B.3 Hamaker Function

B.3.1 Exact Solution for Two Parallel Half-Spaces

The generic term *van-der-Waals forces* refers to any interaction between electrically neutral atoms and molecules which results from the non-uniform distribution of electrons around the atomic nuclei. This non-uniformity may be observable as a permanent polarisation of molecules (e.g. for water), yet it primarily exist on an instantaneous level due to the electron oscillations. Atoms and molecules are, therefore, fluctuating dipoles that interact via Coulombic forces, which eventually produces a net attraction. The strength of these forces diminishes very rapidly with the intermolecular distance r_{ij}; the corresponding energy potential u_{ij} obeys a power-law decay (Kralchevsky et al. 2002, Chap. 5.4.2):

$$u_{ij}(r) = -\frac{\alpha_{ij}}{r^6}, Z \qquad \text{(B.99)}$$

where a_{ij} is the material-specific interaction coefficient. Note that the interaction between fluctuating dipoles—the *dispersion* or *London forces*—is commonly distinguished from the interaction between two permanent dipoles (*orientation* or *Keesom interaction*) as well as that between a permanent and a fluctuating dipole (*induction* or *Debye interaction*) (Israelachvili 1992, pp. 62 and 75).

The van-der-Waals interaction between atomic dipoles leads to corresponding interactions between macroscopic bodies. The effective, macroscopic van-der-Waals forces can be approximated from the pairwise interaction between all atoms (Hamaker-de-Boer theory). Alternatively, one can relate the macroscopic van-der-Waals forces to the fluctuating electromagnetic fields that originate from the macrobodies (Lifshitz theory). These fields arise from the local fluctuations of electron density and can, therefore, be related to the dielectric properties of the material, which are accessible by experiment.

Both approaches merge when the material constant A_{132} at Hamaker-de-Boer (cf. Eq. (3.23)) is replaced by a distance dependent material function. For two parallel half spaces (i.e. thick plates) the Hamaker function $A_{132}(h)$ is computed by:

$$A_{132}(h) = -\frac{3}{2}\cdot \mathrm{k_B}T \cdot \sum_{n=0}^{\infty} 2^{-\delta_{0n}} \int_{x_n}^{\infty} x \cdot \ln[(1 - y_{13}\cdot y_{23}\cdot \mathrm{e}^{-x})\cdot(1 - z_{13}\cdot z_{23}\cdot \mathrm{e}^{-x})]dx, \tag{B.100}$$

where the auxiliary parameters are defined as:

$$y_{ij} = \frac{x\cdot \varepsilon_{i,n} - s_i\cdot \varepsilon_{j,n}}{x\cdot \varepsilon_{i,n} + s_i\cdot \varepsilon_{j,n}}, \quad z_{ij} = \frac{x - s_i}{x + s_i}, \quad s_i = \sqrt{x^2 + x_n^2 \cdot \left(\frac{\varepsilon_{i,n}}{\varepsilon_{3,n}} - 1\right)}, \tag{B.101}$$

$$x_n = \frac{2\cdot \mathrm{h_P}\cdot \xi_n}{\mathrm{c_0}}\cdot \sqrt{\varepsilon_{3,n}}, \; \xi_n = \frac{4\cdot \pi^2 \cdot \mathrm{k_B}T}{\mathrm{h_P}}\cdot n. \tag{B.102}$$

The quantities $\mathrm{c_0}$, $\mathrm{h_P}$, $\mathrm{k_B}$, and T denote the velocity of light in a vacuum, Planck's constant, Boltzmann's constant, and the temperature, respectively. The parameters $\varepsilon_{j,n}$ represent the values of the imaginary dielectric response function $\varepsilon(\mathrm{i}\xi)$ at the imaginary frequencies $\mathrm{i}\cdot\xi_n$, which can be computed from the imaginary part of the dielectric permittivity by means of a Kramers-Kronig relation:

$$\varepsilon(\mathrm{i}\cdot\xi) = 1 + \frac{2}{\pi}\cdot \int_0^{\infty} \frac{x\cdot \varepsilon''(x)}{x^2 + \xi^2}\mathrm{d}x. \tag{B.103}$$

However, the function $\varepsilon(\mathrm{i}\xi)$ is usually approximated by means of a damped oscillator model:

$$\varepsilon(\mathrm{i}\xi) = 1 + \sum_{j=1}^{N} \frac{f_j + h_j\xi}{\omega_j^2 + g_j\xi + \xi^2}, \tag{B.104}$$

where ω_j, f_j and g_j are nonnegative and the sum of h_j is zero (Parsegian and Weiss 1981). The terms in Eq. (B.104) become simplified for the free electron contributions of metals (ω_j, h_j = 0, Drude model), for dielectric substances (h_j = 0, Lorentzian oscillators) and for Debye relaxation (ω_j, h_j = 0). For a lot of substances, the dielectric function is sufficiently described by a few number of relaxation frequencies ω_j in the IR and UV domains. Polar liquids, like water, additionally require a microwave contribution. Values for the parameters ω_j, f_j, g_j and h_j are published for a large variety of substances, including liquids (e.g. Nguyen 2000, Bell and Dimos 2000), dielectric solids (e.g. Bergström 1997) as well as metals and semiconductors (e.g. Parsegian and Weiss 1981, Chen et al. 1991). Note that for vacuum and air, ε = 1 holds true for all frequencies.

B.3.2 Approximation and Special Cases

The shape of the Hamaker function as defined by Eq. (B.100) can be approximated well with a simple three-parameter equation (Viravathana and Marr 2000, cf. Sect. 3.2.1):

$$A_{132}(h) = A_{132,\mathrm{s}} + \left(A_{132,0} - A_{132,\mathrm{s}}\right) \cdot \left(1 + (h/h_{\mathrm{cr}})^{3/2}\right)^{-2/3}, \tag{B.105}$$

in which $A_{132,\mathrm{s}}$ denotes the static (i.e. related to permanent dipoles) contribution to the van-der-Waals interaction, while $A_{132,0}$ defines the non-retarded Hamaker constant, which prevails for a vanishing surface distance ($h = 0$). It can be calculated via integral

$$A_{132,0} = -\frac{3}{2} \cdot \mathrm{k_B}T \cdot \sum_{n=0}^{\infty} 2^{-\delta_{0n}} \int_{x_n}^{\infty} x \cdot \ln(1 - \Delta_{13}\Delta_{23} \cdot \mathrm{e}^{-x})\mathrm{d}x \quad \text{with} \quad \Delta_{ij} = \frac{\varepsilon_{i,n} - \varepsilon_{j,n}}{\varepsilon_{i,n} + \varepsilon_{j,n}}, \tag{B.106}$$

or series representation:

$$A_{132,0} = \frac{3}{2} \cdot \mathrm{k_B}T \cdot \sum_{n=0}^{\infty} 2^{-\delta_{0n}} \sum_{m=1}^{\infty} \frac{(\Delta_{13}\Delta_{23})^m}{m^3}. \tag{B.107}$$

The non-retarded Hamaker constant $A_{132,0}$ reflects the contribution of both static as well as dispersion interactions. At large surface distances h, the latter vanish and the Hamaker function $A_{132}(h)$ tends towards its static part $A_{132,\mathrm{s}}$:

$$A_{132,\mathrm{s}} = -\frac{3}{4} \cdot \mathrm{k_B}T \cdot \int_0^{\infty} x \cdot \ln\left(1 - \Delta_{13}^0 \cdot \Delta_{23}^0 \cdot \mathrm{e}^{-x}\right)\mathrm{d}x \quad \text{with}\ \Delta_{ij}^0 = \frac{\varepsilon_{i,0} - \varepsilon_{j,0}}{\varepsilon_{i,0} + \varepsilon_{j,0}}. \tag{B.108}$$

In contrast to the dispersion interaction, the static van-der-Waals interaction can be repulsive (i.e. $A_{132,\mathrm{s}} < 0$). That occurs exactly if the static permittivity $\varepsilon_{3,0}$ of the medium lies between those of the surfaces 1 and 2.

Note that Eq. (B.107) agrees with classic theory to first approximation:

$$A_{132,\mathrm{s}} \approx \frac{3}{4} \cdot \mathrm{k_B}T \cdot \Delta_{13}^0 \cdot \Delta_{23}^0. \tag{B.109}$$

It is, therefore, possible to estimate the static part for dissimilar surfaces with:

$$A_{132,s}^2 \approx A_{132,s} \cdot A_{232,s}. \tag{B.110}$$

This kind of approximation also applies to the dispersive part:

$$A_{132,d} = A_{132,0} - A_{132,s} = A_{132}(0) - A_{132}(\infty) \approx \sqrt{A_{131,d} \cdot A_{232,d}}. \tag{B.111}$$

The frequently employed approximations:

$$A_{132} = A_{12} - A_{13} - A_{23} + A_{33} \quad \text{and} \quad A_{131} = A_{11} - 2 \cdot A_{13} + A_{33}. \tag{B.112}$$

immediately result from the independency of the pairwise interactions within the Hamaker-de-Boer-theory; they are not valid in a strict sense (deviation up to −30 %).

The quantity h_{cr} from Eq. (B.105) typically amounts to several nanometres; it is best derived by fitting the real Hamaker function (B.100).

B.3.3 Hamaker-Functions for Selected Materials

Bergström (1997) provided comprehensive dielectric data of many substances for the computation of Hamaker functions. They were used to calculate the parameters $A_{132,0}$, $A_{132,s}$, and h_{cr} for three oxides which are frequently produced as pyrogenic powders.

Table B.1 Parameters of the Hamaker function acc. to Eq. (B.105), symmetric pairs

Materials	$A_{131,0}$, J	$A_{131,s}$, J	h_{cr}, nm
$SiO_{2(am)}$–water–$SiO_{2(am)}$	0.455×10^{-20}	0.29×10^{-20}	9
α-Al_2O_3–water–α-Al_2O_3	3.67×10^{-20}	0.2×10^{-20}	7
$TiO_{2(rutile)}$–water–$TiO_{2(rutile)}$	5.35×10^{-20}	0.011×10^{-20}	23

Table B.2 Parameters of the Hamaker function acc. to Eq. (B.105), asymmetric pairs

Materials	$A_{132,0}$, J	$A_{132,s}$, J	h_{cr}, nm
$SiO_{2(am)}$–water–$TiO_{2(rutile)}$	0.687×10^{-20}	-0.052×10^{-20}	16
$SiO_{2(am)}$–water–α-Al_2O_3	0.974×10^{-20}	0.24×10^{-20}	8
$TiO_{2(rutile)}$–water–α-Al_2O_3	3.11×10^{-20}	-0.044×10^{-20}	17.5

References

N. Bell, D. Dimos, Calculation of Hamaker constants in nonaqueous fluid media. MRS Proc. **624** (1), 275–280 (2000). doi:10.1557/PROC-624-275.

L. Bergström, Hamaker constants of inorganic materials. Adv. Colloid Interface Sci. **70,** 125–169 (1997). doi:10.1016/S0001-8686(97)00003-1

X.J. Chen, A.C. Levi, E. Tosatti, Hamaker-constant calculations and surface melting of metals, semimetals and semiconductors. Nuovo Cimento D, **13**(7), 919–937 (1991). doi:10.1007/BF02457179

J.N. Israelachvili, *Intermolecular and Surface Forces* (Academic Press, London, 1992). ISBN 0-12-375181-0

P.A. Kralchevsky, K.D. Danov, N.D. Denkov, Chemical physics of colloid systems and interfaces, in *Handbook of Surface and Colloid Chemistry*, ed. by K.S. Birdi, 2nd edn., Chapter 5 (CRC Press, New York, 2002), pp. 137–344. ISBN 0-8493-1079-2

A.V. Nguyen, Improved approximation of water dielectric permittivity for calculation of Hamaker constants. J. Colloid Interface Sci. 229, 648–651 (2000). doi:10.1006/jcis.2000.7010

V.A. Parsegian, G.H. Weiss, Spectroscopic parameters for computation of van der Waals forces. J. Colloid Interface Sci. **81**(1), 285–289 (1981). doi:10.1016/0021-9797(81)90325-8

P. Viravathana, D.W.M. Marr, Optical trapping of titania/silica core-shell colloidal particles. J. Colloid Interface Sci. **221**, 301–307 (2000). doi:10.1006/jcis.1999.6603

Appendix C
Mathematical Basics

C.1 Expressing and Solving Wave Phenomena in Spherical Coordinates

C.1.1 Spherical Coordinates

Spherical coordinates describe the position of a point P in space by means of its distance r to the origin O and two angular coordinates. The latter are defined with respect to Cartesian coordinates (x, y, z). There are two major conventions for the angular coordinates:

Geographic coordinates (r, λ, φ) employ the latitude φ, which is the angle between the line segment OP and x–y-plane, and the azimuth (or longitude) λ, which is the angle between the x-axis and the projection of OP onto the x–y-plane. The latitude is positive for $z > 0$ and ranges between $-\pi/2$ and $\pi/2$, whereas the azimuthal angle covers the complete cycle from 0 to 2π.

Scattering coordinates (r, ϕ, θ) replace the latitude φ by the polar or scattering angle θ, which is the angle between the z-axis and the line segment OP. It covers the range from 0 to π and is related to the latitude via $\theta = \pi/2 - \varphi$. The second angle, ϕ, is the azimuth and, therefore, identical to λ.

The former convention was traditionally employed by mathematicians, while the latter is used in physics, in particular for the description of scattering phenomena. This book employs the spherical scattering coordinates.

Scattering coordinates can be transformed to Cartesian coordinates by a linear operation:

$$\begin{pmatrix} f_x \\ f_y \\ f_z \end{pmatrix} = S \cdot \begin{pmatrix} f_r \\ f_\phi \\ f_\theta \end{pmatrix} = \begin{pmatrix} \cos\phi\,\sin\theta & -\sin\phi & \cos\phi\,\cos\theta \\ \sin\phi\,\sin\theta & \cos\phi & \sin\phi\,\cos\theta \\ \cos\theta & 0 & -\sin\theta \end{pmatrix} \cdot \begin{pmatrix} f_r \\ f_\phi \\ f_\theta \end{pmatrix}. \quad \text{(C.1)}$$

F. Babick, *Suspensions of Colloidal Particles and Aggregates*,
Particle Technology Series 20, DOI 10.1007/978-3-319-30663-6

C.1.2 Laplace's and Helmholtz Differential Equation

A fundamental equation for various physical phenomena is Laplace's equation, which describes the spatial distribution of a scalar or vectorial quantity (ψ or $\mathbf{F}$):

$$\Delta\psi = 0 \tag{C.2}$$

$$\Delta\mathbf{F} = \mathbf{0}. \tag{C.3}$$

Laplace's equation applies e.g. to Newton's energy potential (like gravitation) or to the potential flow of incompressible fluids. Any function that satisfies this differential equation is said to be harmonic or to be a potential function.

There are several differential equations that are related to Laplace's equation, e.g. the Poisson equation for the distribution of electric potential in the presence of electric charges, the wave equation for the propagation of a disturbance or the Helmholtz differential equation for the time-invariant distribution of harmonic fields. The latter is of particular relevance for scattering phenomena; it has the form:

$$\left(\Delta + k^2\right)\psi = 0 \tag{C.4}$$

$$\left(\Delta + k^2\right)\mathbf{F} = \mathbf{0} \tag{C.5}$$

where k is the wave number.

These equations can be expressed for different types of coordinates. For spherical scattering coordinates one obtains for Eq. (C.4):

$$\frac{\partial^2\psi}{\partial r^2} + \frac{2}{r}\cdot\frac{\partial\psi}{\partial r} + \frac{1}{r^2}\cdot\frac{\partial^2\psi}{\partial r^2} + \frac{\cos\theta}{r^2\sin\theta}\cdot\frac{\partial\psi}{\partial r} + \frac{1}{r^2\sin^2\theta}\cdot\frac{\partial\psi}{\partial\phi} + k^2\psi = 0. \tag{C.6}$$

Solutions of this equation can be obtained by separation of variables. For spherical scattering coordinates such a separation may be defined as:

$$\psi(r,\theta,\phi) = f(r)\cdot g(\theta)\cdot h(\phi), \tag{C.7}$$

which allows to express the partial differential equation (C.6) in terms of ordinary differential equation for each spherical coordinate:

$$r^2\cdot f'' + 2r\cdot f' + \left(k^2r^2 - n(n+1)\right)\cdot f = 0, \tag{C.8}$$

$$g'' + \frac{\cos\theta}{\sin\theta}\cdot g' + \left(n(n+1) - \frac{m^2}{\sin^2\theta}\right)\cdot g = 0, \tag{C.9}$$

$$h'' + m^2\cdot h = 0, \tag{C.10}$$

where n and m are arbitrary integers.

The first of these equations defines the spherical Bessel functions of radial coordinate r, the second leads to the Legendre polynomials for scattering angle θ, while the last defines sinus functions of the azimuthal angle ϕ. Note that for rotational symmetry ($h(\phi) = 1$) the constant m vanishes ($m = 0$).

C.1.3 Spherical Bessel-Functions

The radial dependency of harmonic wave functions in spherical coordinates is described by the differential equation (C.8):

$$x^2 \cdot f''(x) + 2x \cdot f'(x) + \left(x^2 - n(n+1)\right) \cdot f(x) = 0 \quad \text{with} \quad x = kr. \tag{C.11}$$

Its solution are the spherical Bessel functions (also: Heine-Sommerfeld functions), which can be described as (cylindrical) Bessel functions of half-integer order. The spherical Bessel functions of first and second kind (j_n and y_n) are thus defined:

$$j_n(x) = \sqrt{\frac{\pi}{2x}} \cdot J_{n+\frac{1}{2}}(x) = (-1)^n x^n \left(\frac{1}{x}\frac{\mathrm{d}}{\mathrm{d}x}\right)^n \frac{\sin x}{x}, \tag{C.12}$$

$$y_n(x) = \sqrt{\frac{\pi}{2x}} \cdot Y_{n+\frac{1}{2}}(x) = (-1)^{n+1} x^n \left(\frac{1}{x}\frac{\mathrm{d}}{\mathrm{d}x}\right)^n \frac{\cos x}{x}, \tag{C.13}$$

$$\text{with} \left(\frac{1}{x}\frac{\mathrm{d}}{\mathrm{d}x}\right)^n = \frac{1}{x} \cdot \frac{\mathrm{d}}{\mathrm{d}x}\left(\frac{1}{x}\frac{\mathrm{d}}{\mathrm{d}x}\right)^{n-1} = \frac{1}{x^n}\sum_{k=0}^{n-1} \frac{(n-1+k)!}{k!(n-1-k)!} \cdot \left(\frac{-1}{2x}\right)^k \cdot \frac{\mathrm{d}^{n-k}}{\mathrm{d}x^{n-k}}. \tag{C.14}$$

They are complemented by the Hankel functions of first and second kind (spherical Bessel function of third and fourth kind)

$$h_n^{(1)}(x) = j_n(x) + \mathrm{i} \cdot y_n(x) = \mathrm{i}^{-n-1}\frac{\mathrm{e}^{\mathrm{i}x}}{x}\sum_{k=0}^{n} \frac{(n+k)!}{k!(n-k)!} \cdot \left(\frac{\mathrm{i}}{2x}\right)^k, \tag{C.15}$$

$$h_n^{(2)}(x) = j_n(x) - \mathrm{i} \cdot y_n(x) = \mathrm{i}^{n+1}\frac{e^{-\mathrm{i}x}}{x}\sum_{k=0}^{n} \frac{(n+k)!}{k!(n-k)!} \cdot \left(\frac{-\mathrm{i}}{2x}\right)^k. \tag{C.16}$$

For the general description of wave fields it is important to note that the spherical Bessel functions of first kind j_n diverges at small arguments and vanishes at infinity, while the opposite applies to the spherical Hankel functions of first kind $h_n^{(1)}$.

The computation of the spherical Bessel functions may be enhanced by using the recurrence relations:

$$f_n = \frac{2n-1}{x} f_{n-1} - f_{n-2}, \tag{C.17}$$

$$x \cdot f_n' = n \cdot f_n - x \cdot f_{n+1} = x \cdot f_{n-1} - (n+1) \cdot f_n. \tag{C.18}$$

For real arguments we further find:

$$h_n^{(1)} = \left(h_n^{(2)}\right)^*, \tag{C.19}$$

$$j_n \cdot \left(h_n^{(1)}\right)' - h_n^{(1)} \cdot j_n' = \frac{\mathrm{i}}{x^2}, \tag{C.20}$$

$$y_n \cdot \left(h_n^{(2)}\right)' - h_n^{(2)} \cdot y_n' = -\frac{1}{x^2}, \tag{C.21}$$

$$h_n^{(1)} \cdot \frac{\mathrm{d}\left(h_n^{(1)}\right)^*}{\mathrm{d}x} - \left(h_n^{(1)}\right)^* \cdot \frac{\mathrm{d}h_n^{(1)}}{\mathrm{d}x} = -\frac{2\mathrm{i}}{x^2}, \tag{C.22}$$

$$h_n^{(2)} \cdot \frac{\mathrm{d}\left(h_n^{(2)}\right)^*}{\mathrm{d}x} - \left(h_n^{(2)}\right)^* \cdot \frac{\mathrm{d}h_n^{(2)}}{\mathrm{d}x} = \frac{2\mathrm{i}}{x^2}. \tag{C.23}$$

Asymptotes for large arguments are:

$$\lim_{|x|\to\infty} j_n(x) = \frac{(-1)^n}{x} \sin\left(x + n \cdot \frac{\pi}{2}\right), \tag{C.24}$$

$$\lim_{|x|\to\infty} y_n(x) = \frac{(-1)^{n+1}}{x} \cos\left(x + n \cdot \frac{\pi}{2}\right), \tag{C.25}$$

$$\lim_{|x|\to\infty} h_n^{(1)}(x) = \mathrm{i}^{-n-1} \cdot \frac{\mathrm{e}^{\mathrm{i}x}}{x}, \tag{C.26}$$

$$\lim_{|x|\to\infty} h_n^{(2)}(x) = \mathrm{i}^{n+1} \cdot \frac{\mathrm{e}^{-\mathrm{i}x}}{x}. \tag{C.27}$$

Additionally it may be useful to employ the Taylor expansion for very small arguments:

$$\lim_{|x|\to 0} j_n(x) = x^n \Big/ \prod_{k=0}^{n} (2k+1) - x^{n+2} \Big/ 2 \prod_{k=0}^{n} (2k+3) (|x| \leq 0.1\, n), \tag{C.28}$$

$$\lim_{|x|\to 0} y_n(x) = x^{-n-1} \cdot \prod_{k=0}^{n} (2k-1) - x^{-n+1} \cdot \frac{1}{6} \prod_{k=0}^{n} (2k-3) (|x| \leq 0.1\, n). \tag{C.29}$$

The spherical Bessel functions of low degree are:

$$j_0(x) = \frac{\sin x}{x}, \tag{C.30}$$

$$j_1(x) = \frac{1}{x}\left(\frac{\sin x}{x} - \cos x\right), \tag{C.31}$$

$$j_2(x) = \frac{1}{x}\left(\left(\frac{3}{x^2} - 1\right)\sin x - \frac{3}{x}\cos x\right), \tag{C.32}$$

$$y_0(x) = -\frac{\cos x}{x}, \tag{C.33}$$

$$y_1(x) = -\frac{1}{x}\left(\frac{\cos x}{x} + \sin x\right), \tag{C.34}$$

$$y_2(x) = -\frac{1}{x}\left(\left(\frac{3}{x^2} - 1\right)\cos x + \frac{3}{x}\sin x\right), \tag{C.35}$$

$$h_0^{(1)}(x) = -\mathrm{i}\frac{\mathrm{e}^{\mathrm{i}x}}{x}, \tag{C.36}$$

$$h_1^{(1)}(x) = -\frac{\mathrm{e}^{\mathrm{i}x}}{x}\left(1 + \frac{\mathrm{i}}{x}\right), \tag{C.37}$$

$$h_2^{(1)}(x) = \mathrm{i}\frac{\mathrm{e}^{\mathrm{i}x}}{x}\left(1 + \frac{3\mathrm{i}}{x} - \frac{3}{x^2}\right), \tag{C.38}$$

$$h_0^{(1)}(x) = \mathrm{i}\frac{\mathrm{e}^{-\mathrm{i}x}}{x}, \tag{C.39}$$

$$h_1^{(1)}(x) = -\frac{\mathrm{e}^{-\mathrm{i}x}}{x}\left(1 - \frac{\mathrm{i}}{x}\right), \tag{C.40}$$

$$h_2^{(1)}(x) = -\mathrm{i}\frac{\mathrm{e}^{-\mathrm{i}x}}{x}\left(1 - \frac{3\mathrm{i}}{x} - \frac{3}{x^2}\right). \tag{C.41}$$

C.1.4 Ricatti-Bessel-Function

The computation of light scattering frequently employs the Ricatti-Bessel functions, which are related to the spherical Bessel functions:

$$\psi_n(x) = x \cdot j_n(x), \tag{C.42}$$

$$\chi_n(x) = -x \cdot y_n(x), \tag{C.43}$$

$$\xi_n(x) = \psi_n(x) + \mathrm{i} \cdot \chi_n(x), \tag{C.44}$$

where ψ_n and ξ_n are the Ricatti-Bessel functions of first and third kind.

C.1.5 Legendre Polynomial

In case of rotational symmetry ($m = 0$) one can re-write the differential equation (C.9) for the scattering angle by replacing θ with its cosine:

$$\left(1 - x^2\right) \cdot g''(x) - 2x \cdot g'(x) + n(n+1) \cdot g(x) = 0 \quad \text{with}\, x = \cos\theta. \tag{C.45}$$

This is the Legendre differential equation, which has the following general solution for $n \in \mathrm{N}$:

$$y = C_1 \cdot P_n(x) + C_2 \cdot Q_n(x) \tag{C.46}$$

with P_n denoting the Legendre polynomial and Q_n the Legendre function of second kind. The Legendre polynomial can be calculated from the Laplace integral:

$$P_n(\cos\theta) = \frac{1}{\pi} \int_0^{\pi} (\cos\theta + \mathrm{i} \cdot \sin\theta \cos\varphi)^n \mathrm{d}\varphi \tag{C.47}$$

or via Rodrigue's representation:

$$\begin{aligned} P_n(x) &= \frac{1}{2^n n!} \cdot \frac{\mathrm{d}^n}{\mathrm{d}x^n} \left(x^2 - 1\right)^n \\ &= \frac{1 \cdots (2n-3)(2n-1)}{n!} \left(x^n - \frac{n}{2} \cdot \frac{n-1}{2n-1} x^{n-2} + \frac{n(n-1)(n-2)(n-3)}{2 \cdot 4 \cdot (2n-1)(2n-1)} x^{n-4} - \ldots \right). \end{aligned} \tag{C.48}$$

The Legendre polynomials form an orthogonal function system with:

$$\int_{-1}^{1} P_m(x) P_n(x) \mathrm{d}x = \frac{2}{2n+1} \delta_{mn}, \tag{C.49}$$

where δ_{mn} denotes the Kronecker delta.

The Legendre polynomials of low degree are:

$$P_0(x) = 1, \tag{C.50}$$

$$P_1(x) = x, \tag{C.51}$$

$$P_2(x) = \frac{1}{2}\left(3x^2 - 1\right), \tag{C.52}$$

$$P_3(x) = \frac{1}{2}\left(5x^3 - 3x\right), \tag{C.53}$$

$$P_4(x) = \frac{1}{8}\left(35x^4 - 30x^2 + 3\right), \tag{C.54}$$

$$P_5(x) = \frac{1}{8}\left(63x^5 - 70x^3 + 15x\right). \tag{C.55}$$

C.1.6 Associated Legendre Polynomial

The Legendre polynomials define the angular dependency of spherical harmonics for rotational symmetry. Yet for the general case we have to refer to Eq. (C.9) with non-zero constant m, which leads to the associated Legendre differential equation,

$$\left(1 - x^2\right) \cdot g''(x) - 2x \cdot g'(x) + \left[n(n+1) - \frac{m}{1 - x^2}\right] \cdot g(x) = 0 \quad \text{with}\, x = \cos\theta, \tag{C.56}$$

where n is a positive integer and $m = -n \ldots n$.

Its solution are the associated Legendre polynomials $P_n^m(x)$ (of degree n and order m) order and $P_n^{-m}(x)$, which are defined as:

$$P_n^m(x) = (-1)^m \cdot \left(1 - x^2\right)^{m/2} \frac{\mathrm{d}^m}{\mathrm{d}x^m} P_n(x) \tag{C.57}$$

$$\text{or} \quad P_n^m(x) = \frac{(-1)^m}{2^n n!} \cdot \left(1 - x^2\right)^{m/2} \frac{\mathrm{d}^{m+n}}{\mathrm{d}x^{m+n}} \left(x^2 - 1\right)^n \tag{C.58}$$

$$\text{with} \quad P_n^{-m}(x) = (-1)^m \cdot \frac{(n-m)!}{(n+m)!} P_n^m(x) \tag{C.59}$$

$$\text{and} \quad P_0^m(x) = \begin{cases} 1 & \text{for } m = 0 \\ 0 & \text{otherwise} \end{cases}. \tag{C.60}$$

Note, that this definition includes the Condon-Shortley phase $(-1)^m$, which is sometimes left for the sake of convenience (e.g. Bohren and Huffman, p. 90; Spiegel 1995, pp. 243–244), but has then to be included at other places of the spherical harmonics.

The associated Legendre polynomials obey the following orthogonality relation:

$$\int_{-1}^{1} P_n^m(x) P_{n'}^m(x) \mathrm{d}x = \frac{2}{2n+1} \frac{(n+m)!}{(n-m)!} \delta_{nn'}. \tag{C.61}$$

Besides there are several recurrence formulae:

$$(n-m) \cdot P_n^m = x \cdot (2n-1) \cdot P_{n-1}^m - (n+m-1) \cdot P_{n-2}^m, \tag{C.62}$$

$$P_{n+1}^m = P_{n-1}^m - (2n+1) \cdot \sqrt{1-x^2} \cdot P_n^{m-1}, \tag{C.63}$$

$$\frac{\partial}{\partial \theta} P_n^m(\cos\theta) = \frac{1}{\sin\theta} \left(n \cot\theta \cdot P_n^m - (n+m) P_{n-1}^m \right). \tag{C.64}$$

Note that $\theta \in [0, \pi]$, hence $\sin\theta \in [0, 1]$ and thus $\sin\theta = \sqrt{1-\cos^2\theta}$.

For the limiting case of $\sin\theta = 0$ the following solutions apply:

$$P_n^0(\cos\theta)\big|_{\sin\theta=0} = (\cos\theta)^n, \tag{C.65}$$

$$\left.\frac{P_n^1(\cos\theta)}{\sin\theta}\right|_{\sin\theta=0} = -\frac{n(n+1)}{2} \cdot (\cos\theta)^{n+1}, \tag{C.66}$$

$$\left.\frac{P_n^{-1}(\cos\theta)}{\sin\theta}\right|_{\sin\theta=0} = \frac{1}{2} \cdot (\cos\theta)^{n+1}, \tag{C.67}$$

$$\left.\frac{P_n^m(\cos\theta)}{\sin\theta}\right|_{\sin\theta=0} = 0 \quad \text{for } |m| \geq 2. \tag{C.68}$$

C.1.7 Angular Functions for Light Scattering

In calculation of light scattering with generalised Mie theory employs angular functions π_{mn} and τ_{mn}, which are derived from the associated Legendre polynomial:

$$\pi_{mn}(\theta) = (-1)^m \frac{m}{\sin\theta} P_n^m(\cos\theta), \tag{C.69}$$

$$\tau_{mn}(\theta) = (-1)^m \frac{\mathrm{d}}{\mathrm{d}\theta} P_n^m(\cos\theta) = (-1)^{m+1} \sin\theta \frac{\mathrm{d}}{\mathrm{d}\cos\theta} P_n^m(\cos\theta). \tag{C.70}$$

Note that in this definition the Legendre polynomials are defined with Condon-Shortley phase (see above).

For forward scattering ($\theta = 0$) and backscattering ($\theta = \pi$) these functions yield (cf. to Eqs. (C.65)–(C.68); Xu 1995):

$$\pi_{mn}(\cos\theta = \pm 1) = \frac{(\pm 1)^{n+1}}{2} \times \begin{cases} 1, & m = -1 \\ n(n+1), & m = 1 \\ 0, & \text{otherwise} \end{cases}, \tag{C.71}$$

$$\tau_{mn}(\cos\theta = \pm 1) = \frac{(\pm 1)^{n}}{2} \times \begin{cases} -1, & m = -1 \\ n(n+1), & m = 1 \\ 0, & \text{otherwise} \end{cases}. \tag{C.72}$$

The derivative functions τ_{mn} can be computed from the associated Legendre polynomials via a recurrence expression:

$$\tau_{mn}(\theta) = (-1)^m \left(m \frac{\cos\theta}{\sin\theta} P_n^m(\cos\theta) + P_n^{m+1}(\cos\theta) \right). \tag{C.73}$$

For single sphere scattering one just requires solutions for $m = 1$:

$$\pi_n = \pi_{1n} = -\frac{1}{\sin\theta} P_n^1(\cos\theta), \tag{C.74}$$

$$\tau_n = \tau_{1n} = -\frac{\mathrm{d}}{\mathrm{d}\theta} P_n^1(\cos\theta). \tag{C.75}$$

with the low order solutions:

$$\pi_0 = 0, \quad \tau_0 = 0, \tag{C.76}$$

$$\pi_1 = 1, \quad \tau_1 = \cos\theta, \tag{C.77}$$

$$\pi_2 = 3\cos\theta, \quad \tau_2 = 6\cos^2\theta - 3. \tag{C.78}$$

C.1.8 Spherical Harmonics

Spherical harmonics $Y_n^m(\theta, \phi)$ are the angular contribution to the solution of Laplace's equation (or Helmholtz differential equation) in spherical coordinates (i.e. Eqs. (C.9) and (C.10)). They are hence the product of the associated Legendre polynomial of $\cos\theta$ and the general sine of the azimuth ϕ:

$$Y_n^m(\theta,\phi)=\sqrt{\frac{2n+1}{4\pi}\frac{(n-m)!}{(n+m)!}}\cdot P_n^m(\cos\theta)\cdot \mathrm{e}^{\mathrm{i}m\phi}(\mathrm{m}=-\mathrm{n}\ldots\mathrm{n}). \tag{C.79}$$

The computation may be facilitated by using interrelation and recurrence equations:

$$Y_n^m=(-1)^{-m}\left(Y_n^{-m}\right)^*, \tag{C.80}$$

$$Y_n^{-n}=\frac{1}{2^n\cdot n!}\sqrt{\frac{(2n+1)!}{4\pi}}\sin^n\theta\mathrm{e}^{-\mathrm{i}n\phi}, \tag{C.81}$$

$$Y_n^m\cos\theta=\sqrt{\frac{(n+1)^2-m^2}{4(n+1)^2-1}}Y_{n+1}^m+\sqrt{\frac{n^2-m^2}{4n^2-1}}Y_{n-1}^m, \tag{C.82}$$

$$Y_n^m=\sqrt{\frac{4n^2-1}{n^2-m^2}}\left(Y_{n-1}^m\cos\theta-\sqrt{\frac{(n-1)^2-m^2}{4(n-1)^2-1}}Y_{n-2}^m\right). \tag{C.83}$$

The spherical harmonics obey the following closure relations:

$$\sum_{m=-n}^{n}\left|Y_n^m(\theta,\phi)\right|^2=\frac{2n+1}{4\pi}, \tag{C.84}$$

$$\sum_{n=0}^{\infty}\sum_{m=-n}^{n}Y_n^m(\theta_1,\phi_1)\left(Y_n^m(\theta_2,\phi_2)\right)^*=\frac{1}{\sin\theta_1}\delta(\theta_1-\theta_2)\delta(\phi_1-\phi_2). \tag{C.85}$$

C.1.9 Vector Spherical Harmonics

The spherical harmonics can be generalized to vector spherical harmonics by looking for a solution of the vectorial Laplace equation (C.5). This solution can be expressed in terms of vector field, which has a radial and two transverse components:

$$\mathbf{Y}_n^m=Y_n^m\cdot\mathbf{e}_r \tag{C.86}$$

$$\mathbf{\Psi}_n^m=r\cdot\operatorname{grad}Y_n^m, \tag{C.87}$$

$$\mathbf{\Phi}_n^m=\mathbf{e}_r\times\operatorname{grad}Y_n^m. \tag{C.88}$$

C.1.10 Spherical Vector Wave Functions

The spherical vector wave functions (SVWF) are the general solution of the vectorial Helmholtz differential equation in spherical coordinates (Xu 1995):

$$\mathbf{M}_{mn}^{(j)} = \left(\mathrm{i} \cdot \pi_{mn}(\cos\theta) \cdot \mathbf{e}_\theta - \tau_{mn}(\cos\theta) \cdot \mathbf{e}_\phi\right) \cdot z_n^{(j)}(kr) \cdot \mathrm{e}^{\mathrm{i}m\phi}, \tag{C.89}$$

$$\begin{aligned}\mathbf{N}_{mn}^{(j)} &= \mathrm{i} \cdot n(n+1) P_n^m(\cos\theta) \cdot \frac{z_n^{(j)}(kr)}{kr} \cdot \mathrm{e}^{\mathrm{i}m\phi} \cdot \mathbf{e}_r \\ &+ \ldots \left(\tau_{mn}(\cos\theta) \cdot \mathbf{e}_\theta + \mathrm{i} \cdot \pi_{mn}(\cos\theta) \cdot \mathbf{e}_\phi\right) \cdot \frac{1}{kr}\frac{\mathrm{d}}{\mathrm{d}r}\left[r \cdot z_n^{(j)}(kr)\right] \cdot \mathrm{e}^{\mathrm{i}m\phi},\end{aligned} \tag{C.90}$$

where $z_n^{(j)}$ denotes the spherical Bessel function of *j*th kind (j = 1–4).

The following interrelations hold true:

$$k \cdot \mathbf{N} = \mathrm{rot}\,\mathbf{M}, \tag{C.91}$$

$$k \cdot \mathbf{M} = \mathrm{rot}\,\mathbf{N}. \tag{C.92}$$

The SVWF are required for the scattering of vectorial waves. The incident and internal field then needs to be expressed by means of $\mathbf{M}_{mn}^{(1)}$ and $\mathbf{N}_{mn}^{(1)}$, because they do not diverge at the origin (unlike the other SVWF). In contrast, the scattered field is described by $\mathbf{M}_{mn}^{(3)}$ and $\mathbf{N}_{mn}^{(3)}$, as they vanish at infinity. In the far field (i.e. at large distance from the origin) the following asymptotic relations apply:

$$\lim_{kr\to\infty} \mathbf{M}_{mn}^{(3)} = (-\mathrm{i})^n \cdot \left(\pi_{mn}(\cos\theta) \cdot \mathbf{e}_\theta + \mathrm{i} \cdot \tau_{mn}(\cos\theta) \cdot \mathbf{e}_\phi\right) \cdot \frac{\mathrm{e}^{\mathrm{i}kr}}{kr} \cdot \mathrm{e}^{\mathrm{i}m\phi}, \tag{C.93}$$

$$\lim_{kr\to\infty} \mathbf{N}_{mn}^{(3)} = (-\mathrm{i})^n \cdot \left(\tau_{mn}(\cos\theta) \cdot \mathbf{e}_\theta + \mathrm{i} \cdot \pi_{mn}(\cos\theta) \cdot \mathbf{e}_\phi\right) \cdot \frac{\mathrm{e}^{\mathrm{i}kr}}{kr} \cdot \mathrm{e}^{\mathrm{i}m\phi}. \tag{C.94}$$

C.1.11 Solid Harmonics

Linear combinations of spherical harmonics

$$\sum_m A_n^m P_n^m(\cos\theta) \mathrm{e}^{\mathrm{i}m\phi} \tag{C.95}$$

are called surface harmonics (because the radial coordinate r is not varied).

Regular solid harmonics are surface harmonics of degree n multiplied by a factor r^n:

$$v_{mn} = r^n P_n^m(\cos\theta) e^{im\phi}. \tag{C.96}$$

They vanish at the origin and are finite at large distances r. This is in contrast to irregular solid spherical harmonics:

$$u_{mn} = \frac{1}{r^{n+1}} P_n^m(\cos\theta) e^{im\phi}, \tag{C.97}$$

which are singular at the origin and vanish at large distance r. Hence, they can be used to describe the perturbations of flow fields around spherical shaped objects (cf. Sect. 4.4.2). The computation of viscous stresses in such flow fields requires the velocity gradients, i.e. the spatial derivatives of u_{mn}:

$$\frac{\partial u_{mn}}{\partial r} = -\frac{n+1}{r^{n+2}} P_n^m e^{im\phi} = -\frac{n+1}{r} \cdot u_{mn}, \tag{C.98}$$

$$\frac{\partial u_{mn}}{\partial \phi} = \frac{i \cdot m}{r^{n+1}} P_n^m e^{im\phi} = i \cdot m \cdot u_{mn}, \tag{C.99}$$

$$\frac{\partial u_{mn}}{\partial \theta} = \frac{e^{im\phi}}{r^{n+1}} \cdot \frac{\partial}{\partial \theta} P_n^m = n \cot\theta \cdot u_{mn} - \frac{n+m}{r \sin\theta} u_{mn-1}. \tag{C.100}$$

References

M.A. Abramowitz, I.A. Stegun (Eds.), *Handbook of Mathematical Functions*, 8th Dover printing (Dover Publication, New York, 1972), pp. 332–341, 437–456. ISBN: 486-61272-4

C.F. Bohren, D.R. Huffman, *Absorption and Scattering of Light by Small Particles* (Wiley, New York, 1983). ISBN: 0-471-05772-X

W. Greiner. *Elektrodynamik*, 5th edn. In series: W. Greiner (ed.) Theoretische Physik Bd. 3. (Verlag Harri Deutsch, Frankfurt/M., 1991), pp. 83–92.

J. Naas, H.L. Schmid, *Mathematisches Wörterbuch, Bd. I u. II.*, 3rd edn. (Akademie Verlag Berlin und B.G. Teubner Stuttgart, 1984), Bd. I, p. 715, Bd. II, pp. 37–42, 45–48.

M.R. Spiegel. *Höhere Mathematik für Ingenieure und Naturwissenschaftler* (McGraw-Hill Book Company Europe, London, 1991).

E.W. Weisstein, in *CRC Concise Encyclopaedia of Mathematics* (Chapman & Hall/CRC, Boca Raton, 1999), pp. 1057–1060, 1688–1689, 1694–1696. ISBN 0-8493-9640-9

Y.-L. Xu, Electromagnetic scattering by an aggregate of spheres. Appl. Opt. **34**(21), 4573–4588 (1995). doi:10.1364/AO.34.004573

C.2 Singular Fields for Electric Potential Distribution

The electric double layer (EDL) of a particle is determined by the mechanisms of surface charging and ion adsorption in the Stern layer as well as by the balance of Coulombic and osmotic forces on the ions within the diffuse part of EDL. The latter is described by the nonlinear Poisson-Boltzmann equation (PBE, Eq. (3.16)), which needs to be solved numerically for particle aggregates. Additionally, one needs a model for the charge and potential regulation of overlapping EDLs which is inevitable in the case of aggregates and approaching particles.

The Sect. 4.6.2 introduces a numerical scheme for calculating the potential distribution in the EDL of arbitrary clusters of spherical particles: the singularity method (Phillips 1995). This approach considers the total electric field in EDL as superposition of the electric fields of point charges, which are located inside the particles. Such a field is easily described by:

$$\psi = q \frac{\mathrm{e}^{-\kappa a r}}{r}. \tag{C.101}$$

The singularity method further assumes that the linearised PBE (Eq. (3.17)) holds true and that a linear charge regulation model (Eq. (3.27)) can be applied. Most equations in Sect. 4.6.2 are given for normalised expressions of the double layer parameters. These are defined as follows:

$$\psi^* = \frac{\psi}{\psi_0^{\mathrm{iso}}}, \tag{C.102}$$

$$r^* = \frac{r}{a}, \tag{C.103}$$

$$q^* = \frac{q}{a\psi_0^{\mathrm{iso}}}, \tag{C.104}$$

$$K_{\mathrm{reg}}^* = \frac{aK_{\mathrm{reg}}}{\varepsilon_{\mathrm{m}}}, \tag{C.105}$$

$$S_{\mathrm{reg}}^* = \frac{aS_{\mathrm{reg}}}{\varepsilon_{\mathrm{m}}\psi_0^{\mathrm{iso}}}. \tag{C.106}$$

Note that the equations in Sect. 4.6.2 refer to the general case of linear charge regulation. However, for the limiting case of constant potential the boundary condition (4.120) has to be replaced by

$$\psi_0^* = 1, \tag{C.107}$$

and Eqs. (4.124) and (4.125) are reduced to:

$$A_{jl} = \frac{e^{-\kappa a \cdot r_{lj}^*}}{r_{lj}^*}, \tag{C.108}$$

$$b_j = 1 - \sum_{k=1}^{N} \frac{e^{-\kappa a \cdot r_{kj}^*}}{r_{kj}^*} q_c^*. \tag{C.109}$$

Moreover, for constant potential the interaction energy is defined by Eq. (3.32) instead of Eq. (3.31). That means for the singularity method that Eq. (4.127) has to be substituted by:

$$\begin{aligned} V_{\mathrm{DL}}(h) = &-\frac{1}{2}\varepsilon_{\mathrm{m}}\psi_1^{\mathrm{iso}} \sum_{m=1}^{N_1} \int \left(\sum_{k_2=1}^{N_2} \nabla\psi_{\mathrm{c},k_2} \bullet \mathbf{n} + \sum_{l=1}^{\alpha N} \nabla\psi_{\mathrm{oc},l} \bullet \mathbf{n} - \sum_{l_1=1}^{\alpha N_1} \nabla\psi_{\mathrm{oc},l_1}^{\mathrm{iso}} \bullet \mathbf{n} \right) dA_m \\ &-\frac{1}{2}\varepsilon_{\mathrm{m}}\psi_2^{\mathrm{iso}} \sum_{n=1}^{N_2} \int \ldots dA_n. \end{aligned} \tag{C.110}$$

This involves the solution of the surface integral for the gradient of the potential field, which can be transformed into a volume integral:

$$\int \nabla\psi \bullet \mathbf{n}\, \mathrm{d}A = -\int \Delta\psi\, \mathrm{d}V. \tag{C.111}$$

The solution of this integral depends on the distance R between the singularity k or l and the primary particle m and n:

$$\int \nabla\psi \bullet \mathbf{n}\, \mathrm{d}A = 4\pi q \begin{cases} (1+\kappa a)\mathrm{e}^{-\kappa a}, & \text{internal source at centre: } R = 0 \\ \frac{(1+\kappa a)\mathrm{e}^{-\kappa a}}{\kappa R} \sinh(\kappa R), & \text{internal source, offcentre: } a \geq R > 0 \\ \frac{\mathrm{e}^{-\kappa R}}{\kappa R} (\sinh(\kappa a) - \kappa a \cdot \cosh(\kappa a)), & \text{external source: } R > a \end{cases}. \tag{C.112}$$

Equations (C.108)–(C.112) define the double layer and interaction energy of aggregates for a constant potential regulation.

References

R.J. Phillips, Calculation of multisphere linearized Poisson-Boltzmann interactions near cylindrical fibers and planar surfaces. J. Colloid Interface Sci. **175**(2), 386–399 (1995). doi:10.1006/jcis.1995.1469

K. Schießl, F. Babick, M. Stintz, Calculation of double layer interaction between colloidal aggregates. Adv. Powder Technol. **23**(2), 139–147 (2012). doi:10.1016/j.apt.2011.01.005

It is wrong to assume that the positive amount of data and information – which grows tremoundously these days – makes theory obsolet, or that the correlation of data displaces models. Theory – as negativity *– is superior to positive data and information, even superior to models. The data-based positive science […] lacks the negativity of decision, which selects what is or should be.*

Es ist ein Irrtum, anzunehmen, dass die positive Daten- und Informationsmasse, die heute ins Ungeheure wächst, die Theorie überflüssig mache, dass der Abgleich von Daten die Modelle ersetze. Die Theorie als Negativität ist vor positiven Daten und Informationen, auch vor den Modellen angesiedelt. [Der] datenbasierte[n] Positivwissenschaft […] fehlt die Negativität der Dezision, die erst entscheidet, was ist oder zu sein hat.

Byung-Chul Han, *Transparenzgesellschaft.*
Matthes & Seitz, Berlin, 2012

MIX
Papier aus verantwortungsvollen Quellen
Paper from responsible sources
FSC® C105338

If you have any concerns about our products,
you can contact us on
ProductSafety@springernature.com

In case Publisher is established outside the EU,
the EU authorized representative is:
Springer Nature Customer Service Center GmbH
Europaplatz 3, 69115 Heidelberg, Germany

Printed by Libri Plureos GmbH
in Hamburg, Germany